Alan Schilow

OXFORD CLASSIC TEXTS IN THE PHYSICAL SCIENCES

WHETHER the *friction of the heavens* makes a sound or no: . . . in the case of smooth bodies the friction does not create sound, and it would happen in a similar manner that there would be no sound in the contact or friction of the heavens. And if these heavens are not smooth at the contact of their friction it follows that they are full of lumps and rough, and therefore their contact is not continuous, and if this is the case the vacuum is produced, which it has been concluded does not exist in nature. We arrive, therefore, at the conclusion that the friction would have rubbed away the boundaries of each heaven, and in proportion as its movement is swifter towards the centre than towards the poles it would be more consumed in the centre than at the poles; and then there would not be friction any more, and the sound would cease, and the dancers would stop, except that the heavens were turning one to the east and the other to the north.

LEONARDO DA VINCI (1452–1519)
Notebooks, translated into English by Edward MacCurdy; Jonathan Cape, London, 1938.

LE grand usage que tous les Arts sont obligez de faire des machines, est une preuve convainquante de leur absoluë nécessité; ainsi sans perdre temps à établir autrement cette verité, on se contentera de dire icy, que si le nom de Machine est quelquefois pris en mauvaise part, & s'il devient quelquefois méprisable, ce n'est en partie qu'à cause que le peu de régles que nous avons dans les Mécaniques ne suffisent pas toûjours pour prévoir certainement l'effet que les Machines qu'on projette doivent produire dans leur execution ce qui fait bien souvent, que plusieurs personnes qui les ignorent se croyent bien fondées à ne s'en pas instruire, & tombent par là dans les absurdités étranges.

AMONTONS, *Memoires de L'Academie Royale des Sciences*, 1699.

THE FRICTION AND LUBRICATION OF SOLIDS

BY

F. P. BOWDEN†

Formerly Professor of Surface Physics, Cavendish Laboratory, University of Cambridge

AND

D. TABOR

Emeritus Professor of Physics, Cavendish Laboratory, University of Cambridge

CLARENDON PRESS • OXFORD

This book has been printed digitally and produced in a standard specification in order to ensure its continuing availability

OXFORD
UNIVERSITY PRESS

Great Clarendon Street, Oxford OX2 6DP

Oxford University Press is a department of the University of Oxford.
It furthers the University's objective of excellence in research, scholarship, and education by publishing worldwide in

Oxford New York

Auckland Cape Town Dar es Salaam Hong Kong Karachi
Kuala Lumpur Madrid Melbourne Mexico City Nairobi
New Delhi Shanghai Taipei Toronto

With offices in

Argentina Austria Brazil Chile Czech Republic France Greece
Guatemala Hungary Italy Japan South Korea Poland Portugal
Singapore Switzerland Thailand Turkey Ukraine Vietnam

Oxford is a registered trade mark of Oxford University Press in the UK and in certain other countries

Published in the United States by Oxford University Press Inc., New York

Reprinted 2008

ISBN 978-0-19-850777-2

Printed and bound by CPI Antony Rowe, Eastbourne

FRANK PHILIP BOWDEN
1903–1968

Frank Philip Bowden was born in Tasmania in 1903 and came to Cambridge on an 1851 Scholarship in 1927 to work with E. K. (later Sir Eric) Rideal in the Physical Chemistry department. There he continued his studies on electrode potentials for which he obtained the Cambridge Ph.D. In 1930, Rideal was made head and Professor of the newly-established department of Colloid Science. Bowden remained in Physical Chemistry and soon began to develop his interests in surface problems associated with friction and lubrication beginning with a joint paper in 1931 with Stewart Bastow on the contact of smooth surfaces. The work on friction expanded and gradually replaced his earlier interest in electrode chemistry.

In 1939, he went on a lecture tour to the USA and decided to return to Britain via Australia in order to visit friends and relatives there. His wife and child had meanwhile arrived direct from England. While Bowden was in Australia, World War II broke out and he was invited to set up a research laboratory on friction, lubrication and bearings to assist the war effort. (I had been a research student of his in Cambridge and joined him in Melbourne at the beginning of 1940.) Bowden's laboratory in Melbourne was extremely successful in both basic and applied research, and after the war acquired the title of Tribophysics as a division of the Commonwealth Scientific and Industrial Research Organisation (CSIRO). With the passage of time, the work of that division gradually changed and the name was finally dropped in 1978.

Bowden returned to Britain in 1945 and began to rebuild his research team, at first in the department of Physical Chemistry and, after 1957, in the physics department (the Cavendish Laboratory) where he was appointed to an *ad hominem* chair in surface physics in 1960. His research interests brought him into contact with many industrialists and in 1954 he was appointed scientific adviser to, and for a short period director of, a new laboratory set up by Tube Investments (TI) at Hinxton with the aim of carrying out first-class research and applying it to industrial problems. The Hinxton laboratory and the University research group flourished under his guidance, and are still thriving.

In assessing Philip Bowden's remarkable achievements, there are several features which are outstanding. First, he pioneered the application of scientific methods to the study of the unfashionable and relatively 'messy' field of friction and lubrication. Secondly, he recognized at once that in this task he would need in his research team physicists, chemists, engineers, metallurgists, physical chemists, and (occasionally) mathematicians. Thus, over forty years ago, although he was a staff member of a specific university department, he created an interdisciplinary research group. This approach was far ahead (and still is) of its time. Thirdly, he continued to show flexibility in his interests and it is significant that one of his wartime commitments led, in collaboration with A. D. Yoffe and others, to some of the best and most original research on the initiation of explosion in liquids and solids. Finally, in his lifetime he established not one but three extremely successful research groups, in Cambridge, in Melbourne, and in TI. Indeed, in spite of many distractions and outside commitments, his deepest and most sustained interest throughout his life was the 'lab' and the challenge and excitement of scientific research.

This short account cannot deal with many of the facets of his personality and career. But this quotation from the address* I gave at the Memorial Service held in Caius College Chapel on 20 October 1968 gives something of the flavour of the man and his work:

> One of the qualities that most completely describes so much of Philip Bowden's many-sided personality was his keen appreciation of the aesthetic side of life. . . Those of us who worked with him felt the influence of this quality in his scientific research. The choice of subject, the elegant original and direct experimental approach, the forthright language, the uncluttered conclusions—all these reflect his aesthetic sense. His best scientific work is at once art and science and bridges many cultures in an effortless and unselfconscious manner. . . All of us who knew him, in one capacity or another, will miss his wisdom, his charm and his humanity.

These sentiments are as true today as when they were first written.

D.T.

* See *Biographical Memoirs of Fellows of the Royal Society*, 1969, **15**, 1–38.

PREFACE TO PAPERBACK EDITION

This reprint of a book published over 35 years ago is testimony to the perception, originality, and scientific nous of Philip Bowden who directed the researches described in it.

Bowden's approach to the study of friction was based on the recognition that all the relevant tribological processes take place in and around the regions of real contact. A clear idea of this area was thus of primary concern. For this reason much of the original work involved the contact between a small curved slider and a flat surface, for in this configuration the area of true contact, if the surfaces are reasonably smooth, is close to the geometric area. With metals this arrangement, in turn, readily leads to plastic flow in the contact region. As a result the experiments and the interpretation tended to over-emphasize the role of plastic deformation. We now know more about the conditions under which elastic, elasto-plastic, and plastic deformation occur. During the last two decades, surface profilometry, the statistical analysis of surface topography, and the science of contact mechanics have indeed combined to provide a more balanced picture of the way in which surfaces deform when they are brought into contact. Nevertheless the basic conclusions still remain valid: the area of true contact for extended surfaces is roughly proportional to the applied load and in general constitutes only a very small fraction of the geometric area. And with metals the initial contact usually involves some plastic deformation of surface asperities. Of course when sliding takes place the combined normal and tangential stresses modify the area of contact especially in the plastic regime. Slip-line field theory provides a solution to this but the *direct* determination of the true area of contact between surfaces in static and sliding contact still remains a fundamental challenge to the experimental scientist.

The idea of adhesion at the interface also tended to emphasize plastic flow and welding although by the time Part I (the present volume) of *The Friction and Lubrication of Solids* was published it was recognized that welding did not necessarily require high temperatures and the term 'cold welding' was employed. During the last twenty years the adhesion of metals has been studied in considerable detail and various modern spectroscopic techniques have been applied to characterize the contacting surfaces in ultra-high vacuum or in controlled environments. These researches have greatly added to our knowledge of the influence of crystal orientation, of metallic bond-type, of micro-deformation characteristics, and of surface contamination. Broadly speaking, the results confirm the early work carried out when such elegant facilities were not available; clean metals stick together very strongly and, apart from one or two remarkable exceptions, a monolayer of contaminant greatly reduces the adhesion. The later work has also confirmed the role of released elastic stresses and of limited ductility in weakening the interfacial junctions: these factors probably account for the poor adhesion of hard solids and ceramics.

Bowden's researches attributed the frictional process to two main factors: first, the shearing of junctions formed by interfacial adhesion at the regions of real contact, and second, the ploughing of the softer surface by asperities on the harder. We now know that this is an oversimplification. The two processes are not independent and there are now slip-line field solutions which cover the situation when the surface interactions are dominated by plastic deformation. These analyses show how easily high shear strains are generated in the surface layers, leading readily to surface fatigue, surface fragmentation, and other forms of failure. But the detailed behaviour of systems in which plastic deformation is unimportant raises many problems still unresolved concerning the nature of the interfacial shear process. The role of oxides and their breakdown is another challenging problem of great importance, in sliding mechanisms as well as in the performance of electrical contacts.

The major advances in tribological studies during the last thirty years fall into a number of well-defined areas. Some, though not all, of these were discussed in Part II of *The friction and lubrication of solids*, published in 1964. First, a great deal of effort has been expended on the behaviour of non-metals, especially elastomers, where surface energy and bulk viscoelastic properties determine the adhesion as well as the frictional behaviour; polymers, where the friction and wear are influenced both by chemical structure and molecular morphology; and ceramics, where surface forces determine the friction and brittle fracture or plastic-brittle fracture dominates the wear behaviour. There is indeed a better understanding now of the way in which the mechanical and surface properties of these materials modify the frictional mechanism. Secondly, a whole area of lubrication, unknown in 1950, has been opened up. Known as elastohydrodynamic lubrication (EHL), it constitutes one of the most exciting developments in the field. As its name implies, it is a type of hydrodynamic lubrication which takes into account the elastic deformation of the contacting surfaces. With metal surfaces in a non-conformal configuration this leads to very high contact pressures and these in turn produce an enormous increase in the effective viscosity of the trapped lubricant film. As the sliding-speed (i.e. the shear-rate) is increased, the shear stress in the film increases, until, at some critical stage determined primarily by the contact pressure, a new mode of shear takes place: the shear stress becomes almost independent of further increases in the shear-rate. The film at this stage may, indeed, be regarded as a solid and some workers view this as a genuine phase change. These EH films are of major importance in the functioning of gears and of rolling-element bearings: it is also probable that they are generated between individual surface asperities in sliding systems and thus play a vital role in preventing metal–metal contact. The ultimate cause of film failure is still a subject of research, the two most important factors being unfavourable surface roughnesses and flash temperatures. To some extent elasto-hydrodynamic lubrication has squeezed out boundary lubrication as a separate regime since the shear properties of the EH film closely resemble those of the boundary film. However, the chemical and structural properties of the film formed by boundary lubricants still retain their importance in imparting to the film a resistance to failure superior to that of the conventional EH film.

The third area of increased activity concerns the wear of sliding surfaces. Although in engineering practice the level of friction is important it is the life of the operating parts which often determines whether a mechanical system is economically or functionally viable. For this reason the study of wear has gradually displaced the earlier concentration on the mechanism of friction. Many types of wear processes have been investigated in detail, are reasonably well understood, and their regimes of operation well delineated. The main problem is that, in practical affairs, the wear of a particular system is not the result of a single wear process but of several processes which interact, often in an unpredictable way. For this reason it has often proved useful to apply 'systems analysis' to the practical assessment of wear.

The study of friction, lubrication, and wear received a powerful impetus from the Jost Report, published in 1966. It emphasized the interdisciplinary nature of tribology, an approach which Bowden had practised ever since he entered the field in the 1930s. This reprint represents the results of the first twenty years of his research and the contribution he and his colleagues made to an interdisciplinary understanding of the sliding processes.

Cavendish Laboratory D.T.
January 1986

PREFACE TO ORIGINAL EDITION

To Sir David Rivett and to Sir Ben Lockspeiser whose wise guidance of scientific affairs has done so much to stimulate research in general and this work in particular.

This monograph describes an experimental study of the physical and, to a less extent, of the chemical processes that occur during the sliding of solids—particularly of metals—and an investigation into the mechanism of friction and boundary lubrication. The field covered is somewhat wider than the title might indicate and it deals with a number of the physical properties of solid surfaces. It is not a general text-book, since it deals almost entirely with experimental researches carried out by the writers and their collaborators and colleagues. The work is still in progress and it is abundantly clear that there are many loose ends and that much remains to be done.

The work was begun in Cambridge a few years before the war, was continued (though much mingled with war work) in the Tribophysics Division of the Australian Council for Scientific and Industrial Research, Melbourne, from 1939 to 1944, and since then has been resumed in Cambridge. Collaboration on some aspects of the work is being continued with the Tribophysics laboratory in Melbourne, which is under the direction of Dr. S. H. Bastow.

Our indebtedness to others is very great. To the late Sir William Hardy and to Dr. David Pye, who were members of the pre-war Lubrication Research Committee; to Sir David Rivett and to members of the Executive and Council of the Australian CSIR; to the University of Melbourne and in particular to Professor E. J. Hartung in whose laboratory the Tribophysics Division is still housed; to Sir Henry Tizard, and to the late Sir Ralph Fowler who encouraged us to start up the work again in Cambridge; and finally to Sir Ben Lockspeiser and the Ministry of Supply (Air) for stimulating interest and support, and for generous grants to the University of Cambridge which have made it possible for the work to be continued. Active interest and help has also been given by Sir Edward Appleton and the Department of Scientific and Industrial Research, and by many others.

We also owe a great debt to past and present members of this Laboratory and to Dr. S. H. Bastow and members of the Tribophysics Laboratory for the assistance they have given in the preparation of this monograph. Their names are too numerous to give here, but they are embodied in the text.

LABORATORY ON THE PHYSICS AND
CHEMISTRY OF RUBBING SOLIDS,
DEPARTMENT OF PHYSICAL CHEMISTRY,
CAMBRIDGE.

February 1949

NOTE

The experimental study of problems described in this monograph has been continued. The book as a whole is brought up to date by accounts of recent work and relevant references given in the addenda, pp. 328–63. Reference to these addenda is indicated in the text by [*A*]. Corrections and small alterations are made in the text itself.

November 1953

F.P.B.
D.T.

PUBLISHER'S NOTE (1986)

A second volume of *The friction and lubrication of solids*, by the same authors, was issued in 1964 as a companion to the present work, which thereafter was known as Part I. Part II is currently out of print.

CONTENTS

Plates fall between pages 14 and 15

INTRODUCTION

THE two basic laws of friction, that the frictional resistance is proportional to the load and that it is independent of the area of the sliding surfaces, have been known for a long time. It would seem that Leonardo da Vinci (1452–1519) with his astonishing practical genius and insight was clear about them and had verified them experimentally, for he writes: 'Friction produces double the amount of effort if the weight be doubled', and 'The friction made by the same weight will be of equal resistance at the beginning of the movement although the contact may be of different breadths or lengths.' The rediscovery of these laws by the French engineer Amontons in 1699 was received with some surprise and scepticism by the Académie Royale des Sciences:

'Dans le Discours que fit M. Amontons sur son Moulin à fęu, il avança seulement en passant, que c'étoit une erreur de croire, comme l'on fait communément, que le frottement de deux corps qui se meuvent en s'appliquant l'un contre l'autre, soit d'autant plus grand, que les surfaces qui frottent sont plus grandes. Il dît qu'il avoit reconnu par experience que le frottement n'augmente que selon que les corps sont plus pressés l'un contre l'autre, & chargés d'un plus grand poids. Cette nouveauté causa quelque étonnement à l'Academie.'

These observations were verified by Coulomb in 1781, who made a clear distinction between static friction, the force required to start sliding, and kinetic friction, the force required to maintain it. He showed that kinetic friction could be appreciably lower than static friction and made the observation (which is sometimes quoted as a third law) that the kinetic friction is nearly independent of the speed of sliding. Coulomb considered the possibility that friction might be due to molecular adhesion between the surfaces, but rejected it on the grounds that if this were so, the friction should be proportional to the area of the sliding bodies whereas it was found to be independent of it. He concluded that the friction was due to the interlocking of the surface asperities and represented in the main the work of lifting the load over the summits of these asperities. It would seem that he had certain reservations about this theory, and it would also appear that the Académie was not prepared to stand any nonsense from its members, for he writes:

'Je ne m'étendrai pas davantage sur cette théorie; elle paroît expliquer avec facilité tous les phénomènes du frottement; mais l'Académie ne demande aujourd'hui que des recherches qui puissent être utiles: ainsi il seroit dangereux de trop se livrer à un systême qui pourroit peut-être influer sur la manière de rendre compte des expériences qui nous restent à faire.'

In recent times important experimental investigations in friction, particularly of the static friction of solids lubricated with very thin films of hydrocarbon compounds, were made by Sir William Hardy. In particular, Hardy showed the important part which is played by a single molecular layer of lubricant (the boundary layer). This monolayer which is attached to the surface and oriented has a profound effect on the friction. His own experiments were elegant and simple, and he emphasized the necessity for applying modern physical and chemical concepts and methods to the study of these surface phenomena. It is Hardy's work more than any other that has stimulated a modern approach to the subject.

This book will describe an experimental study of the physical and chemical processes that occur during the contact and sliding of solid surfaces, and an investigation into the mechanism of friction and of boundary lubrication. The first point of inquiry is, What is the real area of contact between solids which are placed together? Experiments show that even the most carefully prepared surfaces contain hills and valleys which are large compared with molecular dimensions. The solids are supported on the summits of the highest of these irregularities, so that the area of intimate contact is very small. The real area of contact is in fact almost independent of the size of the surfaces and is determined by the load, since, under the intense pressure at the localized points of contact, plastic deformation and flow occur until the area is sufficiently great to support the load.

When sliding takes place all the friction occurs over this very small area and we may expect that the surface temperature at these points of rubbing contact would reach a high value. Various experimental methods are used to investigate this, and it is shown that even at moderate speeds of sliding the surface temperatures of metals are indeed high and may easily cause a thermal softening or local melting of the metal at the points of contact. With non-conductors these local high temperatures occur even more readily, and this localized softening or melting plays an important part in polishing and in a number of physical processes.

The third point of inquiry is into the nature of the surface damage. It is shown that with metals there is a real adhesion and welding together at the points of contact. The frictional force is in a large measure the force required to shear these junctions. These observations explain the classical laws of friction, since the area over which the junctions are formed is almost independent of the apparent area of the sliding surfaces and is directly proportional to the applied load. With stationary surfaces or at low speeds of sliding this 'cold welding' is produced by the intense pressure in the region of contact and at higher speeds it is assisted by a high-temperature softening or melting of the metal. It is the shearing, deformation, and plucking away of these metallic junctions which constitute the physical wear of metals, and the various ways in which this can occur are investigated. The smearing of a thin layer of a soft metal over a harder one can be effective in reducing the friction, and the part which this may play in the action of bearing alloys is considered.

The interaction between metal surfaces and the intimacy of contact between them is profoundly changed by the presence of adsorbed films of gas or of oxide and some study of this is made. It is shown that the friction and adhesion between naked metal surfaces is very great. A brief account of the frictional behaviour of some non-metallic solids is included, but the experimental background is incomplete and it is clear that further work is necessary.

We then turn our attention to lubricated surfaces. In general, engineers strive to obtain fluid lubrication, where the surfaces are separated by a relatively thick film of lubricant and the frictional resistance is determined essentially by the hydrodynamic properties of the fluid. The fundamentals of hydrodynamic or fluid lubrication have been well established since the pioneer work of Osborne Reynolds in 1886. Under many conditions of operation, however, it is not possible to maintain fluid lubrication and the surfaces are separated only by a boundary layer which may be of molecular dimensions. It is this aspect of lubrication with which we are mainly concerned here. In particular we consider the general behaviour of metal surfaces lubricated with boundary films of long chain molecules. It is shown that the earlier view that lubrication is due to physically adsorbed oriented monolayers sliding over one another is an oversimplification. The use of radioactive metals and other sensitive physical methods shows that in general some penetration of the boundary film occurs with a consequent localized metallic adhesion. It is shown that the lubrication of metals by fatty acids is most effective only when the acids react with the metal to form the soap. The attack of the metal frequently occurs via the oxide film, so that inert metals or metals which are free from oxide may not be effectively lubricated by fatty acids. The lateral adhesion between the hydrocarbon chain is of primary importance, so that the surface films act as efficient boundary lubricants only when they are 'solid'. Electron diffraction experiments on the structure of these films and on their disorientation or 'melting' as the temperature is raised support this view. The chemical attack of the surface is also of major importance in 'extreme pressure' lubrication, and a study is made of the mechanism of lubrication by compounds containing sulphur and chlorine and other active groups.

Although this book does not deal with full fluid lubrication, some experiments are described on the lubrication between the piston ring and cylinder wall of a running engine and of the lubrication in a journal bearing. It is shown that breakdown of fluid lubrication and momentary contact between the surfaces can occur very readily.

Some account is then given of the impact between colliding solids both in the absence and in the presence of liquid films. It is shown that the forces transmitted through the liquid film may readily cause plastic deformation and damage of metals although no solid contact occurs.

After a discussion of metallic wear an account is given of some experiments on the mechanism of the adhesion of metals and other solids and of the influence of surface films on the adhesion. The adhesion experiments provide direct evidence for the formation of metallic junctions between metal surfaces. Under appropriate conditions the normal force required to separate the surfaces may be of the same order as the tangential force required to slide them over one another; in fact, the 'coefficient of adhesion' may be nearly equal to the coefficient of friction.

The last chapter deals with chemical reaction which is produced by impact and by friction, and discusses the initiation of explosions which are produced in this way. It is shown that the generation of localized frictional hot spots plays an important part in these reactions.

I

AREA OF CONTACT BETWEEN SOLIDS

Measurement of Surface Irregularities and Surface Contour

It is a very difficult matter to prepare surfaces which are really flat. Even on carefully polished surfaces, hills and valleys are present which are large compared with the size of a molecule. If two solids are placed in contact, the upper surface will be supported on the summits of the irregularities, and large areas of the surfaces will be separated by a distance which is great compared with the molecular range of action. Although the techniques of grinding and polishing have advanced in the past few years, it is still a difficult matter to prepare surfaces of appreciable size which are flat to within 100 to 1,000 A. Most of the surfaces used in engineering practice have surface irregularities which are very much greater than this. Since the range of molecular attraction is only a few Ångströms, we may expect that the area of intimate contact, that is, the area over which the surfaces are within molecular range, will, even for carefully prepared surfaces, be quite small. In recent years our knowledge of the surface structure and surface contours of solids has advanced considerably, and various experimental methods have been developed and applied to the measurement of the size and shape of the irregularities which are present on solids. Some of the methods used in the work which we shall describe may be mentioned briefly here.

Stylus methods

Instruments have been developed which amplify the movement of a tracer needle as it passes slowly over the surface and follows its contours. The stylus is usually a conical diamond which may have a radius of curvature at its tip as small as 2×10^{-4} cm. The vertical movement of the stylus is amplified electrically and recorded on a moving paper. Considerable ingenuity and precision has been employed in the design and construction of these instruments, and they can operate successfully with a vertical magnification of 40,000. Certain inherent difficulties arise at high magnification, such as the difficulty of fixing accurately the reference datum level and, if the surface is soft, its damage by the stylus. A great advantage of this type of instrument is that it gives a record of the surface contour more quickly and conveniently than other methods, and it does not involve the destruction of the surfaces. Its ultimate sensitivity is limited by the finite size of the needle which prevents it from penetrating the finest scratches or irregularities. It will, however, under favourable conditions, detect pits or scratches which are only 250 A deep. An excellent account of the underlying theory and of the value and the limitations of stylus methods has been given by Reason, Hopkins, and Garrod (1944) (see also Conference on Surface Finish, 1940). Measurements by the Talysurf instrument of surface irregularities and of the surface contour of solids are described later.

Optical interference [*A*]

Simple two-beam interference methods for measuring surface contour have, of course, been in use for a long time. A familiar example is the use of Newton's rings between a lens and an optical flat to measure the lens curvature. With this simple method the light and dark interference bands are comparatively broad. The variation of intensity from one maximum to the next is a sine squared function and the width of a 'line' at half-maximum intensity is 50 per cent. of the spacing of the lines. This means that detail causing a shift of the maximum of less than 1/5th of an order would be almost invisible. Tolansky (1948) has recently made notable advances by developing interference methods to a high degree of sensitivity. If the surfaces of the optical flat and the lens are coated with a layer of silver having a high reflecting coefficient (85–95 per cent.) and a significant transmission coefficient, the incident light will be reflected back and forth and the interference pattern will be the resultant of all these beams. The effect (which is analogous to increasing the number of lines on a diffraction grating) is to leave the maxima and minima of intensity as before, but to alter the distribution of intensity between one maximum and the next. The position of the maxima is given by the usual formula:

$$n\lambda = 2\mu t \cos\theta,$$

where λ is the wave-length of the light used,
μ is the refractive index of the material between the silvered surfaces,
t is the thickness of the gap between the silvered surfaces,
θ is the angle of incidence measured from the normal,
n is an integer for maxima of illumination.

The intensity distribution is similar to that obtained in a Fabry–Perot interferometer: narrow bright lines in a dark field; the width at half-maximum intensity about 2 or 3 per cent. of the distance between orders. Fine detail of imperfections of surface finish corresponding to variations of thickness of the film 1/100 of that between successive orders is easily visible. Taking $\lambda = 5{,}000$ A, this corresponds to the detection of features 25 A high. Plate I. 1 shows typical interference fringes obtained in this way by Mr. Courtney-Pratt between a very flat surface and a piece of commercial plate-glass. The ripples in the fringes correspond to surface irregularities of the order of 100 A.

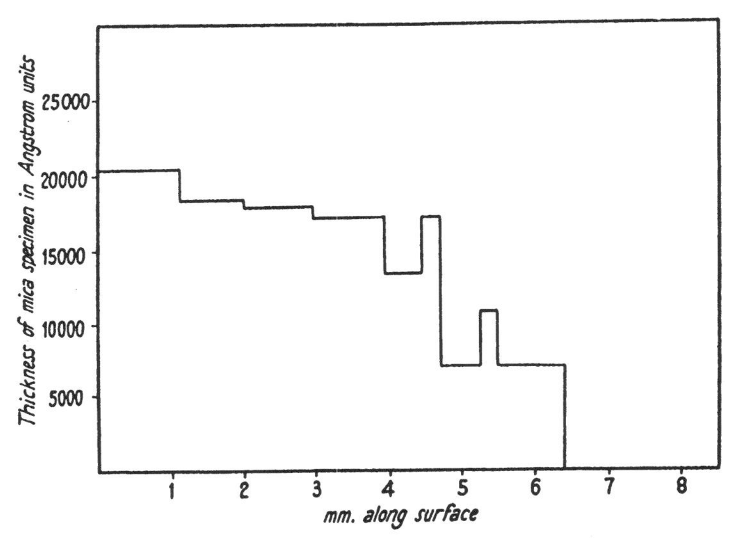

FIG. 1. Thickness of a flake of mica determined by optical interference between the two faces of the flake using the multiple reflection technique. The changes in thickness are accurate to about ± 3 A and are integral multiples of the lattice spacing of the mica.

For this sort of accuracy, considerable attention must be paid to the details of the physical layout of the optical system. Also, the silver must be put on the surfaces very uniformly so that it follows the contours of the surface precisely. Atomic evaporation in a high vacuum and extreme cleanliness of the surfaces are most important; and the silver must be pure or its optical absorption is troublesome. The optical flat must be of high quality and exceptional smoothness at least over small areas. The angle between the reflecting surfaces on either side of the 'wedge' must not be too great and, even more important, the actual separation of the reflecting surfaces must be kept to the absolute minimum.

By a modification of the method and the use of a spectrograph and white light the sensitivity can be increased still further, and steps and similar features in the surface as small as 10 A or 15 A can be easily detected and their magnitude measured to an accuracy of about 2 A or 3 A. Using this method, Tolansky (1945) has shown that the cleavage face of a crystal is usually not flat over any appreciable area but contains steps many hundreds of Ångströms high, which are a definite multiple of the atomic spacing. A typical result obtained by Mr. Courtney-Pratt, showing the contour of the cleavage faces of mica, is shown in Fig. 1. This refined optical method has been applied by Mr. Courtney-Pratt to the measurement of the surface irregularities and the surface damage of sliding solids, and also to the measurement of the frictional interaction between solids at very small displacements.

Electron microscope [*A*]

The electron microscope is a powerful tool for the examination of surface irregularities and surface structure. In its present state of development the instrument is capable of a useful magnification of the order of 100,000 and has a limit of resolution of about 20 A. Since satisfactory 'metallurgical' electron microscopes which would work by reflection have not yet been developed, transmission must be used. It is therefore necessary to prepare thin replicas of the surface which can be studied by transmission, and the preparation of these replicas is frequently troublesome. In order to estimate the height of the surface irregularities, the shadowing method must be used (Williams and Wyckhoff, 1944). The surface of the replica is bombarded with atoms of a heavy metal (such as gold or chromium) from small fragments of the metal heated *in vacuo*. The angle of incidence of the metal atoms is very small and a small projection on the surface therefore throws a long 'shadow'. The height of a surface projection is estimated from the length of the shadow and the angle at which the surface is bombarded. By this means it is possible to estimate the height of a surface asperity of the order of only 30 A.

Some of the later types of electron microscope are fitted with a tilting object stage to enable stereoscopic pairs of electron micrographs to be obtained. Pictures of great beauty have been made in this way, but it is questionable whether this method is quantitatively as accurate as the shadowing technique described above.

Electron microscope pictures of the most carefully polished surfaces will usually reveal surface irregularities or scratches which cannot be detected by the optical microscope but which are very large compared with molecular dimensions. The small hills and pits present on an electrographically polished aluminium surface may be clearly seen in Plate I. 2, where the magnification is 50,000. The height of the little hillocks varies from 100 to 1,000 A. Electron microscope pictures of surface structure and surface damage are given in Chapter IV.

Oblique sectioning

A standard method of studying surface irregularities is to cut a cross-section at right angles to the surface and examine it under the microscope. For small surface irregularities the highest powers must be used and the area of the field is therefore very restricted. Also, the minimum size of irregularity which can be observed is limited by the resolution of the microscope. Provided the surface irregularities are elongated in one direction—this is so, for example, with scratches caused by grinding or abrasion and with the surface damage caused by sliding—both these limitations can, to a considerable degree, be overcome by cutting the section obliquely to the surface. In this way a taper section is obtained with the vertical component of the surface contour greatly magnified relative to the horizontal one. Fig. 2 illustrates diagrammatically the effect of cutting a section at an angle to a surface which contains a V-shaped groove. The ratio of the magnification in the vertical and in the horizontal direction in the final section is the cosecant of the angle α. If the section is cut at an angle $\alpha = 5° 43'$ the ratio of the vertical to the horizontal magnification would be 10:1. It is necessary to protect the surface irregularities from damage by electroplating with a metal of similar hardness. This method of cutting is, of course, not new, but it appears to have first been described by Nelson (1940). It has been further developed by Dr. A. J. W. Moore, and its application to the study of surface damage is described in Chapter IV.

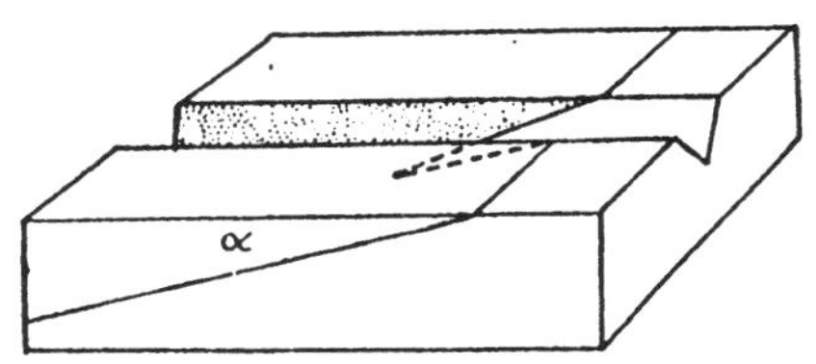

Fig. 2. Sketch showing the principle of the oblique-section technique. For an angle α of about 6°, the vertical magnification is 10 times the horizontal.

Oblique sections prepared in this way by Dr. A. J. W. Moore which show some characteristic contours of surfaces which have been finely scratched, machined, and also lapped with abrasives of varying degrees of fineness are shown in Plate I. 3 *a*, *b*, *c*, and *d*. It is seen that, even with the finest abrasive, the surface irregularities are of the order of 10^{-5} cm. in height.

If the surfaces are polished, the effect, as we shall see in Chapter III, is to cause the summits of the peaks to flow into the valleys so that the contours resemble rolling downs rather than rugged alpine peaks. Again, however, the surfaces will touch on the summits of the hills and the area of intimate contact will be small. This means that even with lightly loaded surfaces, the pressure at the points of contact will be high, and it is these small regions of intimate contact that are responsible for the friction, surface damage, and interaction between the solids. As the load is increased, the irregularities will crush down and distribute the load over a larger area. We may now ask, What are the factors that determine the real area over which the load is supported?

Asperities of spherical shape

We first consider an idealized case. Let us assume that the tips of the asperities are perfectly smooth and spherical in shape. Suppose the

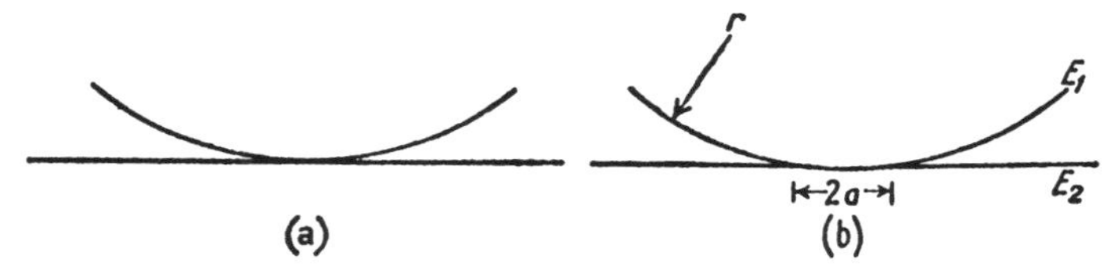

FIG. 3. Contact between a spherical and a flat surface. (*a*) Contact under zero load. (*b*) Contact under a finite load. If the load is not too large the deformation is elastic and the size of the circle of contact is given by the Hertzian equations.

asperity rests on a softer metal and that at the region of contact the surface of the softer metal may be considered to be plane (see Fig. 3 *a*). If the surfaces are pressed together with a load W they will at first deform elastically according to Hertz's classical equations (Hertz, 1886). The region of contact will be bounded by a circle of radius a, where

$$a = 1{\cdot}1\left\{\frac{Wr}{2}\left(\frac{1}{E_1}+\frac{1}{E_2}\right)\right\}^{\frac{1}{3}}, \tag{1}$$

where r is the radius of curvature of the asperity, and E_1, E_2 are Young's moduli for the asperity and the surface respectively (see Fig. 3 *b*). At this stage, therefore, the area of contact $A = \pi a^2$ will be proportional to $W^{\frac{2}{3}}$, whilst the mean pressure over the area of contact $p_m = W/\pi a^2$ will be proportional to $W^{\frac{1}{3}}$. The way in which A and p_m vary with W is shown in Fig. 4. In this range the deformations involved are elastic and are reversible: if the load between the surfaces is removed they return to their original configuration.

As the load W is increased the mean pressure p_m increases until it reaches such a value that at a critical point within the softer material the elastic limit is exceeded. This occurs at the region where the shear stresses are a maximum. The Hertzian analysis shows that this region is situated at a point Z about 0·5a below the centre of the circle of

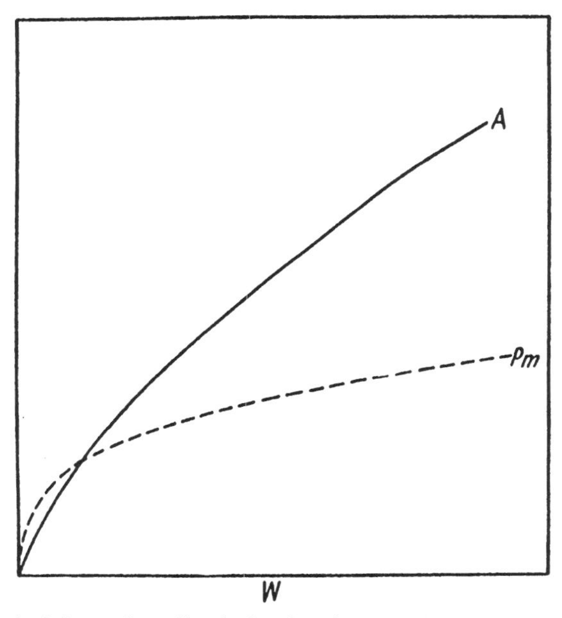

FIG. 4. Elastic deformation. Graph showing the area of contact A and the mean pressure p_m as a function of the load W, for a sphere pressing on a flat surface. A varies as $W^{2/3}$ and p_m as $W^{1/3}$.

contact (Timoshenko, 1934). The elastic limit is just exceeded at this point when

$$p_m = 1{\cdot}1Y, \tag{2}$$

where Y is the elastic limit of the softer metal as found in pure tension (or frictionless compression) experiments. At this stage the metal around Z (Fig. 5a) is plastic and yields irreversibly. The material outside this region has not yet reached the conditions for plasticity and its deformation is still essentially elastic. Consequently, when the load is removed only a very slight amount of residual deformation remains.

If now the load is increased further, the area of contact A and the mean pressure p_m rise in a manner which deviates increasingly from that shown in Fig. 4. The region of plasticity around Z grows rapidly, and a stage is soon reached at which the whole of the material around the region of contact is flowing plastically (Fig. 5*b*). At this stage the theoretical work of Hencky (1923) and of Ishlinsky (1944) shows that

$$p_m = cY, \tag{3}$$

where c has a value of approximately 3.

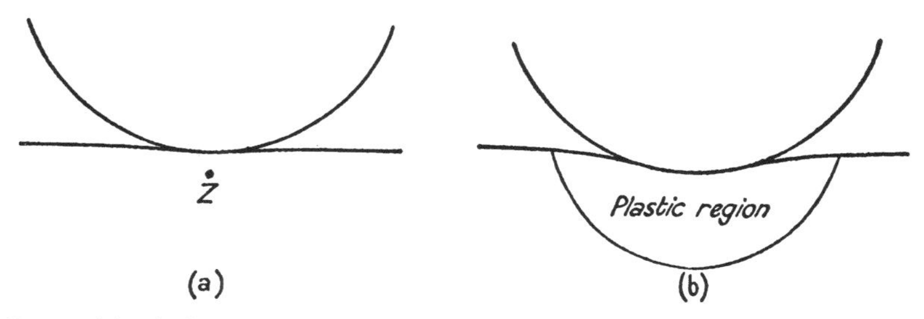

Fig. 5. Plastic deformation of a flat surface by a harder spherical surface. (*a*) The onset of plasticity occurs at the point Z below the surface when the mean pressure $p_m \approx 1{\cdot}1Y$. (*b*) At a later stage the whole of the material around the indentation flows plastically; at this stage $p_m \approx 3Y$.

If the load is still further increased it is found that although the size of the deformed area increases, relation (3) is still valid, provided (i) the deformed area is not too large compared with the size of the specimens, (ii) the elastic limit Y does not increase as a result of the plastic deformation produced, i.e. provided there is no work-hardening. In practice, of course, it is impossible to find a metal that does not work-harden, but a close approximation may be obtained by using metals which have been so highly worked that further deformation produces no appreciable change in their elastic limit.

Some simple experiments along these lines have indeed fully confirmed the general validity of equation (3). A hard steel sphere was pressed into the surface of various metals which had been very highly worked. The mean pressure p_m, which we may call the 'yield pressure', was found to be almost independent of the size of the indentation, and therefore of the load. (There is a slight increase in p_m when the indentations are large, presumably due to the increased confinement of the displaced material (Bishop, Hill, and Mott, 1945), but the effect is small.) Comparison between these values of p_m and the elastic limit of the materials as found in 'frictionless' compression experiments is shown in Table I. It is seen that the constant of proportionality is established for materials ranging in yield pressure from $p_m = 6$ to 100 kg./mm.^{2}

TABLE I

Work-hardened metal	Y kg./mm.²	p_m kg./mm.²	$c = p_m/Y$
Tellurium–lead . .	2·1	6·1	2·9
Copper	31	88	2·8
Steel	65	190	2·8

Reverting to our earlier model, we may now describe graphically the variation of p_m with load for materials which do not work-harden (Fig. 6). The portion OL represents the increase of p_m with W over

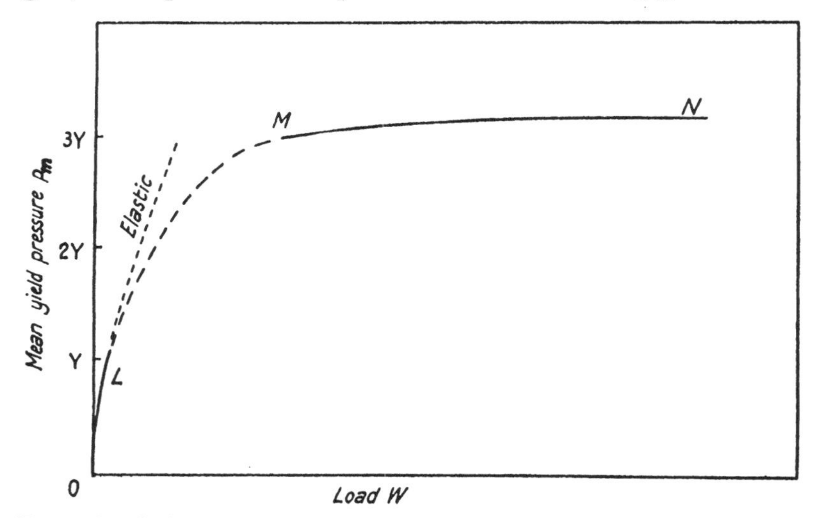

FIG. 6. Graph showing the variation of p_m with W as a hard sphere is pressed into a flat surface. The initial deformation OL is elastic and p_m varies as $W^{\frac{1}{3}}$. Plastic deformation commences at L and continues until the whole of the indentation is flowing plastically. At this stage (MN) the mean pressure p_m is almost independent of W.

the purely elastic range where the deformation is completely reversible. At the point L where p_m reaches a value of about $1{\cdot}1Y$ the onset of plastic deformation commences. There is a gradual increase in p_m and a value is reached at about $p_m = 2{\cdot}8Y$ where 'full' plasticity occurs. The mean pressure is now more or less independent of the force W and follows the curve MN.

If the deformation beyond L still followed the elastic equations the load at which p_m becomes equal to $2{\cdot}8Y$ would be $(2{\cdot}8/1{\cdot}1)^3$, i.e. about 16 times the load at L. That is to say that full plasticity would be reached at a load about 16 times that at which the onset of plasticity occurs. Experiments show, however, that once plastic flow has commenced the pressure increases somewhat more slowly than this and full plasticity occurs at a load between 100 and 200 times that at which the onset of plastic flow commences.

It is interesting to calculate the range of OL for various metals. Assuming that the onset of plasticity occurs where $p_m = 1{\cdot}1Y$, the corresponding load W_L is found from equation (1) to be

$$W_L = 13{\cdot}1 p_m^3 r^2 \left(\frac{1}{E_1} + \frac{1}{E_2}\right)^2. \tag{4}$$

If we assume that the hard spherical asperities have a Young's modulus E_1 of 2.10^{12} dynes/cm.^{2} (the usual value for steel), the onset of plasticity for various metals occurs at loads given in Table II.

TABLE II

Material	Y kg./mm.^{2}	p_m at L kg./mm.^{2}	E_2 dynes/cm.^{2}	$\left(\frac{1}{E_1}+\frac{1}{E_2}\right)^2$	W_L gm.		
					$r = 10^{-4}$ cm.	$r = 10^{-2}$ cm.	$r = 1$ cm.
Tellurium-lead	2·1	2·3	$0{\cdot}16 \times 10^{12}$	$45{\cdot}5 \times 10^{-24}$	8×10^{-8}	8×10^{-4}	8
Commercial copper	20	22	$1{\cdot}2 \times 10^{12}$	$1{\cdot}78 \times 10^{-24}$	$2{\cdot}5 \times 10^{-6}$	$2{\cdot}5 \times 10^{-2}$	250
Highly worked copper	31	34	$1{\cdot}2 \times 10^{12}$	$1{\cdot}78 \times 10^{-24}$	$9{\cdot}1 \times 10^{-6}$	$9{\cdot}1 \times 10^{-2}$	910
Mild steel	65	71	2×10^{12}	1×10^{-24}	$4{\cdot}7 \times 10^{-5}$	0·47	4,700
Alloy steel	200	220	2×10^{12}	1×10^{-24}	$1{\cdot}4 \times 10^{-3}$	14	$14 . 10^4$

It is clear from these results that for radii of curvature of the order of $r = 10^{-4}$ cm. the load required to bring the deformation up to the onset of plastic flow is extremely small. For example, even with tool steel, the load necessary to initiate plastic deformation is of the order of 10^{-3} gm., whilst 'full' plasticity will be reached at a load less than 0·1 gm. It follows that when metal surfaces are in contact, even under quite small loads, the material around the finer surface asperities is subjected to stresses that readily exceed the elastic limit. Indeed, in most cases the material around the tips of the asperities will be deformed well beyond this range and will be subjected to 'full' plasticity, the mean pressure over the deformed areas being given by the relation $p_m = 3Y$. If the asperities are softer than the surface against which they press, so that the asperities themselves are deformed plastically, similar considerations apply. For metals which do not work-harden Y is a constant, so that p_m is a constant. Consequently the area A over which plastic flow occurs is directly proportional to the load W and inversely proportional to the yield pressure.

Effect of work-hardening

We shall now consider the deformation produced by spherical surfaces when Y increases with the deformation. We again consider the simple

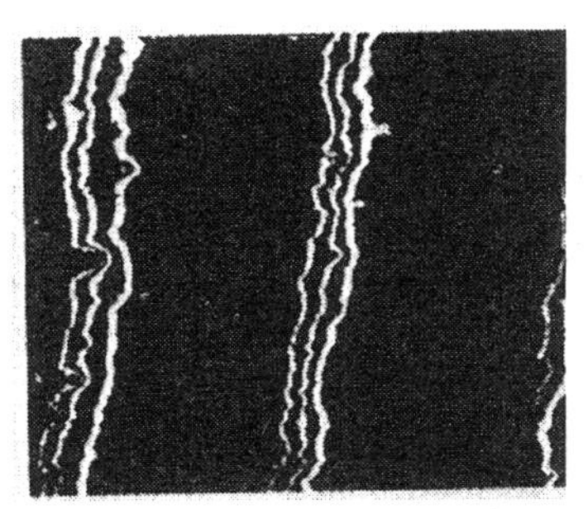

Fig. 1. Fizeau interference fringes sharpened by multiple reflection between optical flat and commercial plate glass. Irregularities about 100 Å high.

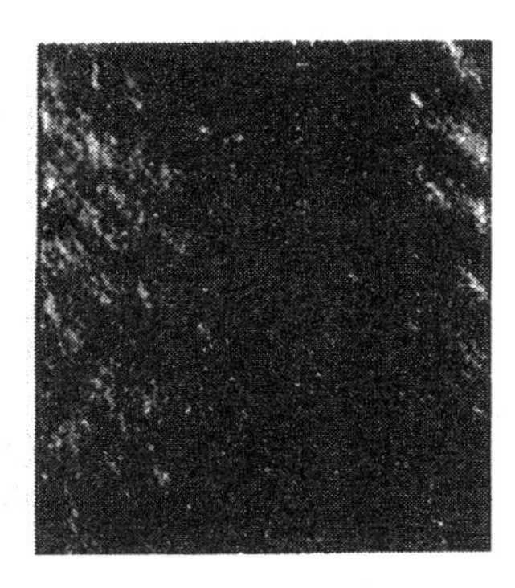

Fig. 2. Electron microscope photograph of electrolytically polished aluminium ($\times$ 50,000). Surface irregularities between 100 and 1000 Å.

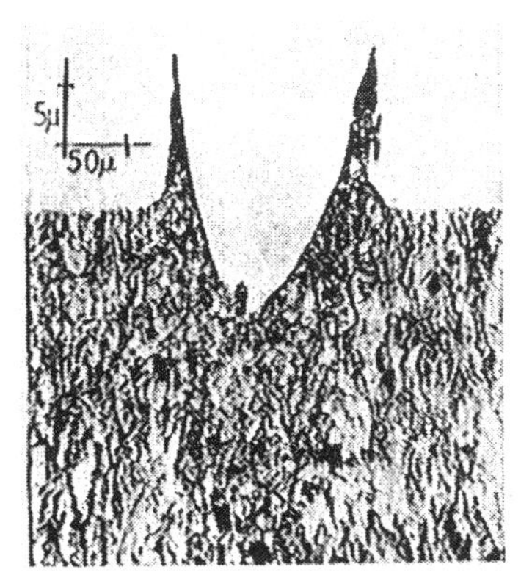

Fig. 3 *a*. Taper section of pin scratch in copper surface. Note built-up edge and work hardening below the scratch.

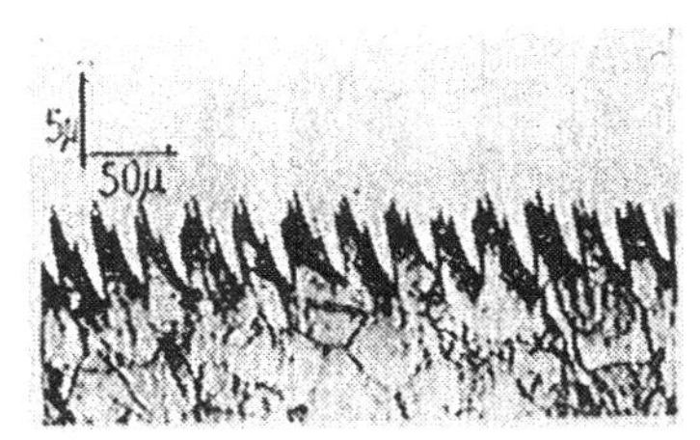

Fig. 3 *b*. Taper section of finely turned copper surface. Irregularities are about 5.10^{-4} cm. or 2.10^{-4} in. high.

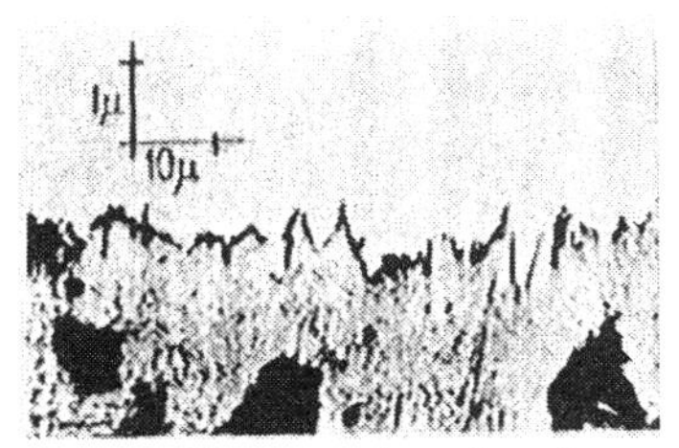

Fig. 3 *c*. Taper section of steel surface lapped with 150 grade carborundum paper. Irregularities are about 5.10^{-5} cm. high.

Fig. 3 *d*. Taper section of steel surface polished in 600 grade carborundum paper. Irregularities are about 10^{-5} cm. high.

PLATE II

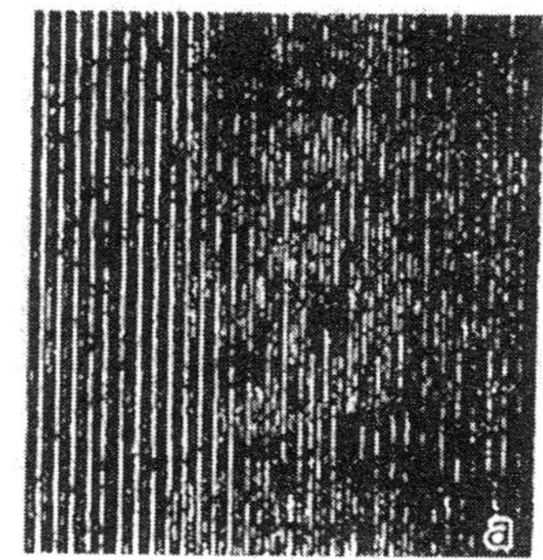

FIG. *a*. Central portion of deep cylindrical indentation in annealed copper showing persistence of grooves in surface.

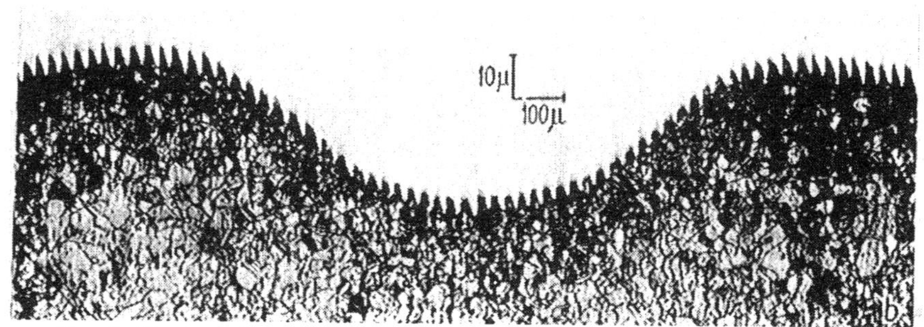

FIG. *b*. Taper section of deep indentation in annealed grooved copper surface showing persistence of grooves after extensive bulk deformation.

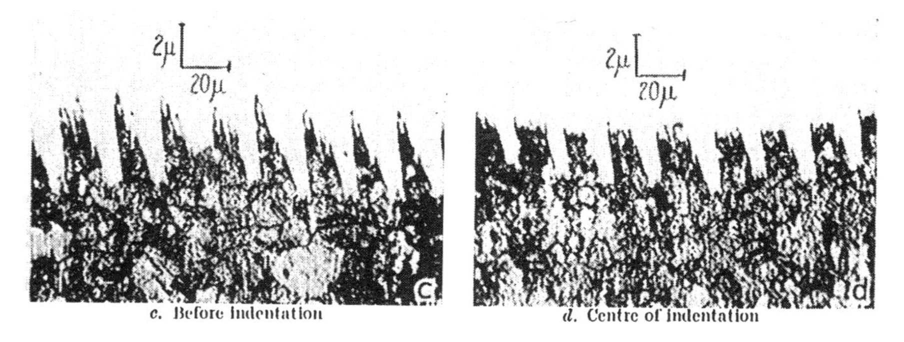

c. Before indentation

d. Centre of indentation

FIG. *c and d*. Comparison of grooves before and after indentation. The work hardening of the asperities increases their yield pressure considerably above that of the underlying bulk material.

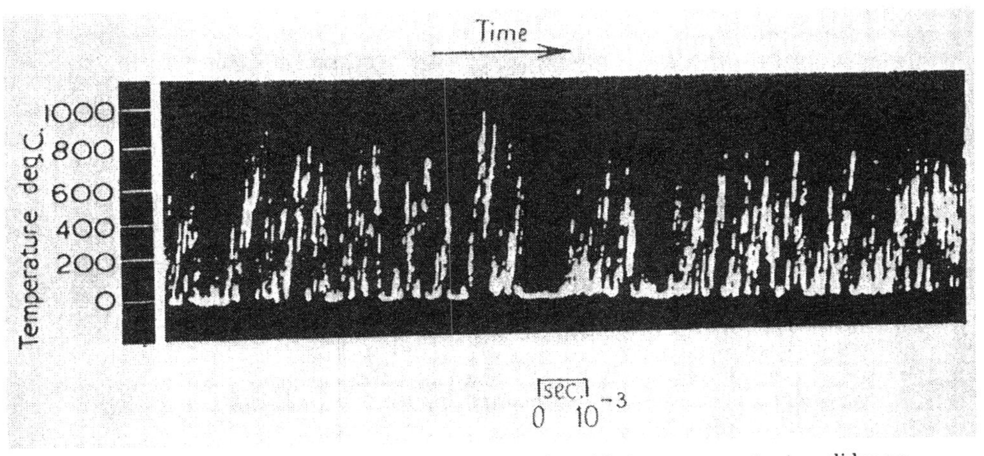

FIG. 1. Cathode ray trace of thermal p.d. developed between a constantan slider on a lapped steel surface. Load 500 gm. Speed of sliding 300 cm./sec. The temperature flashes exceed 700° C. and their duration is less than 10^{-4} sec.

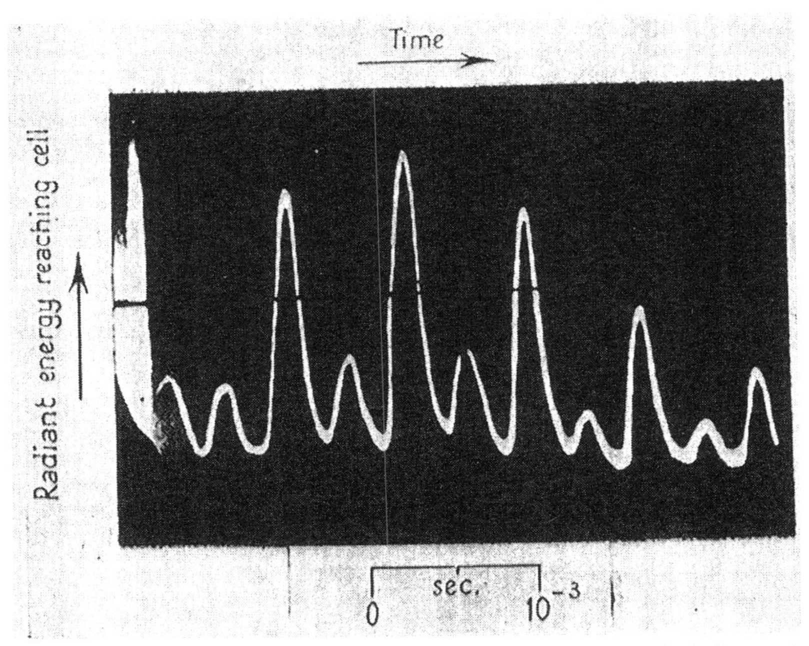

FIG. 2. Cathode ray trace of radiant energy received from a single hot spot developed between a steel slider and a rotating glass disk. Load 350 gm. Speed of sliding 700 cm./sec. The radiant energy is interrupted by a chopper in which alternate segments are covered with a filter. The ratio of the filtered to unfiltered peaks on the trace provides a measure of the temperature. The maximum temperature of the hot spot is about 900° C. and it reaches this value in about 2×10^{-3} sec.

PLATE IV

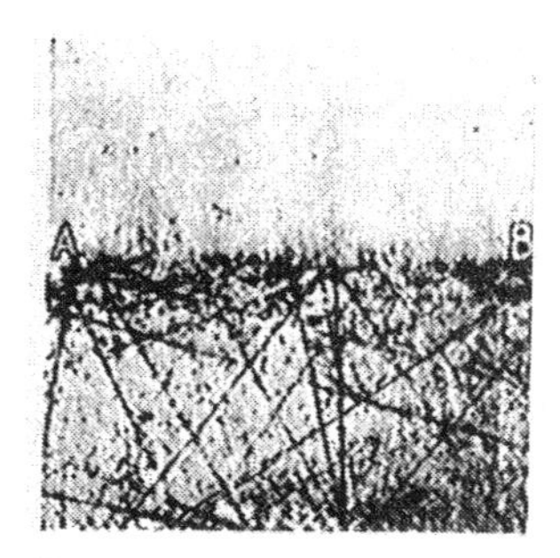

FIG. 1. Mechanically polished copper when etched (below *AB*) reveals original scratches ($\times 525$).

a. Start   *b.* 1 hour   *c.* 2 hours

FIG. 2. Polishing of Wood's alloy (M.P. 69° C.) on camphor (M.P. 178° C.). Surface flow occurs ($\times 160$).

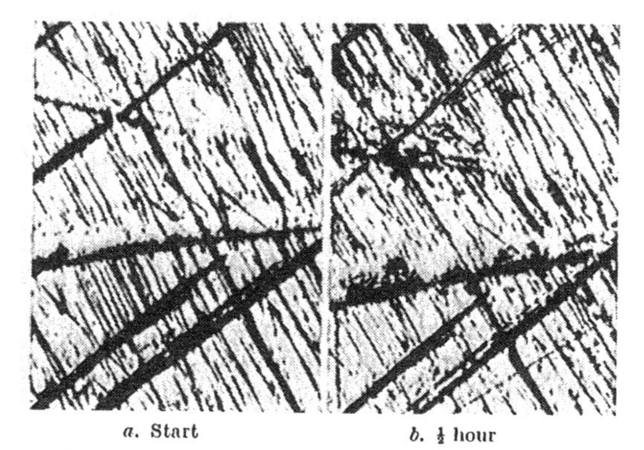

a. Start   *b.* ½ hour

FIG. 3. Rubbing speculum metal (M.P. 745° C.) on oxamide (M.P. 417° C.). No surface flow occurs ($\times 160$).

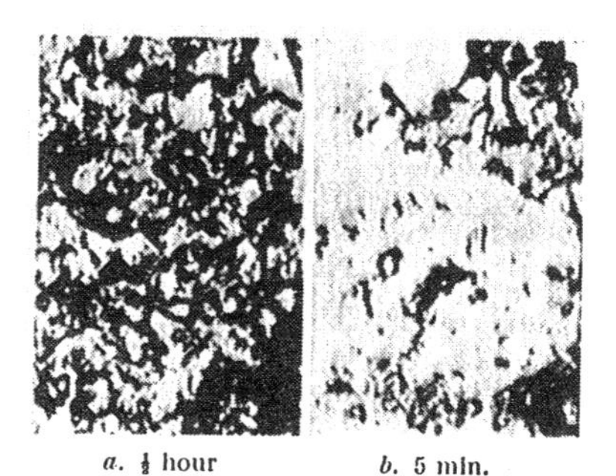

a. ½ hour   *b.* 5 min.

FIG. 4. *a.* Rubbing calcite (M.P. 1333° C.) on cuprous oxide (M.P. 1235° C.) gives inappreciable surface flow: *b.* on zinc oxide (M.P. 1800° C.) gives rapid surface flow ($\times 235$).

a. General view of friction apparatus

b. Detailed view of surfaces

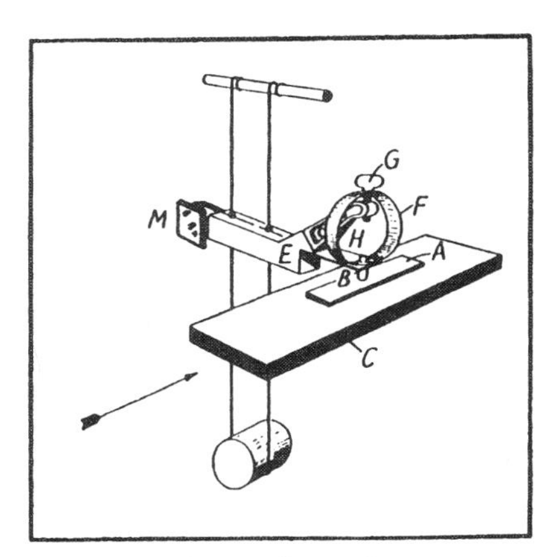

c. Diagram of apparatus

FIG. 1. Bowden-Leben apparatus for measuring friction between sliding surfaces, either dry or in the presence of boundary films.

A, lower surface; *B*, upper surface or slider; *C*, carriage holding lower surface and heating element; *D*, rails; *E*, duralumin arm holding upper surface; *F* and *G*, spring and screw for applying load; *H*, stiffening spring; *K*, massive supports for bifilar suspension; *L*, pulleys and hook for determining normal load; *M*, mirror responding to deflexion of bifilar suspension; *N*, surface thermocouple; *P*, galvanometer for determining surface temperature. Recording camera not shown.

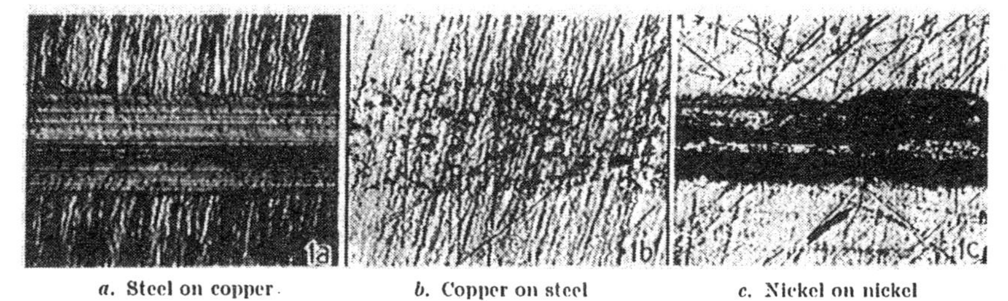

a. Steel on copper *b.* Copper on steel *c.* Nickel on nickel

Fig. 1. Three types of tracks formed during unlubricated sliding (*a*) hard on soft; (*b*) soft on hard; (*c*) similar metals (×31).

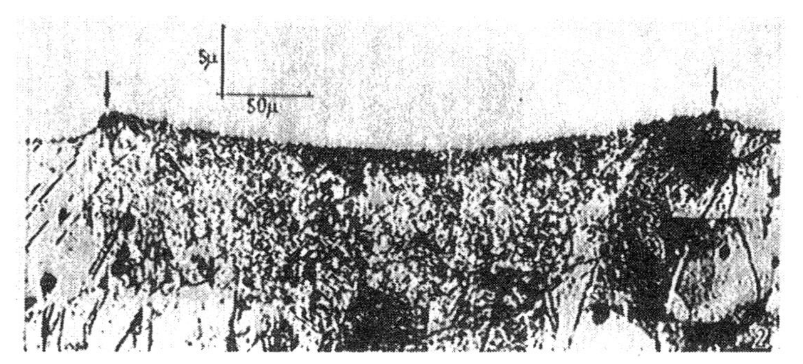

Fig. 2. Taper section of track formed on cold rolled copper when traversed once by a hard hemispherical steel slider (unlubricated). Track width indicated by arrows. The track appears to be relatively smooth but there is heavy deformation below the surface of the track.

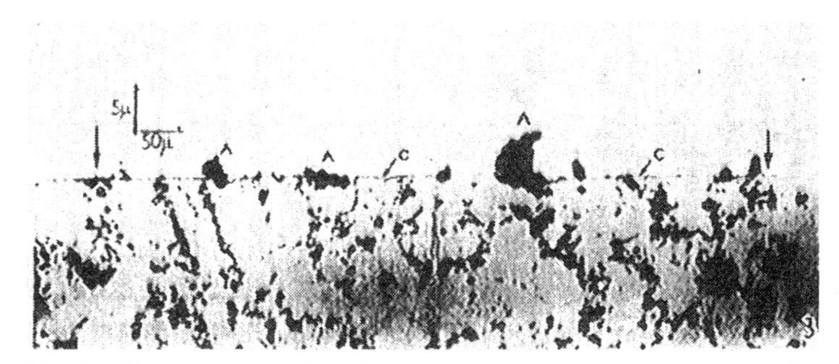

Fig. 3. Taper section of track formed on steel when traversed by a hemispherical copper slider (unlubricated). Track width indicated by arrows. Large particles of copper (*A*) are left adhering to the steel surface showing that when the copper-steel junctions are sheared during sliding the break usually occurs within the bulk of the copper. In some cases, however, small fragments of steel may be plucked out of the steel surface (*C*).

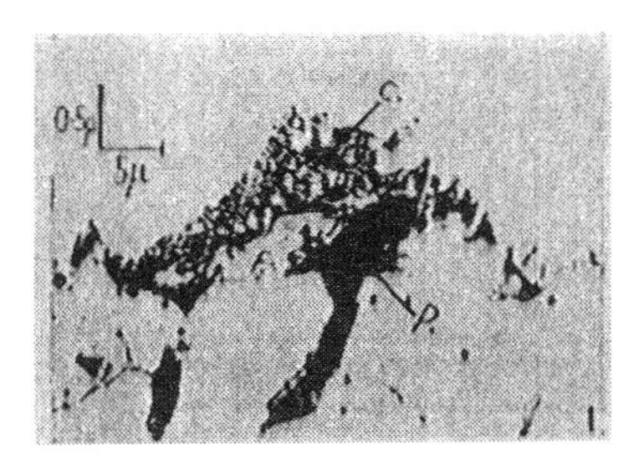

Fig. 1. Copper pick-up on steel in greater detail. Note the rupture and penetration of the steel surface. F = ferrite, P = pearlite, C = copper.

Fig. 2. Welding of copper fragment on steel. The junction is so strong that it has pulled the steel above its original level.

Fig. 3. Taper section of copper surface after a hemispherical copper slider has traversed the surface once (unlubricated). The surface is heavily torn and there is severe distortion below the track.

Fig. 4. Taper section of copper surface after a hemispherical copper slider has passed over it eleven times (unlubricated). A particle of copper has become deeply embedded in the surface.

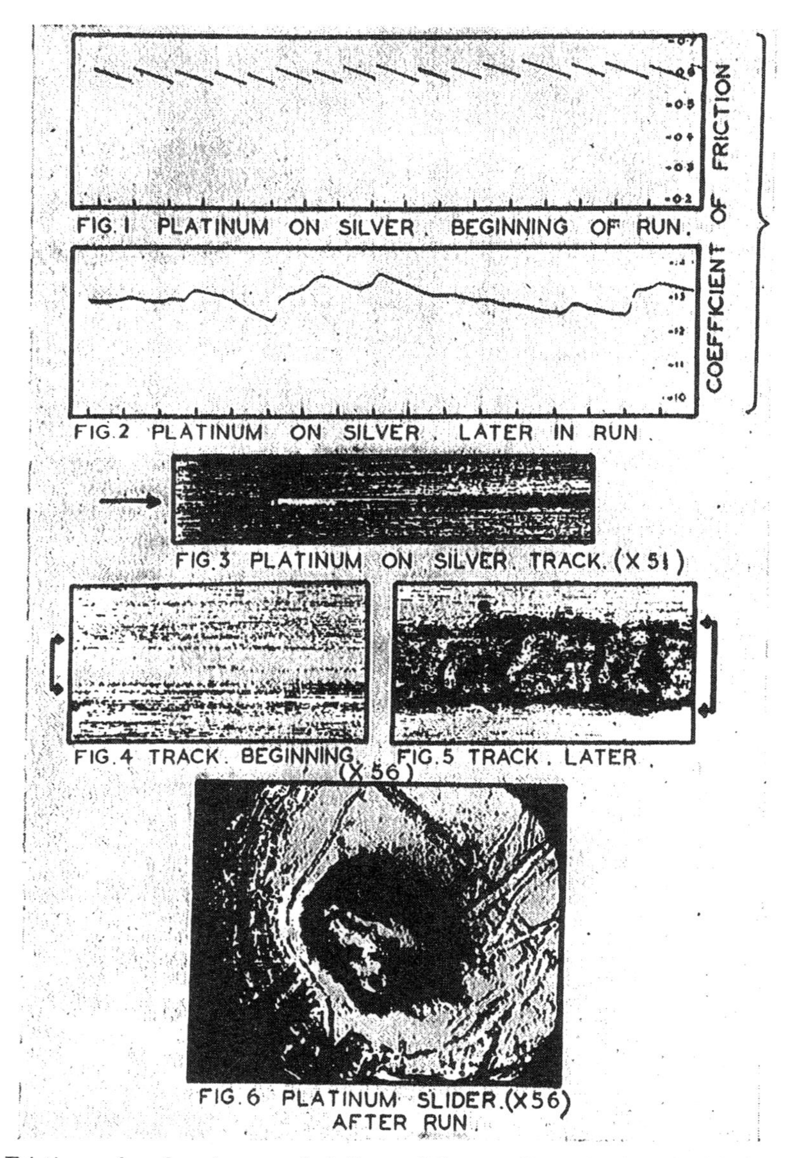

Friction and surface damage of platinum sliding on silver. At first the friction and track correspond to those characteristic of dissimilar metals. After a short time both change over to the behaviour characteristic of similar metals. Silver has been picked up from the lower surface and has formed a welded blob at the tip of the platinum slider.

FIG. 1. Taper section of track formed by sliding a hemispherical steel slider on an unlubricated steel surface. Width of track indicated by arrows. The tearing and plucking of the surface are very marked at some regions. At other regions the surface is scarcely damaged.

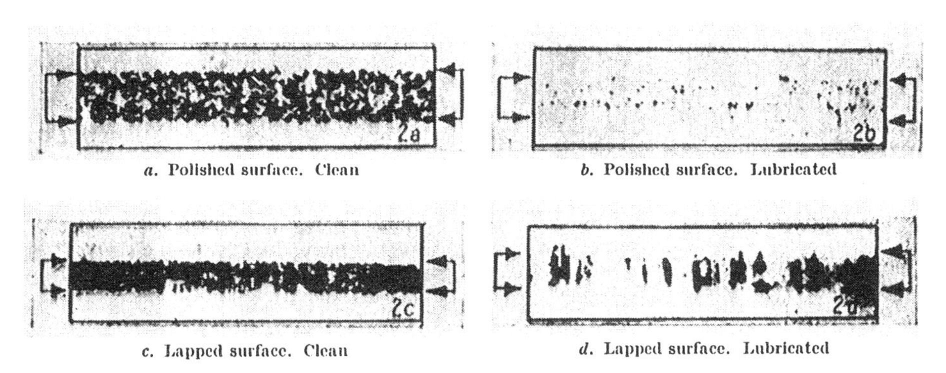

a. Polished surface. Clean

b. Polished surface. Lubricated

c. Lapped surface. Clean

d. Lapped surface. Lubricated

FIG. 2. Electrographic surface analysis showing distribution of copper adhering to a steel surface after traversing the steel surface once (×8). Clean and lubricated surfaces. Although the metallic interaction is greatly reduced by the presence of the lubricant, it is still appreciable.

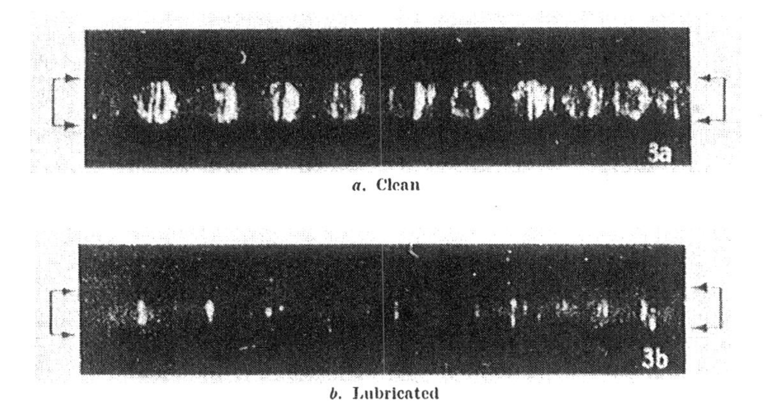

a. Clean

b. Lubricated

FIG. 3. Radioactive autograph of track formed by lead sliding on steel (×8). The lead slider is radioactive and the lead pick-up on the steel surface produces a pattern on the photographic plate. The lubricant reduces the amount of metallic interaction but does not eliminate it entirely.

PLATE X

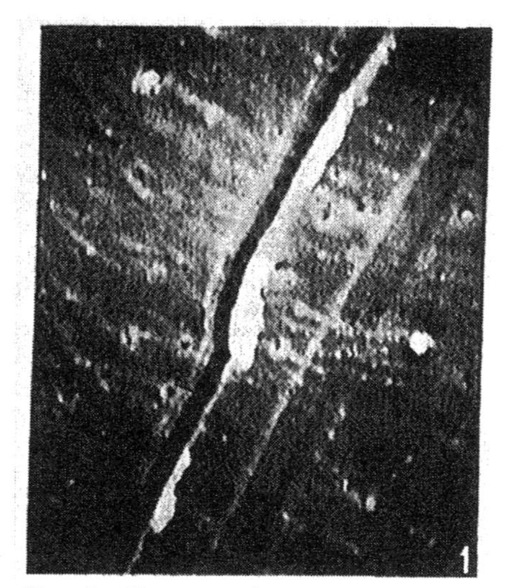

FIG. 1. Electron micrograph of track formed by copper sliding on copper. Clean. Formvar replica ($\times$ 7,000). Load 0·64 gm. μ = 0·5.

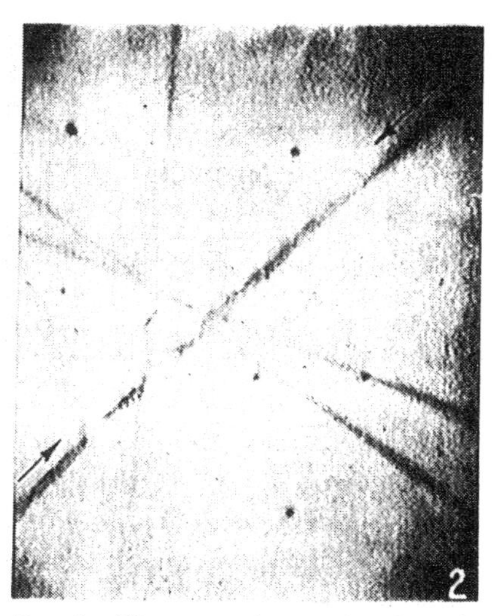

FIG. 2. Electron micrograph of track formed by sapphire sliding on sapphire. Clean. Formvar replica ($\times$ 7,000). Load 0·64 gm. μ = 0·25.

FIG. 3. Electron micrograph of track formed by steel sliding on aluminium. Clean. Oxide replica ($\times$ 14,000). Load 0·34 gm. μ = 1.

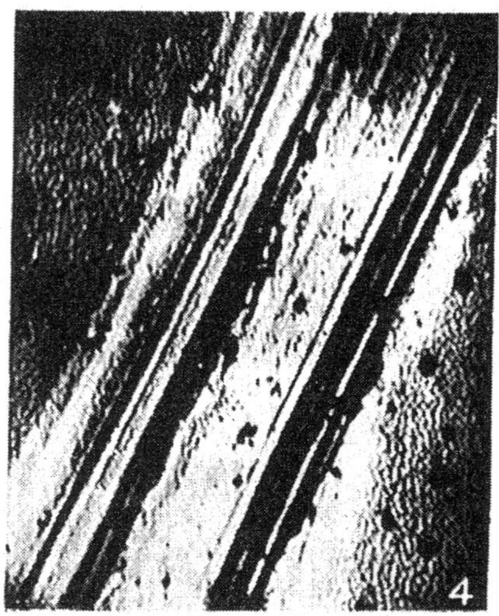

FIG. 4. Electron micrograph of track formed by sliding steel on aluminium. Lubricated. Oxide replica ($\times$ 6,500). Load 0·34 gm. μ = 0·3.

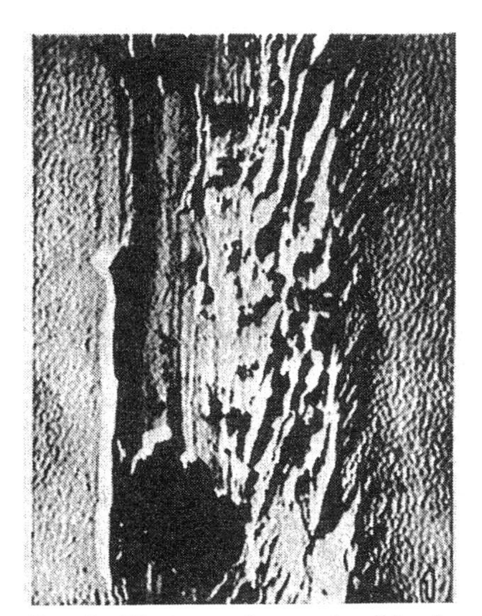

FIG. 1. Electron micrograph of track formed by sliding steel on aluminium. Clean. Shadowed oxide replica (× 8,500). Load 0·34 gm. $\mu \approx 1$.

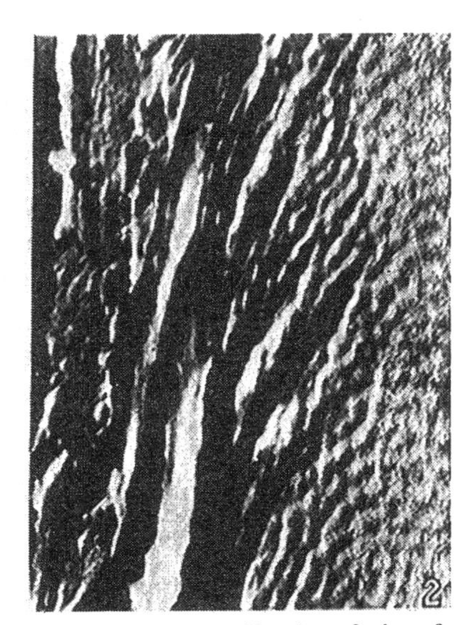

FIG. 2. Larger magnification of edge of track in 1 (× 25,000). The slip lines at the edge of track are between 100 and 1000 Å apart.

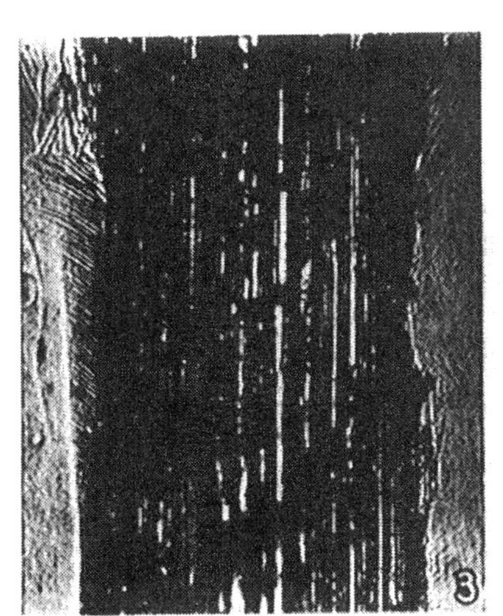

FIG. 3. Optical micrograph of track formed by steel sliding on aluminium. Clean (× 500). Load 20 gm., $\mu \approx 1$. Note heavy surface damage and slip lines at edge of track.

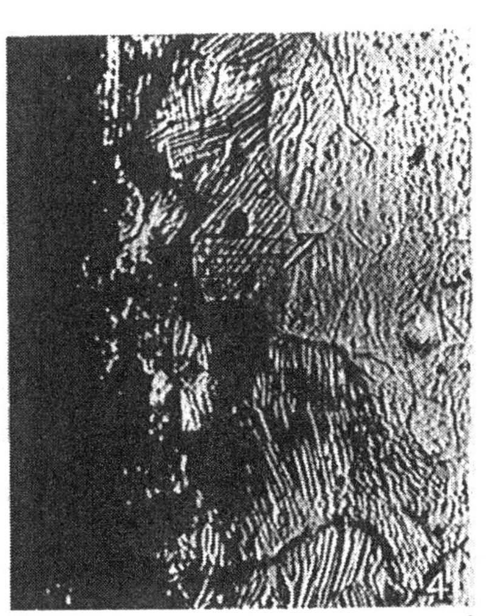

Fig. 4. Edge of track formed by steel sliding on aluminium. Clean (× 500). Load 2 kg., $\mu \approx 1$. The slip lines show that the deformation extends beyond the boundaries of the track itself.

PLATE XII

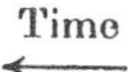

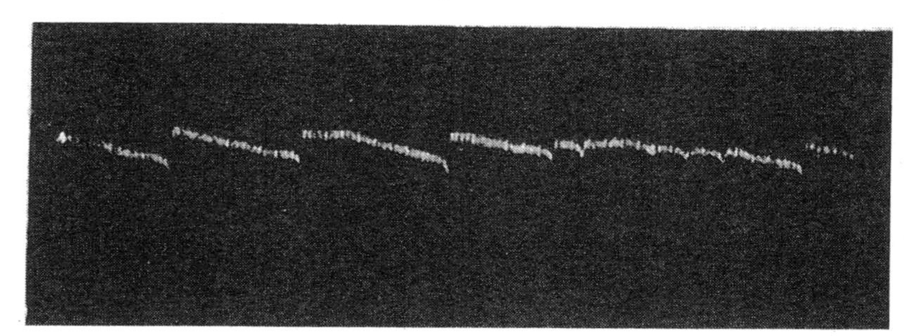

Type 1. Constantan on steel

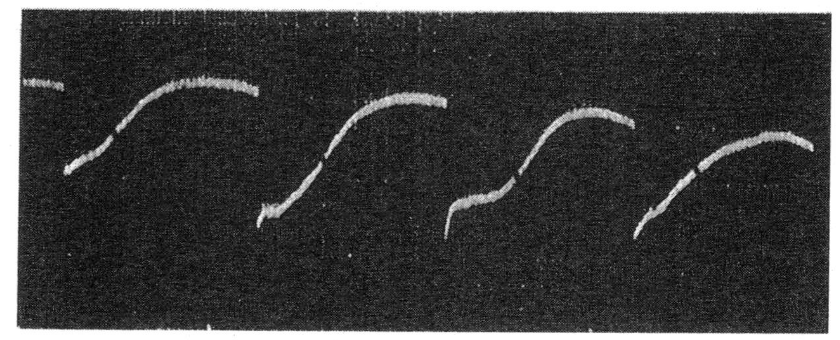

Type 2. Tin on steel

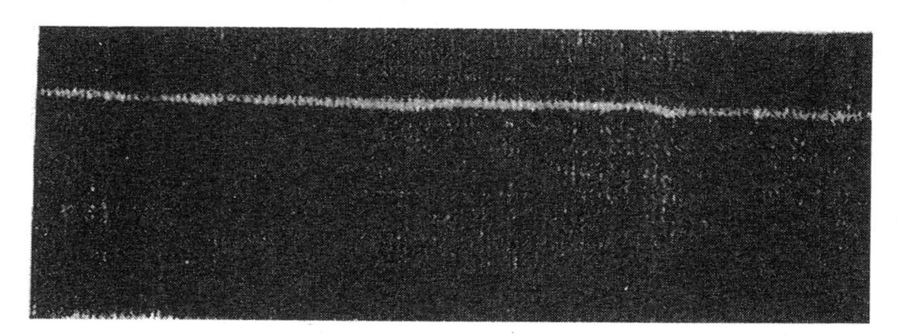

Type 3. Platinum on platinum

Friction and electrical conductance traces for three types of sliding. Dark trace, friction. Light trace, electrical conductance. Type 1, hard metal on soft. Type 2, soft metal on hard. Type 3, similar metals.

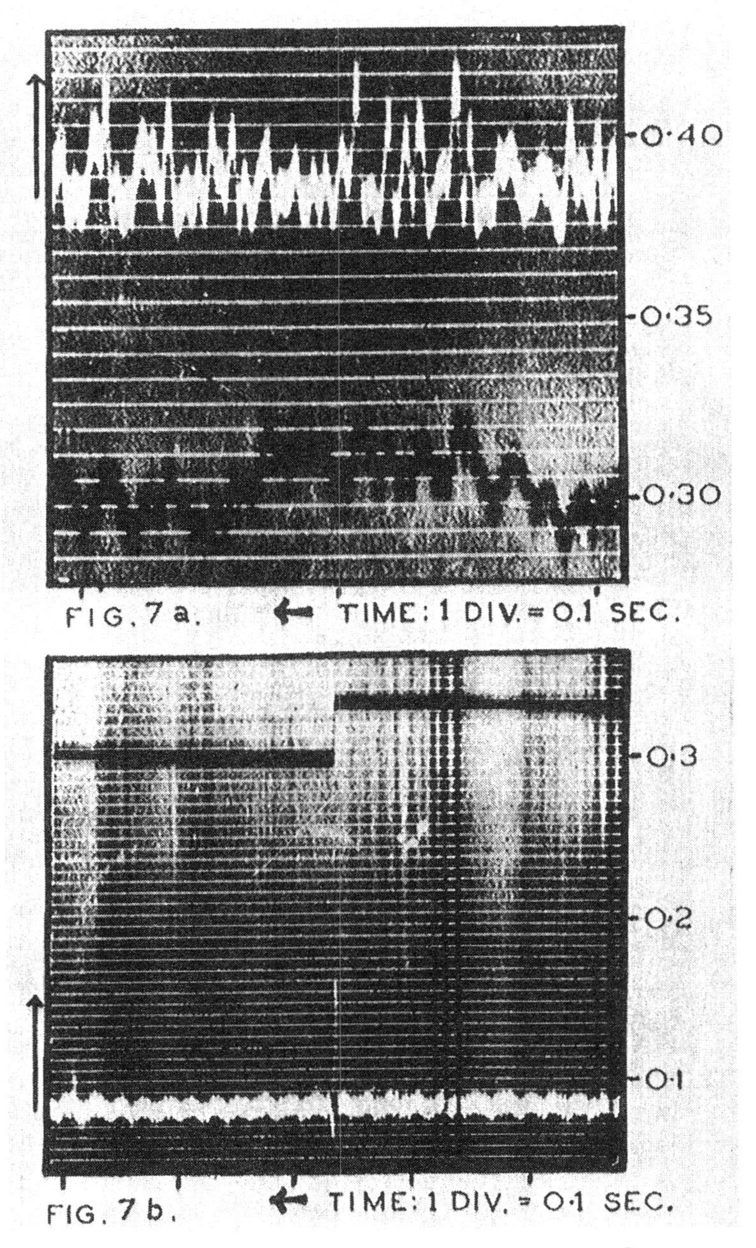

Thermoelectric potential developed between sliding surfaces. Constantan on steel. Light trace, thermal p.d. Dark trace, friction. The surface temperature fluctuates with the friction.

Fig. 1. Frictional behaviour of indium film on steel after repeated traversals of the same track. After seven runs over the same track the friction begins to increase and this becomes very marked by the twentieth run.

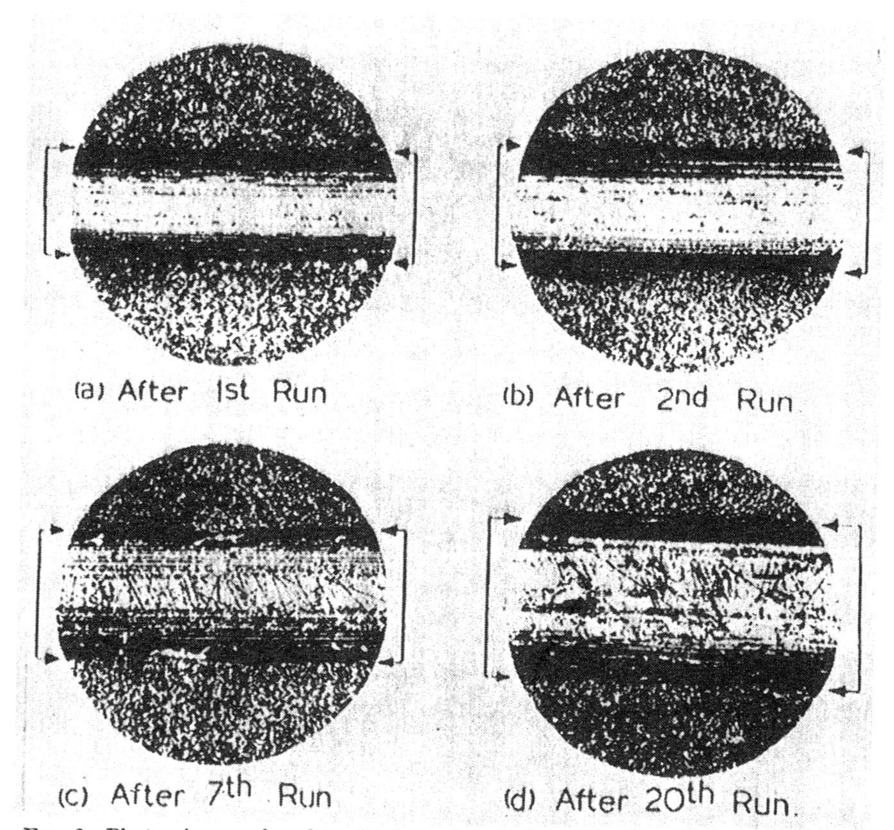

Fig. 2. Photomicrographs of tracks formed on indium film deposited on steel after repeated traversals of the same track ($\times 27$). The wearing away of the indium film leads to increased exposure of the underlying steel surface. This is accompanied by a corresponding increase in friction.

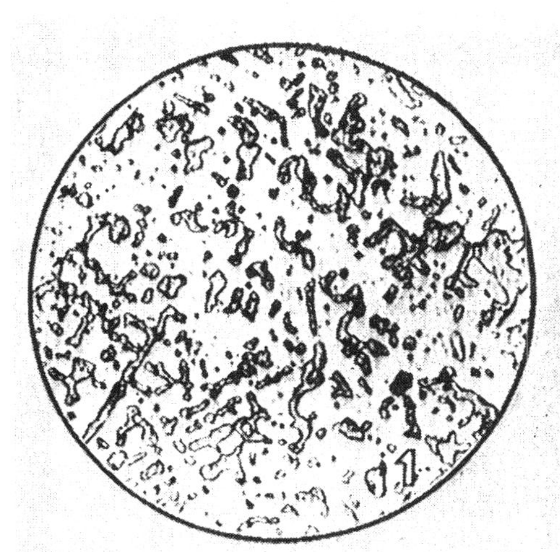

FIG. 1. Non-dendritic copper-lead alloy. The lead (dark phase) is distributed as discrete particles within the harder copper matrix ($\times$ 208).

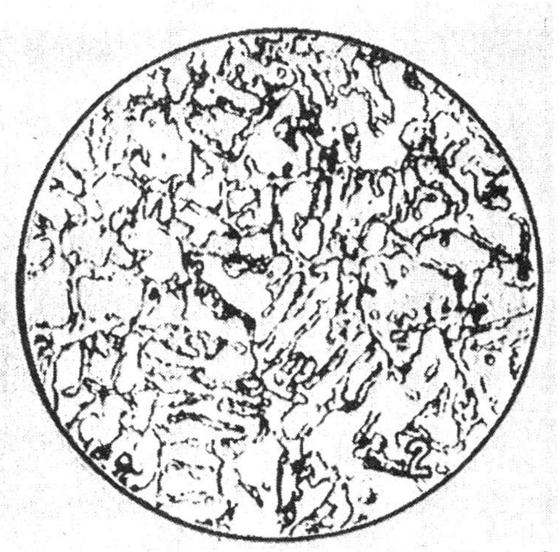

FIG. 2. Dendritic copper-lead alloy. The lead (dark phase) is distributed as a continuous network between the copper dendrites ($\times$ 208).

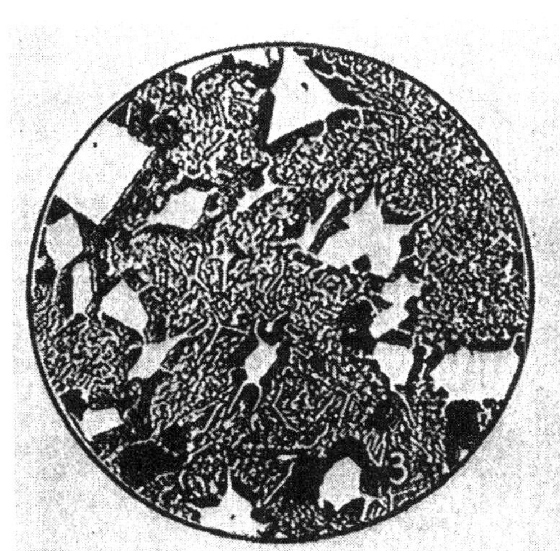

FIG. 3. Lead-base bearing alloy. The matrix is a lead-tin-antimony eutectic. The hard cuboids are a tin-antimony compound and each cuboid is surrounded with a lead-rich solid solution ($\times$ 250).

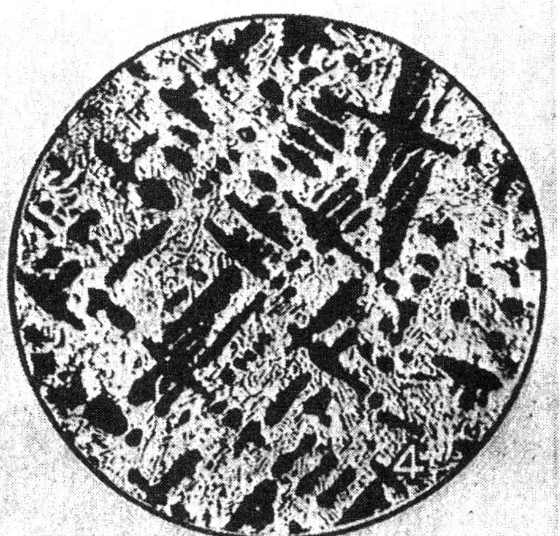

FIG. 4. Lead-base 'matrix' alloy. Dendrites of lead-rich solid solution are set in a background of a lead-tin-antimony eutectic. There are no hard particles and the alloy is similar to the matrix material of the bearing alloy shown in 3 ($\times$ 250).

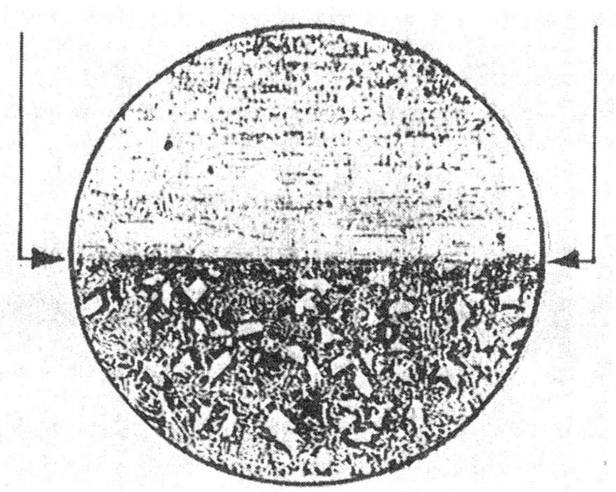

FIG. 1 *a*. Portion of track formed in etched lead-base bearing alloy when a steel slider passes over the surface once. The hard crystallites have been completely obliterated ($\times 90$).

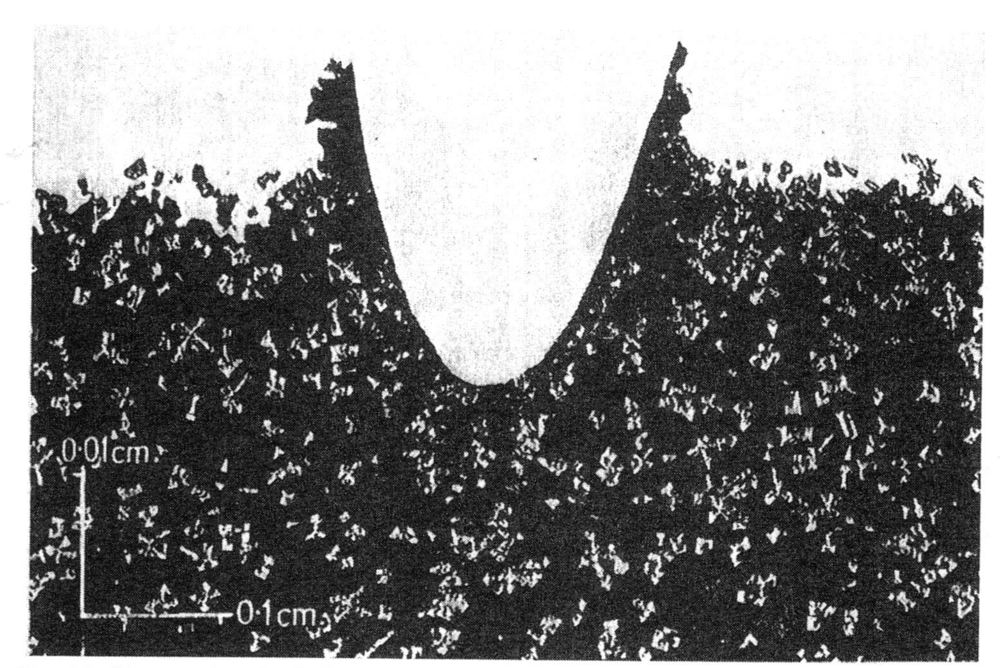

FIG. 1 *b*. Taper section of a similar lead-base bearing alloy after a steel slider has passed once over the surface. The hard crystallites have been pushed under the surface and the matrix material smeared over the surface of the track. (Horizontal magnification $\times 18$; vertical magnification $\times 180$.)

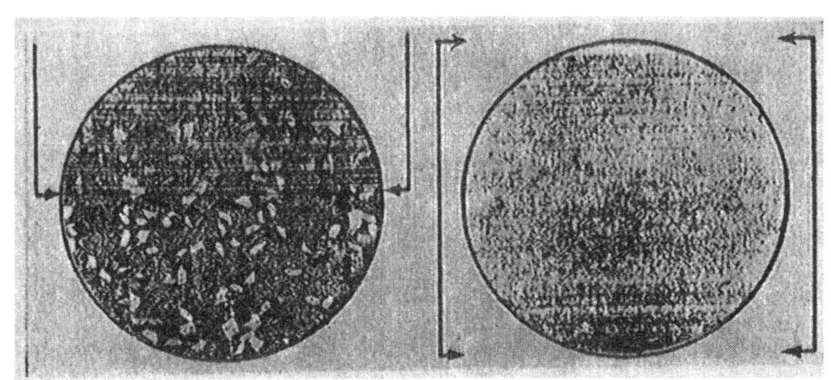

FIG. 1. Lead-base bearing alloy (lubricated). Portion of track formed by steel slider during first run. Partial smearing of matrix over hard particles (× 66).

FIG. 2. Lead-base bearing alloy (lubricated. Central portion of track, hundredth run. Almost complete obliteration of hard particles by smeared matrix material (× 66).

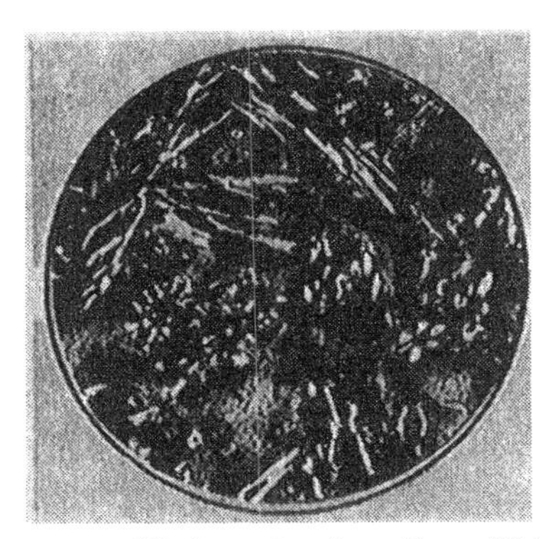

FIG. 3. Tin-base bearing alloy. This consists of a tin-rich solid solution (the matrix) in which are embedded the hard acicular crystallites of copper-tin compound (× 180).

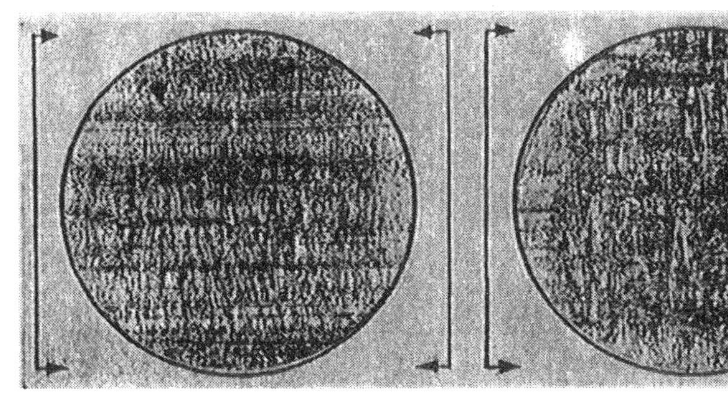

FIG. 4. Track formed on steel by tin-base bearing alloy after traversing the surface 400 times (unlubricated). The lapping marks in the steel surfaces are still visible (× 66).

FIG. 5. Track formed on steel by lead-base bearing alloy after traversing the surface 400 times (unlubricated). The smearing of the alloy is very heavy and completely obscures the lapping marks in the steel surface (× 66).

PLATE XVIII

Fig. 1. Lead-base alloy after 100 cycles. Note marked cracking around the bond *BB* but not in bulk of alloy.

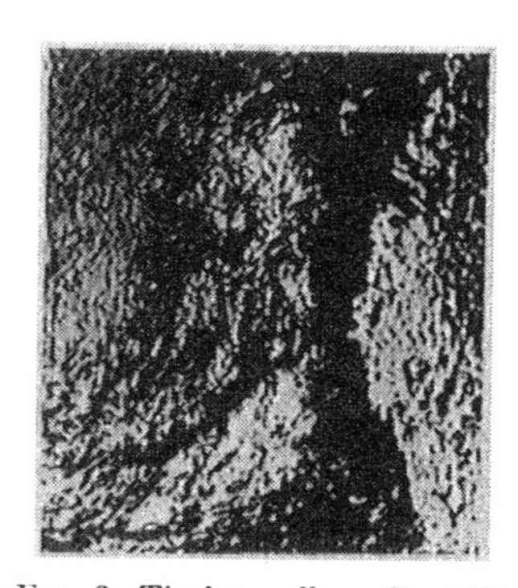

Fig. 2. Tin-base alloy after 100 cycles. Note marked deformation in bulk of alloy.

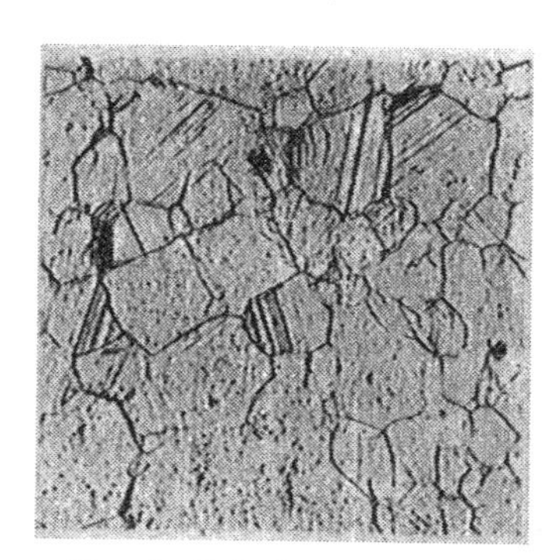

Fig. 3. Cadmium 50 cycles.

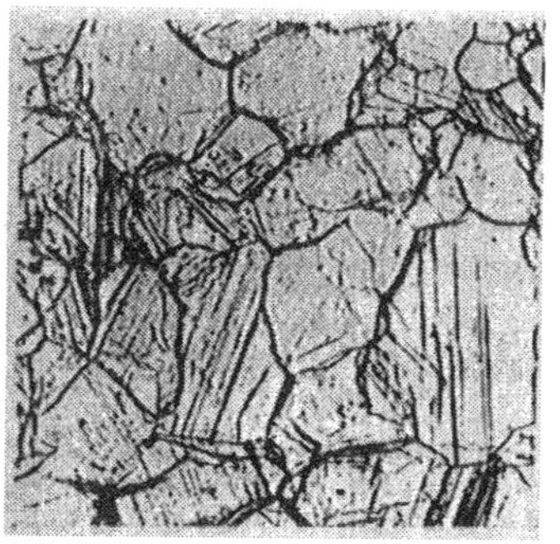

Fig. 4. Cadmium 200 cycles.

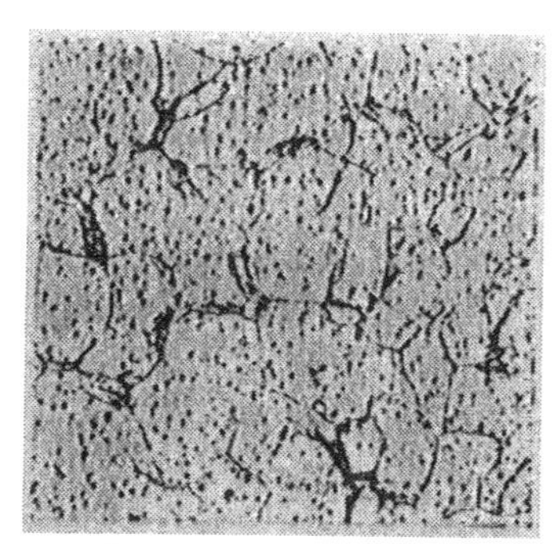

Fig. 5. Tin 50 cycles.

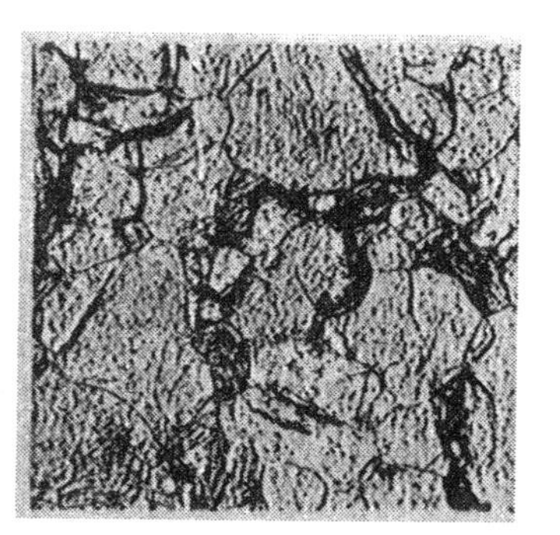

Fig. 6. Tin 200 cycles.

Deformation produced in metals and alloys by cyclic heating and cooling between 30 and 150° C. (all micrographs $\times 100$).

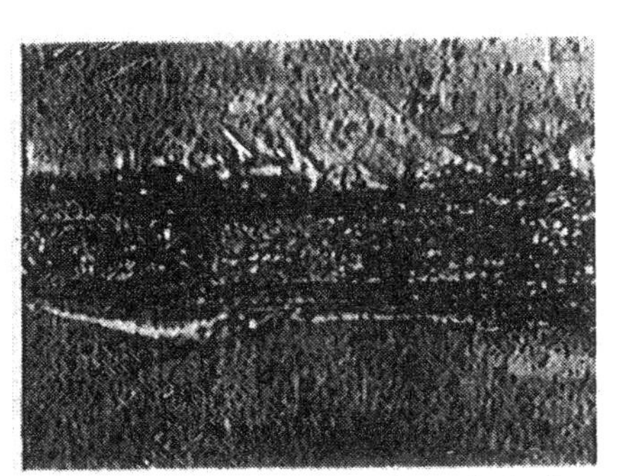

1. Track formed on electrolytically polished copper when traversed once by a copper slider. Load 16 gm., $\mu = 1{\cdot}5$. The surface is heavily torn ($\times 820$).

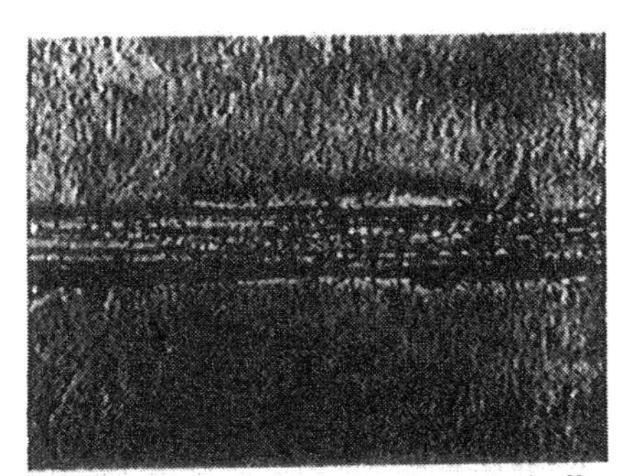

2. Track formed on electrolytically polished copper when traversed once by a copper slider. Load 3 gm., $\mu = 1$. There is still appreciable metallic seizure and tearing ($\times 820$).

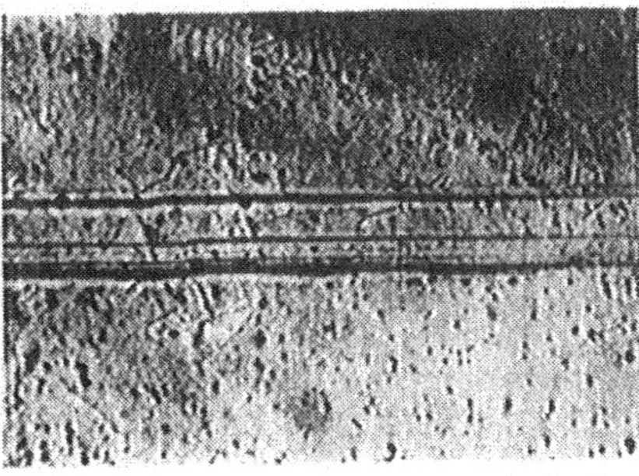

3. Track formed on electrolytically polished copper when traversed once by a copper slider. Load 1 gm., $\mu = 0{\cdot}8$. The surface oxide film is penetrated only slightly and there is little metallic seizure ($\times 820$).

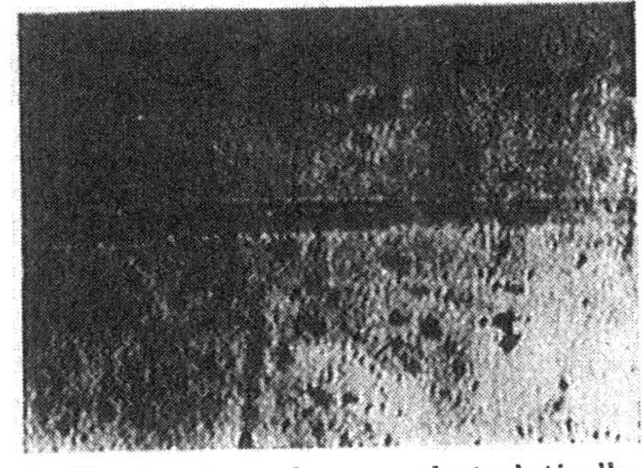

4. Track formed on electrolytically polished copper when traversed once by a copper slider. Load 0·15 gm., $\mu = 0{\cdot}5$. The surface oxide film has not been penetrated and the friction and surface damage are characteristic of the oxide, rather than of the metal ($\times 820$).

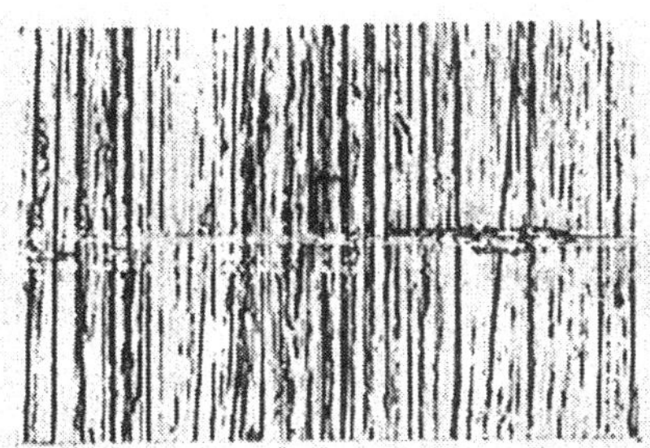

5. Track formed on platinum surface when traversed once by a platinum slider ($\times 110$). Surfaces immersed in N/10 sulphuric acid. Potential 1 volt (hydrogen scale) giving minimum friction (see Fig. 60). Light surface damage.

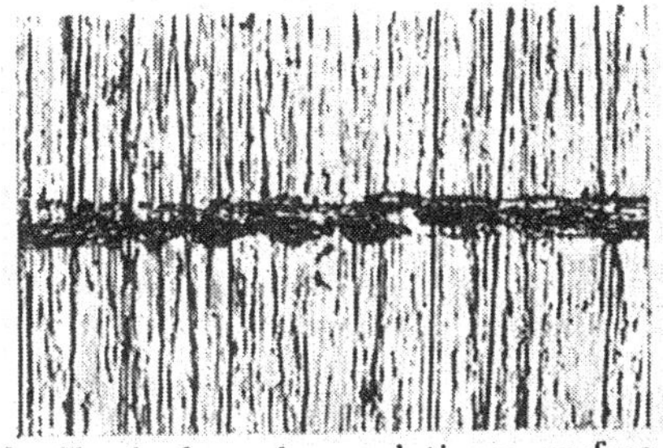

6. Track formed on platinum surface when traversed once by a platinum slider ($\times 110$). Surfaces immersed in N/10 sulphuric acid. Potential 0·3 volts (hydrogen scale) giving maximum friction (see Fig. 60). Heavy surface damage.

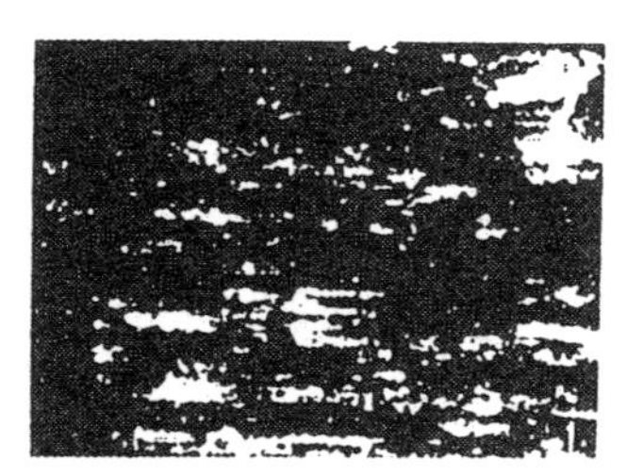

FIG. 1. Glass surface after titanium slider has passed over the surface 3 times (× 130). The smearing and transfer of metal on to the glass is very marked.

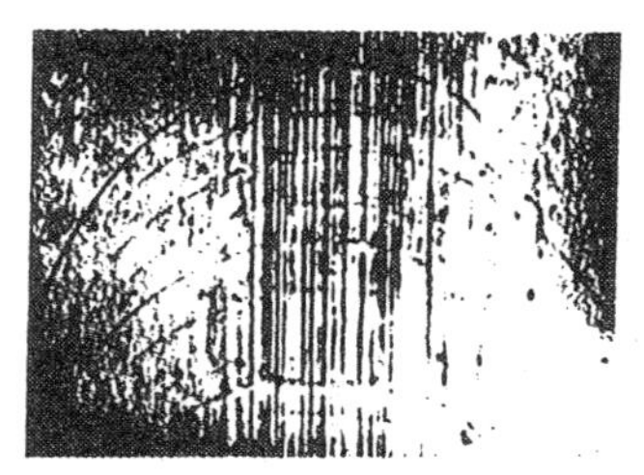

FIG. 2. Glass surface after tungsten carbide slider has passed over the surface once (× 130). Load 4 kg. Direction of sliding from top to bottom. The cracks produced in the surface show the extent of the stresses.

3 *a*. Zinc surface, room temperature

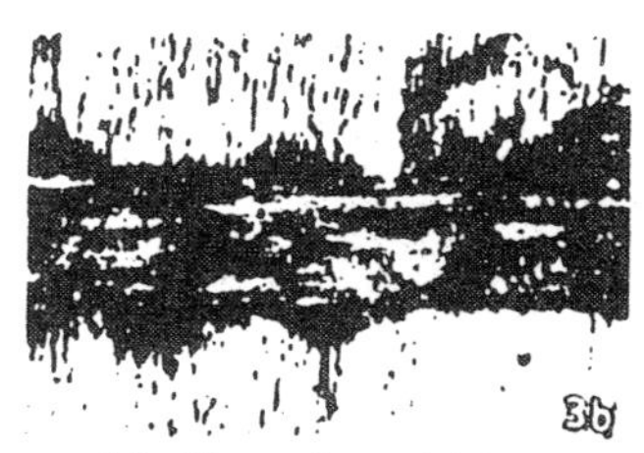

3 *b*. Zinc surface, 130° C.

FIG. 3. Micrographs of track produced when zinc slides on zinc lubricated by a 1% solution of lauric acid (× 15). At room temperature when the friction is low, the surface damage is slight. At elevated temperatures when the friction rises to high values the surface damage is heavy (see Fig. 66).

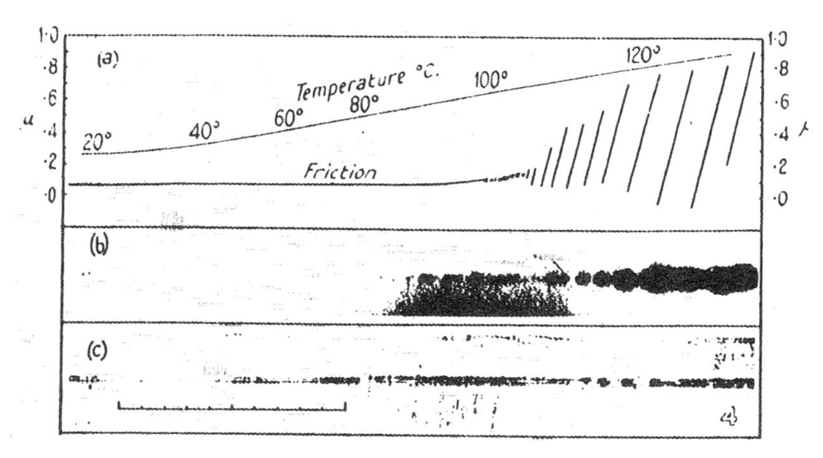

FIG. 4. Friction and surface damage produced when a copper slider (radioactive traverses a copper surface lubricated with palmitic acid (M.P. 63° C.). (*a*) Frictioi trace, shows breakdown of lubrication at about 110° C. (*b*) Radioactive autograpl shows increased metallic interaction and transfer at about 100° C. (*c*) Optica micrograph of track formed in copper surface. Length of track ≈ 3 cm.

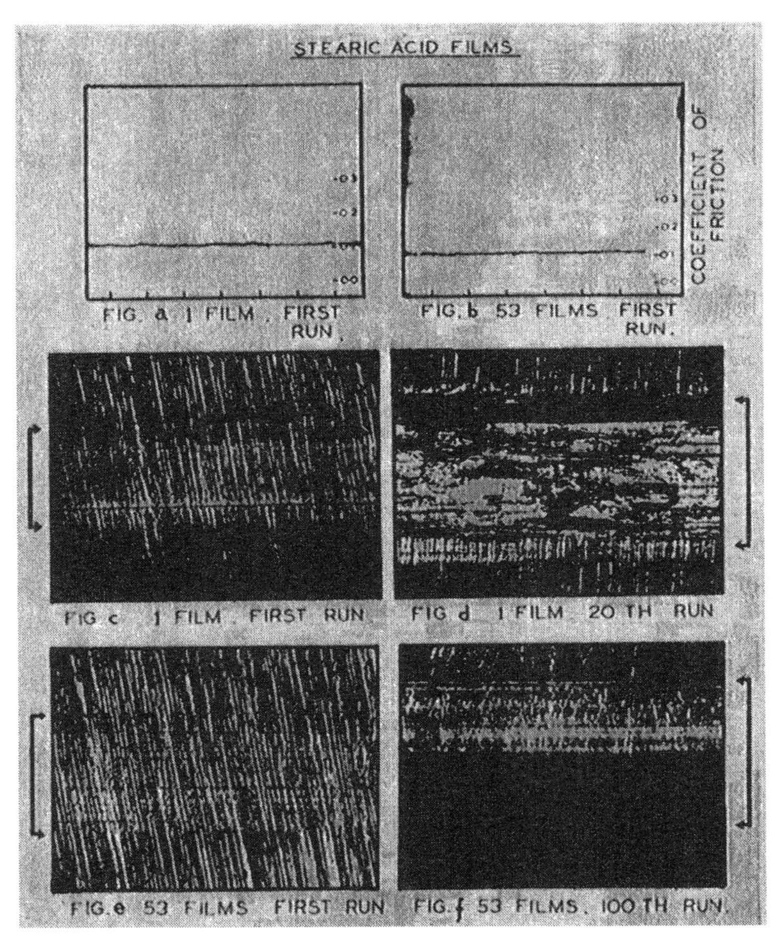

Wear of fatty acid films on stainless steel. Mono- and multi-layers of stearic acid deposited on lower surface only.

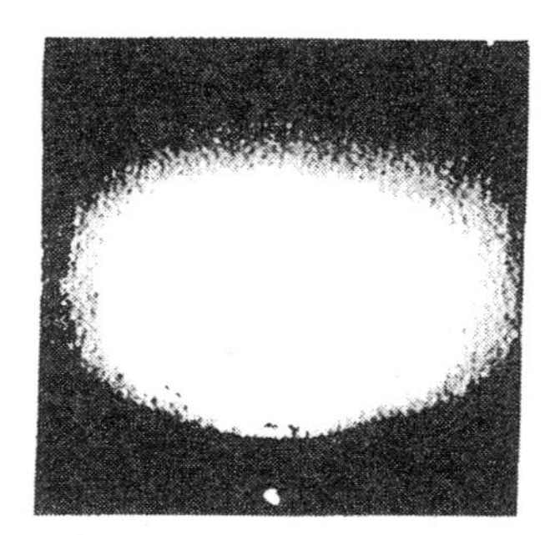

a. Stearic acid (M.P. 69° C.) on platinum at 25° C. There is fairly well-defined orientation of the surface film.

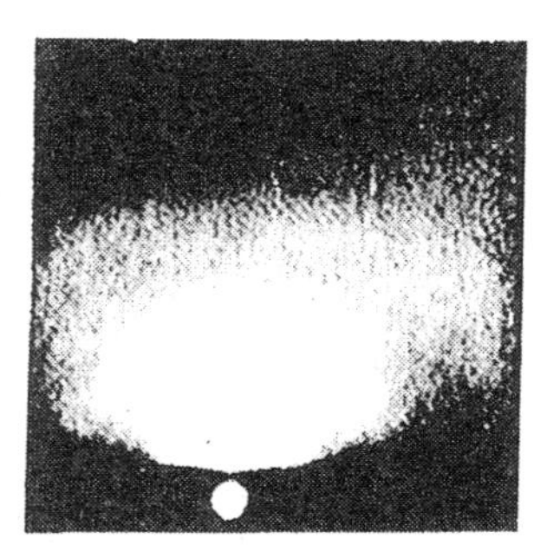

b. Stearic acid on platinum at 65° C. The oriented pattern has disappeared and the blurred rings now visible are those of the underlying polished platinum.

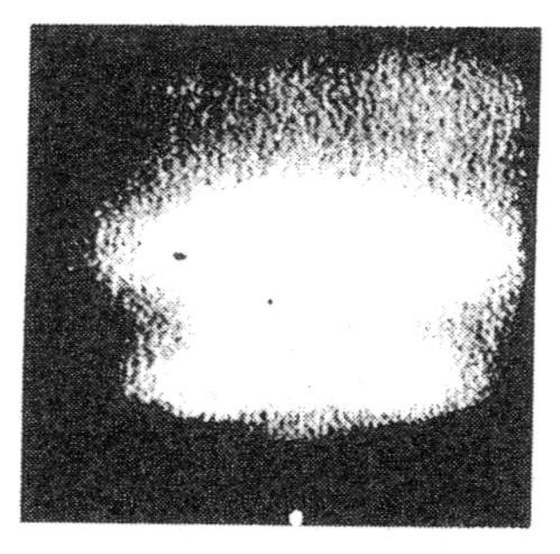

c. Stearic acid (M.P. 69° C.) on cadmium 33° C. Very strong oriented pattern.

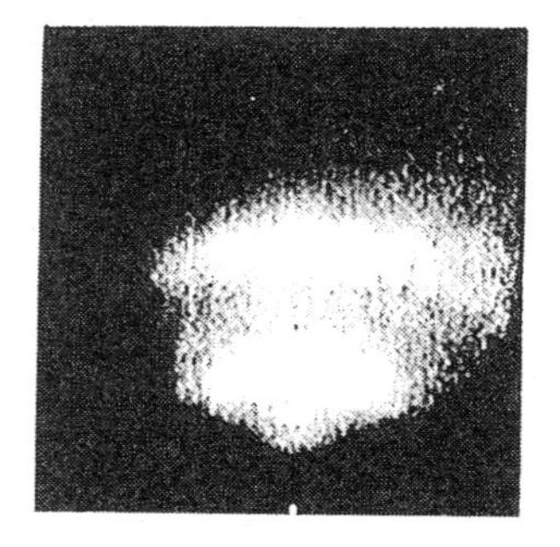

d. Stearic acid on cadmium 106° C. Pattern still shows orientation of surface film at nearly 40° C. above M.P. of acid.

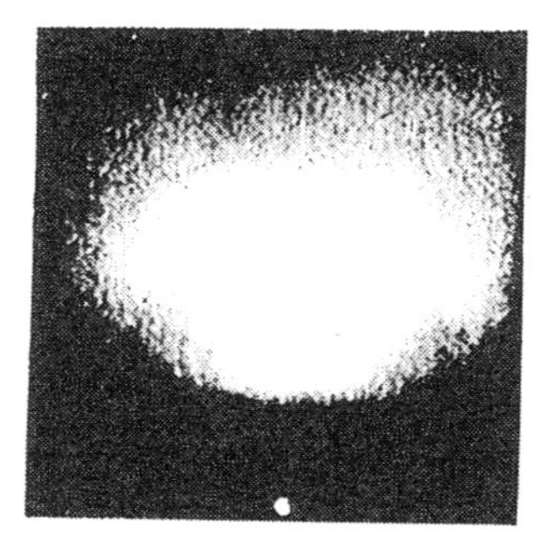

e. Stearic acid on cadmium 131° C. Pattern very weak.

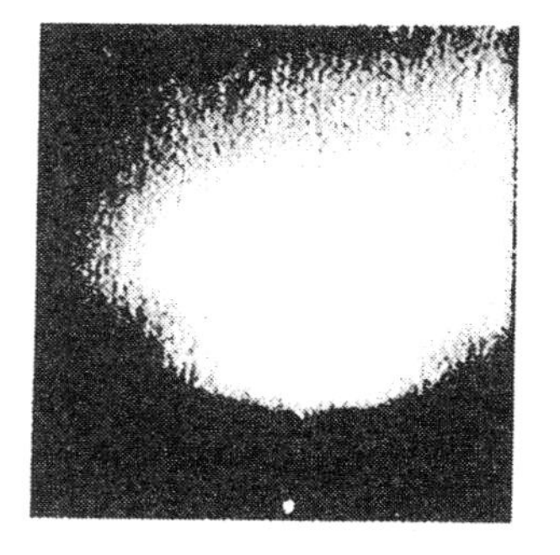

f. Stearic acid on cadmium 139° C. Pattern of surface film has disappeared. Blurred rings are those of underlying cadmium.

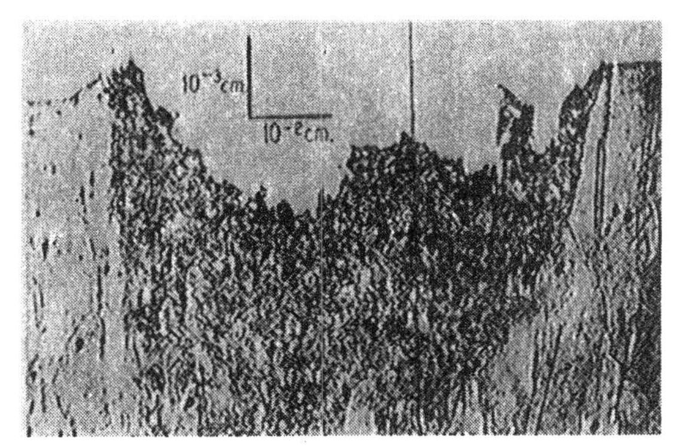

FIG. *a*. Taper section of track produced on a copper surface when a hemispherical copper slider traverses it once (unlubricated). Note the considerable depth of work hardening.

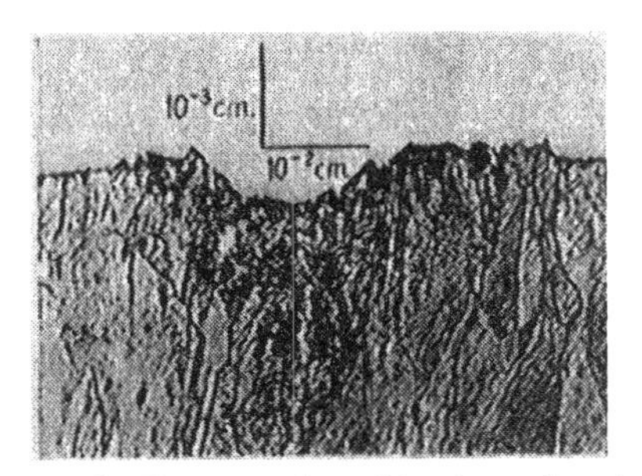

FIG. *b*. Taper section of track produced by a hemispherical copper slider when it traverses a copper surface covered with a sulphide film about 2×10^{-5} cm. thick. The decrease in surface damage and sub-surface deformation is very marked.

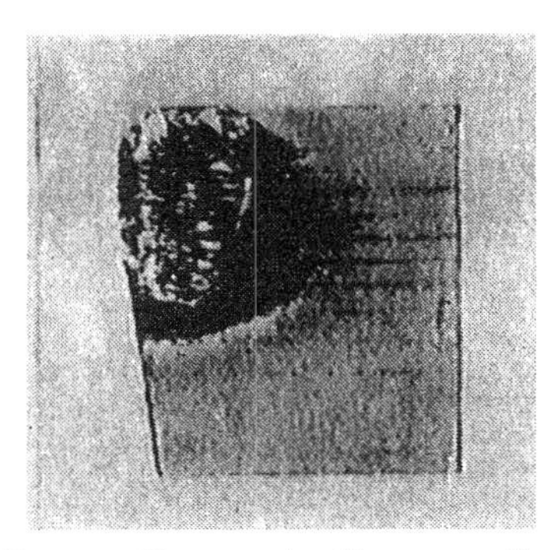

FIG. *c*. Damage in the nose of a lathe tool produced by the breaking away of the built-up edge and a typical wear 'crater' produced by the rubbing of the chip ($\times 5$).

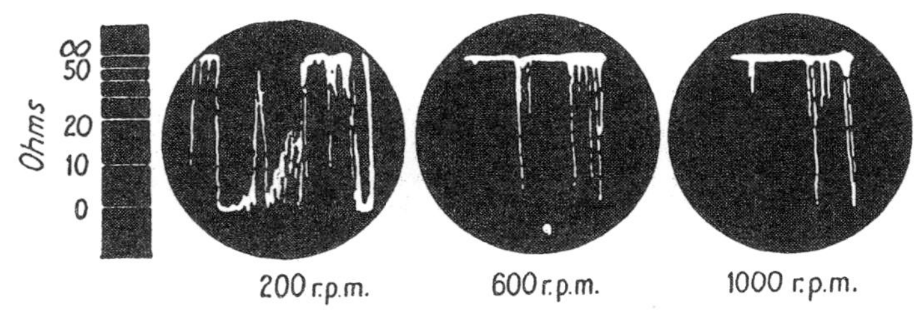

Fig. 1. *Effect of speed.* Oil at 20° C. No compression. The amount of breakdown of the lubricant film is considerably less at higher speeds.

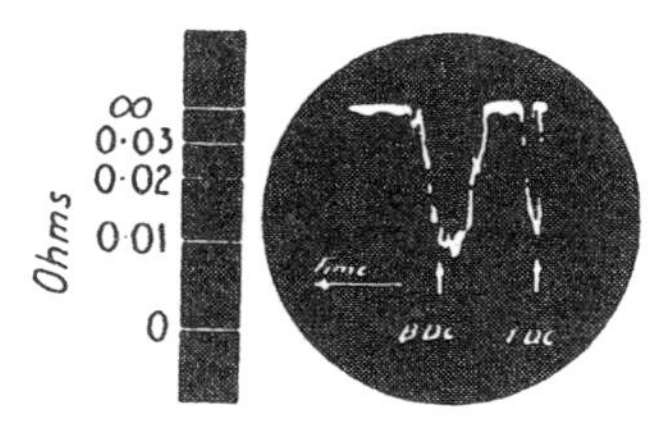

Fig. 2. *Effect of speed variation during one cycle.* Speed 300 r.p.m. Oil at 20° C. There is very marked breakdown of the lubricant film at top and bottom dead centres (T.D.C. and B.D.C. respectively).

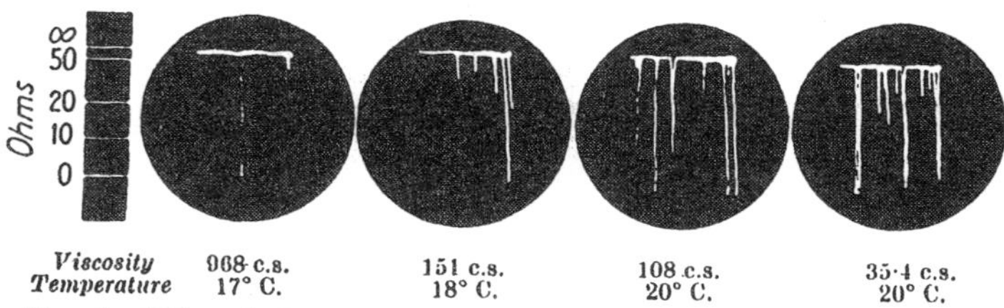

Fig. 3. *Effect of viscosity.* Speed 1000 r.p.m. Oil supply maintained at temperatures given. The amount of non-hydrodynamic lubrication is much greater for oils of low viscosity.

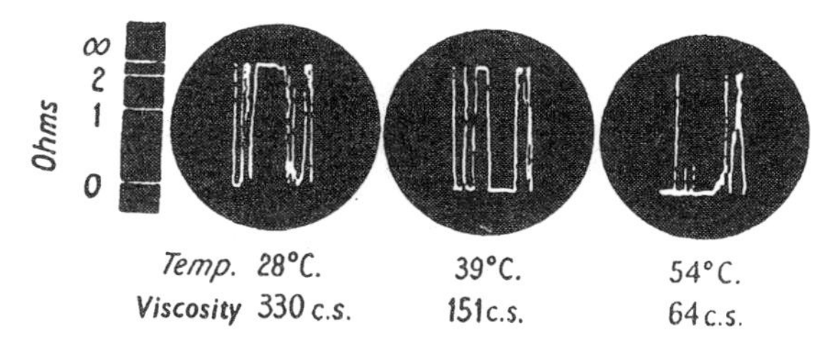

Fig. 4. *Effect of temperature.* Speed 600 r.p.m. Oil supply maintained at temperatures given. At higher temperatures there is marked breakdown of the lubricant film.

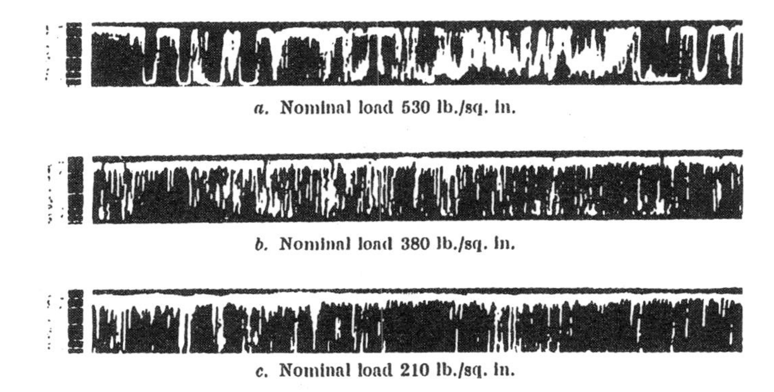

FIG. 1. *Effect of load.* Speed 190 r.p.m. Temperature 20° C. Each trace corresponds to one complete revolution of the shaft. At low loads (*c*) the greater part of the resistance trace is at high resistance values showing predominantly hydrodynamic conditions. At higher loads there is more breakdown of the hydrodynamic film.

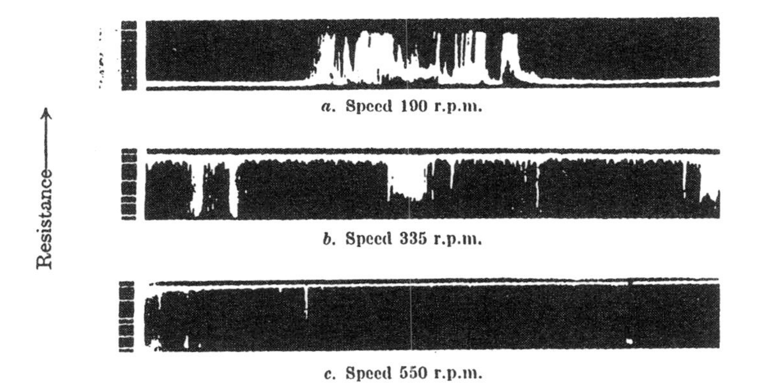

FIG. 2. *Effect of speed.* Nominal load 530 lb./sq. in. Temperature 20° C. Each trace corresponds to one complete revolution of the shaft. At higher speeds the lubrication is more hydrodynamic but there is still some breakdown of the lubricant film.

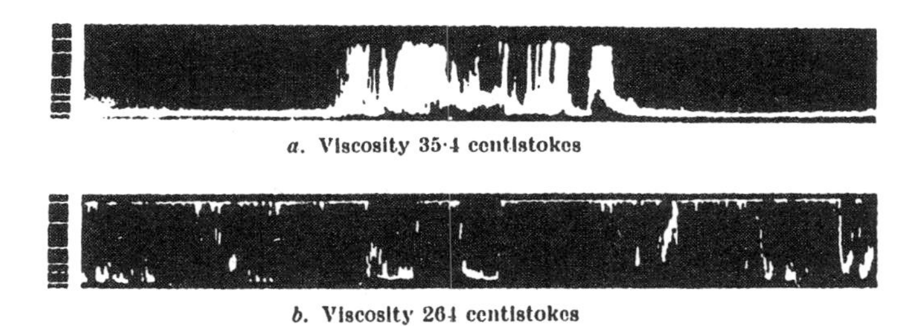

FIG. 3. *Effect of viscosity.* Speed 190 r.p.m. Nominal load 530 lb./sq. in. Temperature 20° C. Each trace corresponds to one complete revolution of the shaft. With the oil of higher viscosity there is considerably more hydrodynamic lubrication.

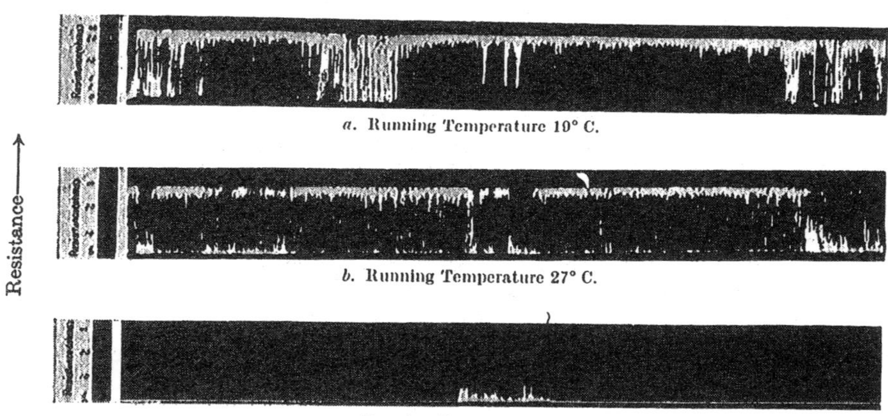

a. Running Temperature 19° C.

b. Running Temperature 27° C.

c. Running Temperature 30° C.

FIG. 1. *Effect of temperature.* Speed 190 r.p.m. Nominal load 380 lb./sq. in. Oil viscosity 36 centistokes at 19° C. Each trace represents one revolution of shaft. The resistance traces show that a temperature rise of 11° C. changes the lubrication from largely hydrodynamic to mainly non-hydrodynamic.

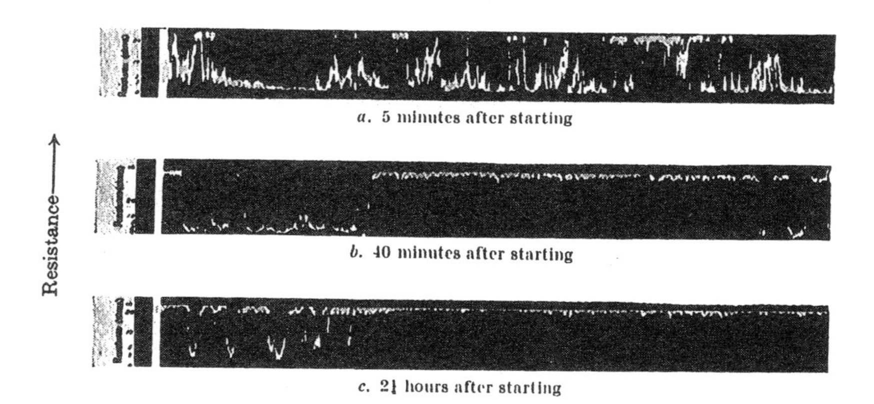

a. 5 minutes after starting

b. 40 minutes after starting

c. 2½ hours after starting

FIG. 2. *Effect of running-in.* Speed 190 r.p.m. Nominal load 460 lb./sq. in. Temperature 18° C. Each trace represents one revolution of shaft. It is seen that after 2 hours' running there is far less breakdown of the hydrodynamic film.

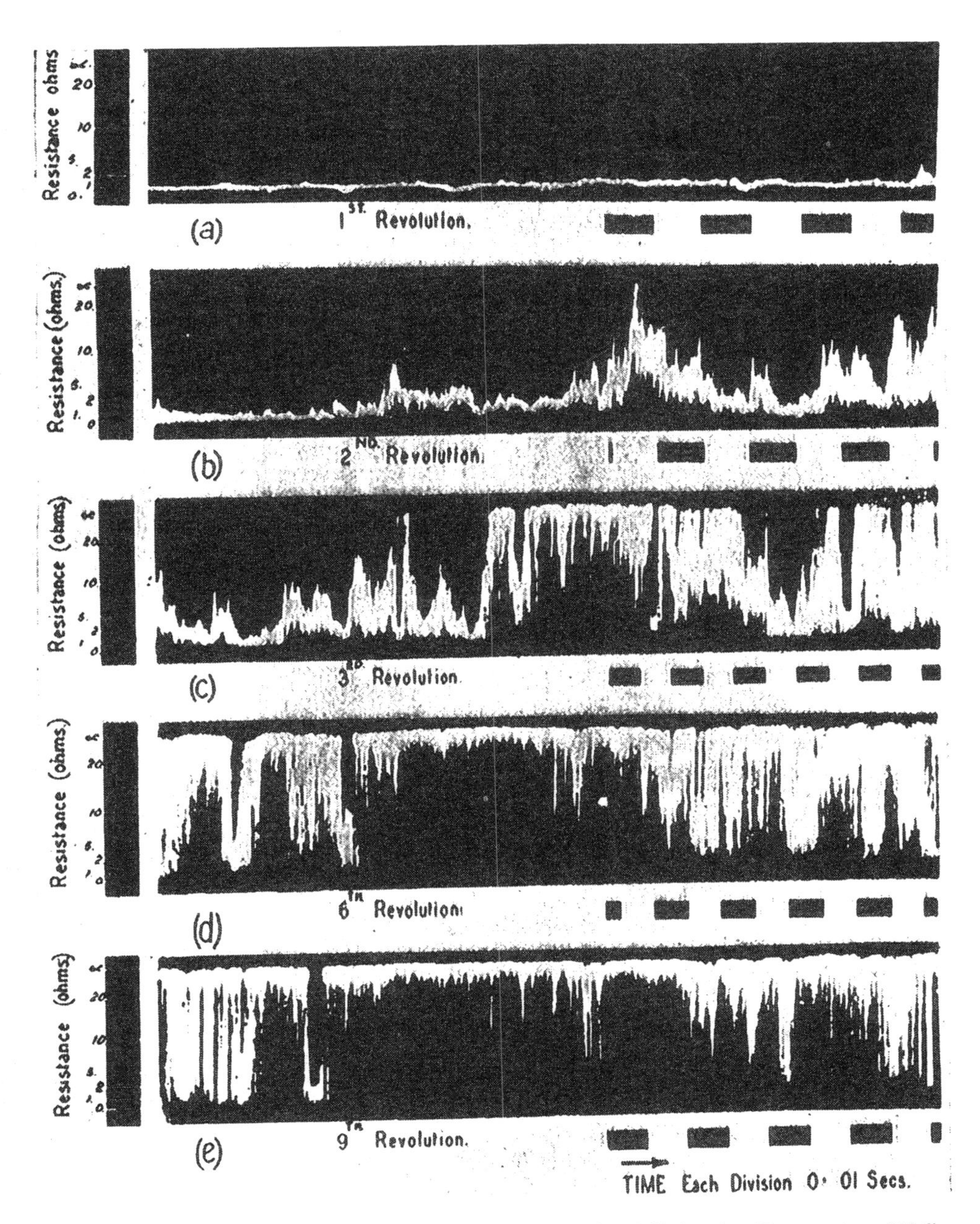

Build up of lubricant film. Speed 335 r.p.m. Nominal load 380 lb./sq. in. Temperature 19° C. Each trace represents one complete revolution of the shaft. The first revolution is almost completely non-hydrodynamic. After 9 revolutions the lubricant film has built itself up and provides considerable hydrodynamic lubrication.

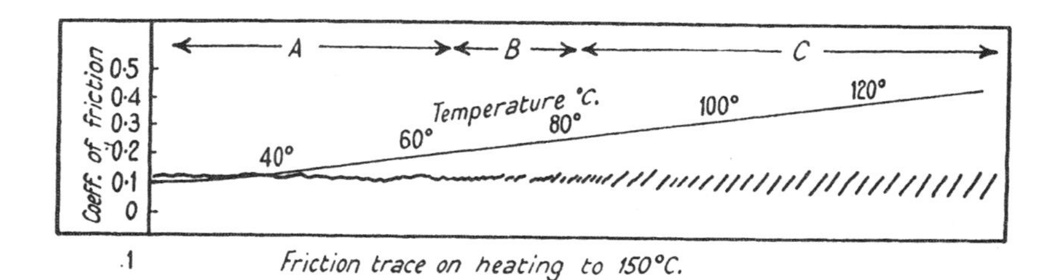

FIG. 1. Steel surfaces lubricated with commercial lubricant. Friction trace as surface is heated to 150° C.

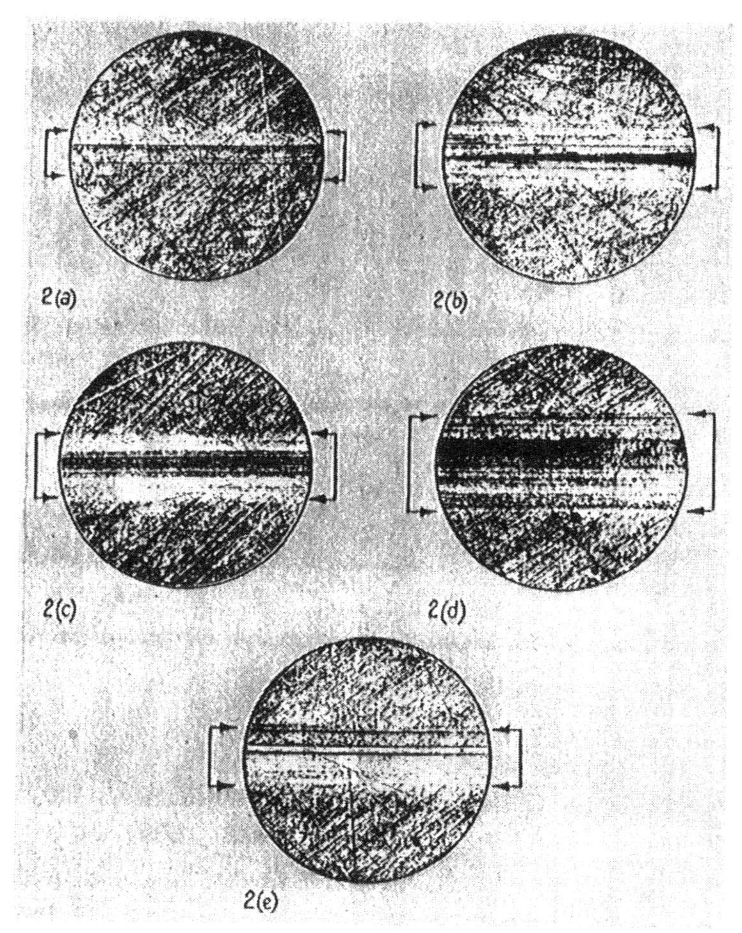

FIG. 2 *a*. Track at room temperature (× 33). Portion *A* of friction trace. Surface damage slight, $\mu = 0{\cdot}13$.
2 *b*. Track at 70° C. (× 33). Transition region *B* of friction trace.
2 *c*. Track between 120 and 150° C. (× 33). Portion *C* of friction trace.
2 *d*. Track at 200° C. (× 33), $\mu > 0{\cdot}35$. Heavy surface damage.
2 *e*. Track on cooling to 20° C. (× 33). Surface damage is slight and friction is again low at $\mu = 0{\cdot}13$.

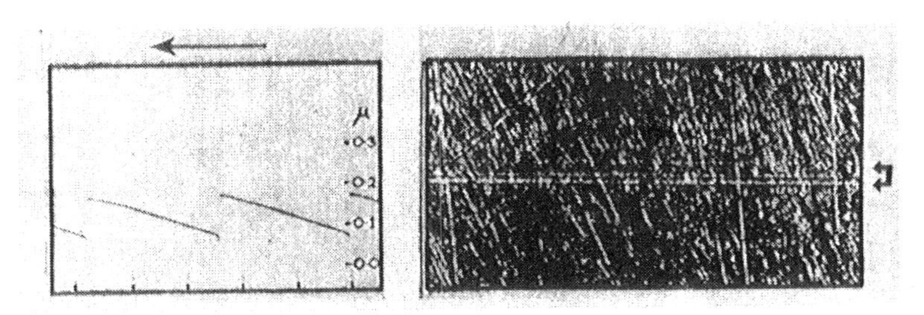

a. Friction record *b.* Track (×60)

FIG. 1. Mineral oil on steel surfaces at room temperature.

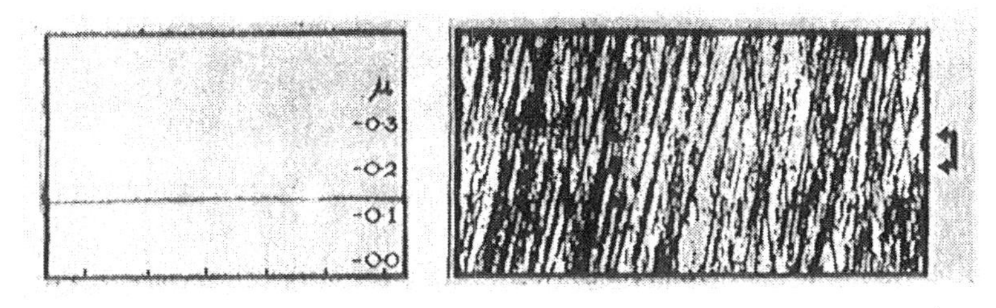

a. Friction record *b.* Track (×60)

FIG. 2. Mineral oil + 1% caprylic acid on steel surfaces at room temperature.

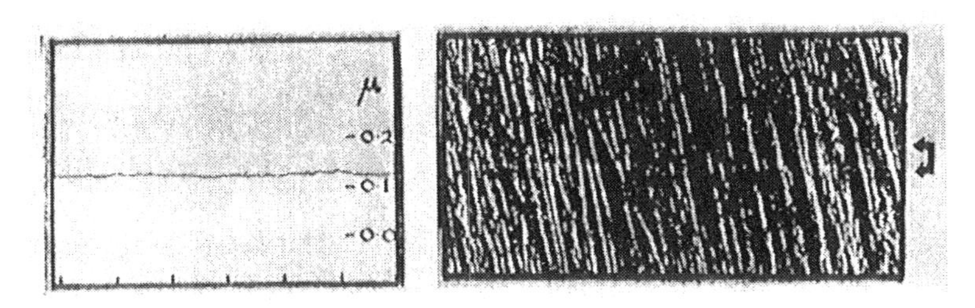

a. Friction record *b.* Track (×60)

FIG. 3. Mineral oil on steel surfaces after heating to 200° C. for 15 minutes.

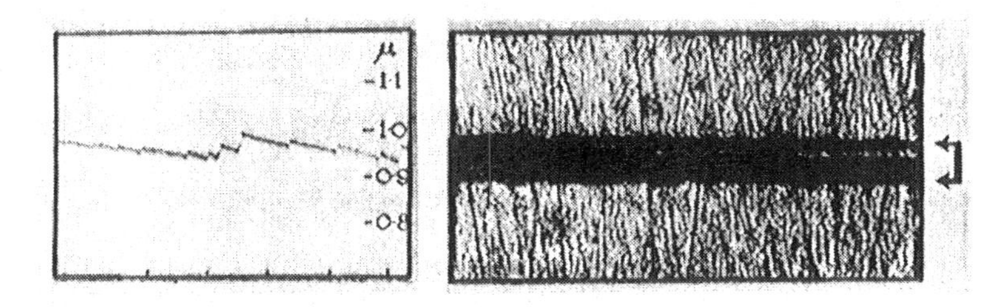

a. Friction record *b.* Track (×60)

FIG. 4. Mineral oil on steel surfaces after heating to 300° C. for 20 minutes.

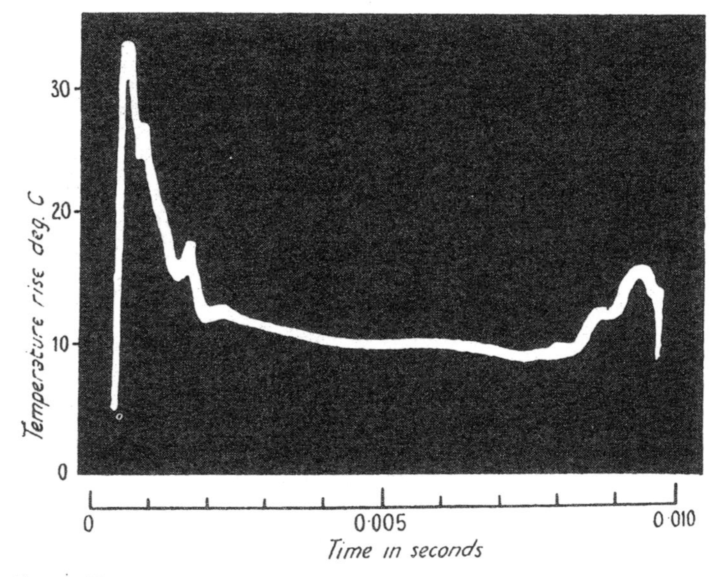

FIG. 1. Thermal e.m.f. generated during impact between a constantan point and a Wood's metal sphere. The temperature rise reaches its maximum in less than $5 \cdot 10^{-4}$ sec. and dies away to half its maximum value in 10^{-3} sec.

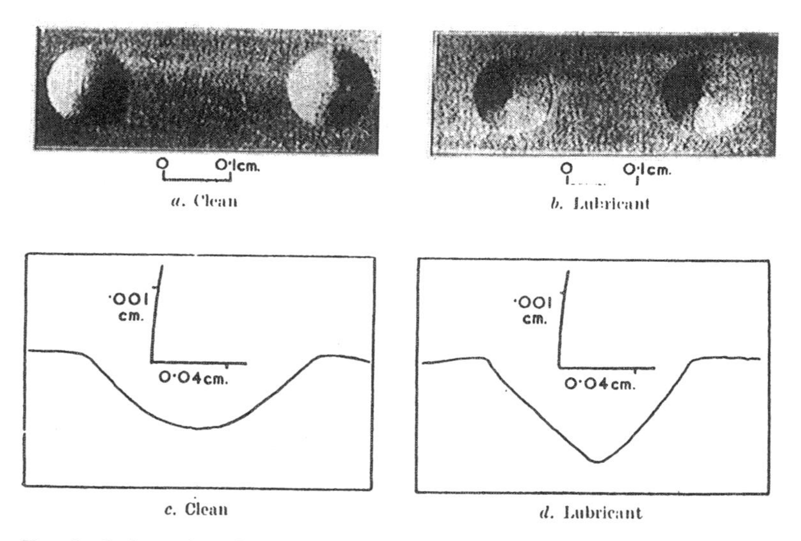

a. Clean

b. Lubricant

c. Clean

d. Lubricant

FIG. 2. Indentations formed in a copper surface when struck by a steel ball of mass 70 gm. (diam. 2·5 cm.) with a velocity of 90 cm./sec. The micrograph (*a*) and the profilometer record (*c*) are for clean surfaces. The micrograph (*b*) and the profilometer record (*d*) are for surfaces covered with an oil film of viscosity 500 centistokes. Electrical resistance measurements showed that in the presence of the oil film no metallic contact occurred between the ball and the surface during the collision. The copper surface has been deformed plastically by the pressure developed in the oil film trapped between the colliding surfaces.

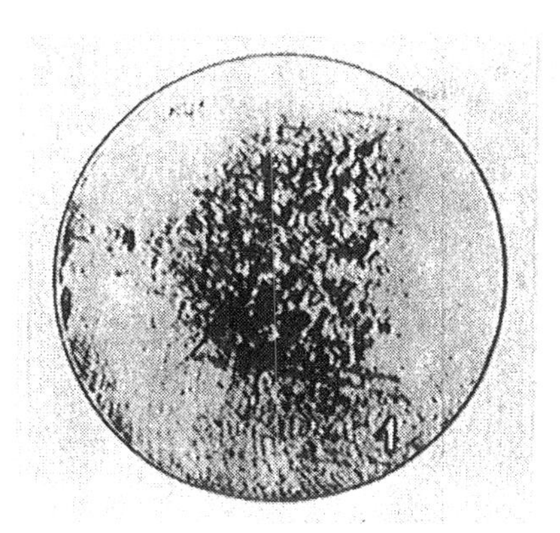

FIG. 1. Worn portion of a hemispherical steel slider (V.D.H. 600) after it has traversed 1000 cm. of track on unlubricated copper (V.D.H. 60). The plucking away of the steel is clearly visible (× 66).

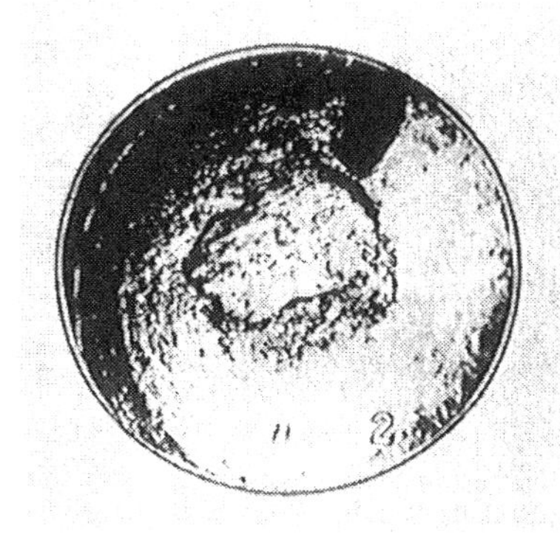

FIG. 2. Steel slider after rubbing on brass surface. Note pick-up of brass (× 30).

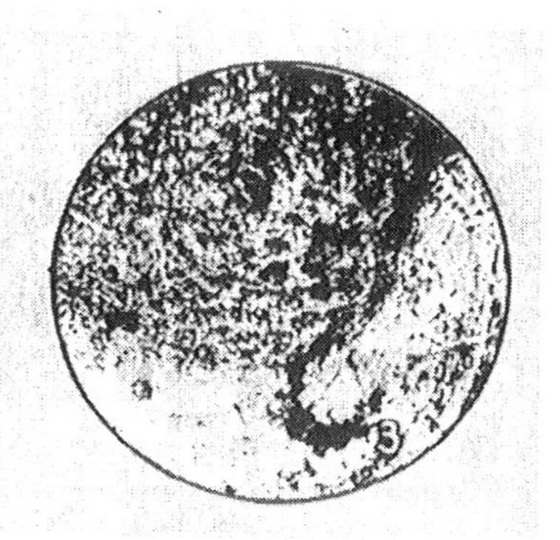

FIG. 3. Portion of 2 at higher magnification after etching away the brass. There is marked plucking of the steel (× 66).

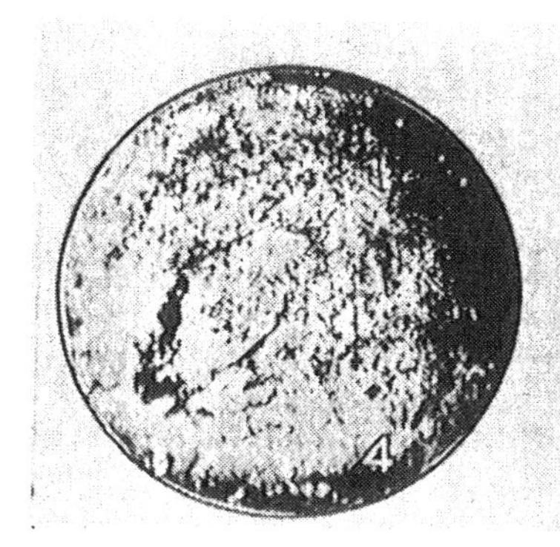

FIG. 4. Rhodium plated brass slider after rubbing on brass. (Thickness of rhodium 5×10^{-4} cm.) Note pick-up of brass (× 30).

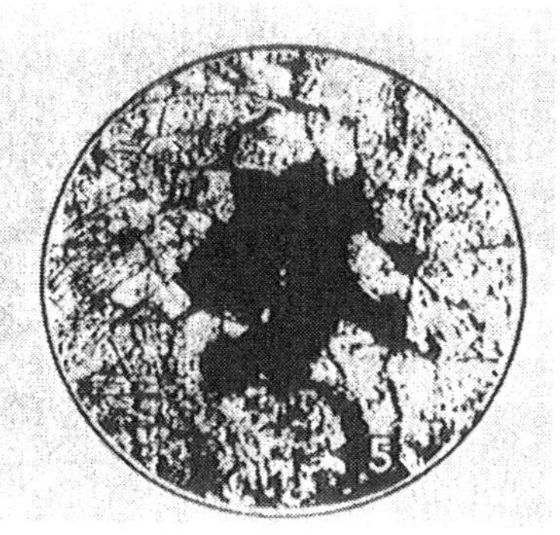

FIG. 5. Portion of 4 at higher magnification after etching away the brass. Note breakdown and cracking of rhodium film (× 66).

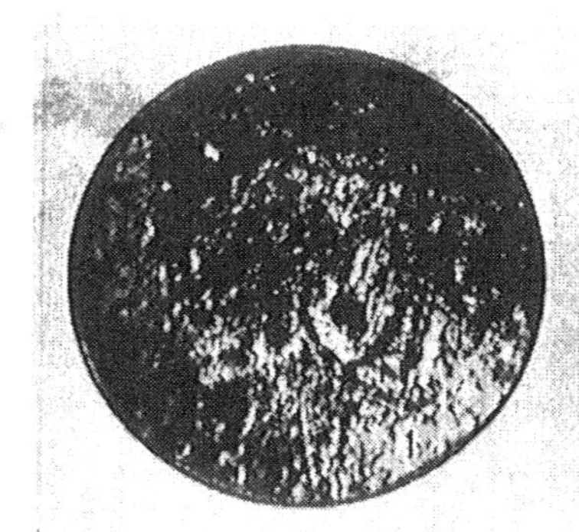

FIG. 1. Rhodium plated brass slider after rubbing on brass. (Thickness of rhodium 10^{-3} cm.) Note pick-up of brass ($\times 30$).

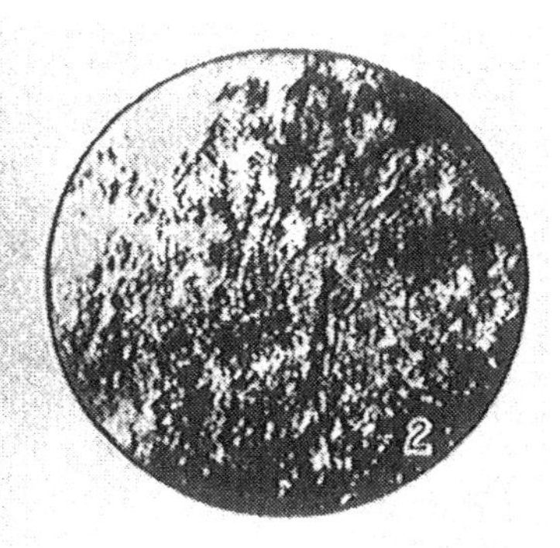

FIG. 2. Portion of 1 at higher magnification after etching away the brass pick-up. There is no sign of damage to the rhodium film ($\times 66$).

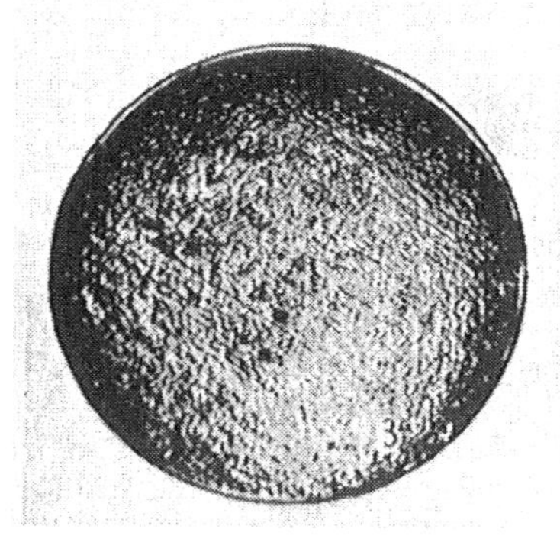

FIG. 3. Chromium plated brass slider after rubbing on brass. (Thickness of chromium $5 . 10^{-3}$ cm.) Note very slight pick-up of brass ($\times 30$).

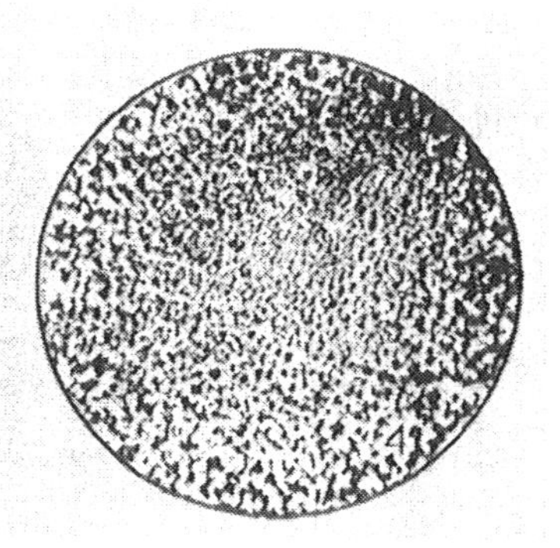

FIG. 4. Portion of 3 at higher magnification after etching away brass. There is no sign of damage to the chromium film ($\times 66$).

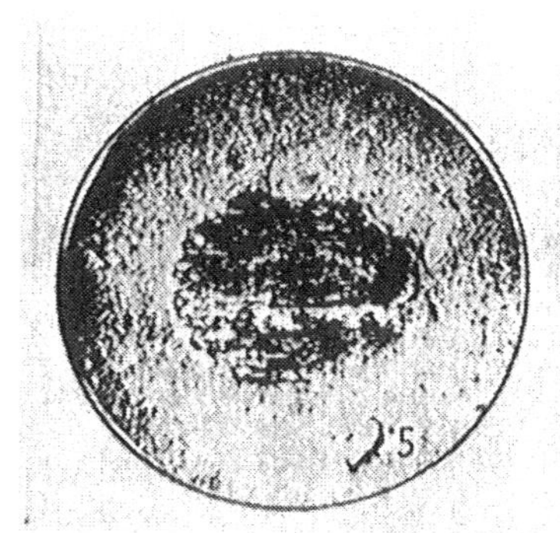

FIG. 5. Cast iron slider, plated with a film of chromium 2×10^{-4} cm. thick after sliding a few cm. on steel. The chromium film has broken down and there is already appreciable wear of the underlying cast iron ($\times 30$).

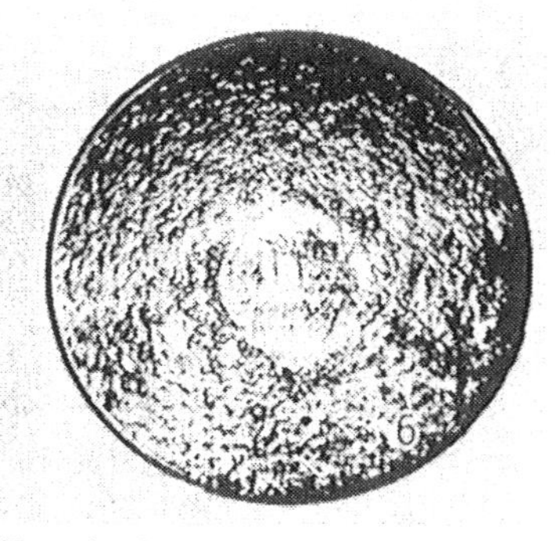

FIG. 6. Cast iron slider plated with a thicker film of chromium (about 5×10^{-4} cm.) after sliding 1200 cm. on steel. It is seen that the wear of the chromium is relatively light ($\times 30$).

model of a hard spherical surface penetrating the surface of a softer metal. Both surfaces are assumed perfectly smooth. If the metal does not work-harden, so that Y is a constant, then, as we saw above, $p_m = 3Y =$ constant. If now the metal is capable of work-hardening, the formation of this indentation itself will produce an increase of the elastic limit Y. Theoretical consideration and practical measurements show that the elastic limit around the indentation will not be constant but will vary from point to point. Nevertheless, we may assume an average or representative value of the elastic limit which is related to the mean pressure p_m by a relation of the same type as equation (3). A detailed experimental investigation shows that, in fact, the elastic limit Y_e at the edge of the indentation may be used for this purpose. If p_m is compared with Y_e we find that over a wide range of indentation sizes $p_m = cY_e$, where c has a value lying between 2·7 and 3. Further, if d is the diameter of the indentation and D the diameter of the indenter, the shape of the indentation is completely defined by the dimensionless ratio d/D, and it is found that the deformation corresponding to Y_e is approximately proportional to the ratio d/D. If we express the deformation δ at the edge as a percentage strain, the factor of proportionality is approximately 20, so that we may write

$$\delta = 20\left(\frac{d}{D}\right). \tag{5}$$

An example will make this clearer. Suppose we use a hard sphere of diameter 1 cm. and apply such a load that an indentation is formed of diameter 5 mm. The ratio d/D is $\frac{1}{2}$ and the effective deformation is 10 per cent. From a stress-strain curve for the metal we may find the elastic limit corresponding to a strain of 10 per cent. This is now the value of the 'representative' elastic limit for an indentation of this size, and it will be found that this value is directly proportional to the yield pressure p_m, the factor of proportionality being about 2·8 (Tabor, 1948).

If we make a series of indentations in the surface at increasing loads we find that p_m increases with the size of the indentation, the plot of p_m against d/D giving the points shown in Fig. 7. We may now convert the d/D axis into equivalent deformations δ, using the relation $\delta = 20d/D$. Then with this as the deformation or strain axis we plot the stress-strain characteristics of the metal, multiplying the stress ordinates by the ratio 2·8. It is seen that this curve coincides with the indentation measurements. Thus the indentation measurements really follow the work-hardening characteristics of the metal.

We may express this variation of p_m with indentation size in a quantitative manner. Suppose W is the load between the sphere and the surface when the diameter of the indentation is d. Then

$$p_m = \frac{4W}{\pi d^2}.$$

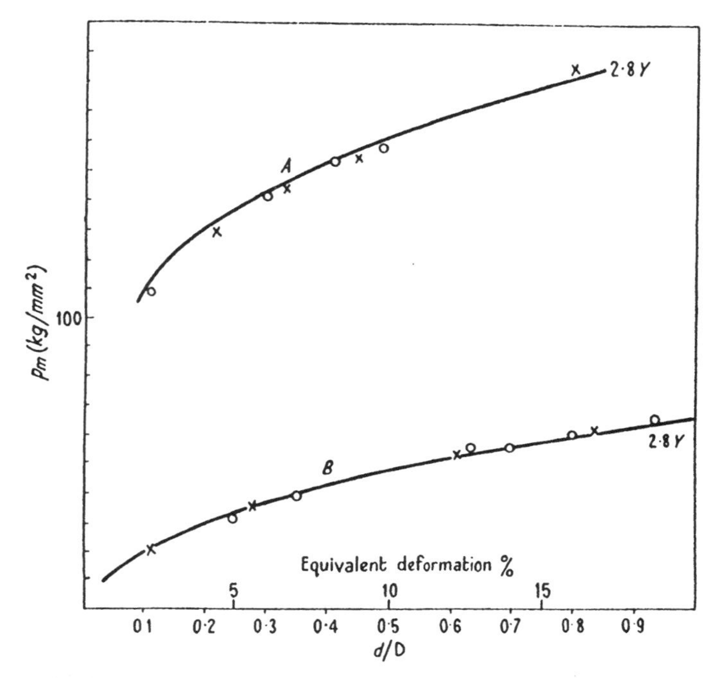

FIG. 7. Plastic deformation of mild steel (curve A) and annealed copper (curve B) by a hard steel sphere of diameter D. The mean pressure p_m is plotted against d/D, where d is the diameter of the indentation. The parameter d/D provides a measure of the work-hardening produced around the indentation and there is close agreement between the indentation measurements (circles and crosses) and the stress-strain curves of the metals (full line).

To a first approximation the elastic limit for many metals is found to be a simple power-function of the deformation or strain, i.e. $Y = b\delta^x$ (Nadai, 1931).

Then $$p_m = cY = cb\delta^x = \text{constant} \times (d/D)^x.$$

Hence $$W = k\frac{d^n}{D^{n-2}}, \tag{6}$$

where $n = x+2$. For many annealed metals x has a value of about $\frac{1}{2}$, so that for a fixed value of D the area over which plastic flow occurs is given by

$$A = \frac{\pi d^2}{4} = k'W^{\frac{8}{9}}. \tag{7}$$

This means that the area of real contact is not quite proportional to the load on account of the work-hardening. If the material is partially work-hardened, as is usually the case, the work-hardening accompanying indentation proceeds less rapidly. A relation similar to equation (6) still holds, but the index n is closer to 2. For example, with ordinary extruded brass or mild steel the value of n is about 2·15. Consequently the area of contact is proportional to $W^{0\cdot 92}$. With fully worked metals, where the elastic limit is not changed by further deformation, the yield pressure is, as we have seen, independent of the size of the indentation, and the area over which plastic flow occurs is directly proportional to W.

Similar relations are found to hold for a soft hemisphere pressing against a flat surface of a harder metal and for two hemispheres of the same metal pressing against one another (O'Neill, 1934). The area of contact for fully annealed materials varies approximately as $W^{\frac{8}{9}}$, whilst for fully worked specimens the area of contact is directly proportional to W, provided the deformed region is small compared with the dimensions of the surfaces.

Asperities of conical and pyramidal shape

Let us now consider the behaviour of asperities which are conical or pyramidal in shape. For simplicity we shall again consider a hard conical or pyramidal indenter penetrating the surface of a softer metal. The surfaces of the indenter and the metal are both assumed to be perfectly smooth.

The tip of the indenter may be considered to be a portion of a sphere of vanishingly small radius of curvature. Consequently, as we see from equation (4), an infinitesimal load will deform the metal surface beyond its elastic limit and a condition of 'full' plasticity will readily be established. As the indenter penetrates the surface it is again found that, for a metal which does not work-harden, the yield pressure p_m is related to the elastic limit of the metal by a relation of the type $p_m = cY$. The factor c is constant for any given indenter, but varies with its angle. The more pointed the indenter the higher the value of c, presumably on account of the greater confinement of the displaced metal. Typical results taken from a paper by Bishop, Hill, and Mott are plotted in Fig. 8 for the indentation of highly worked copper by conical indenters of various angles. It is seen that for semi-angles α ranging from 60° to 90°, c decreases from 3·6 to about 2·9, so that for indenters of large semi-angle the constant of proportionality does not vary rapidly with the angle and has a value of about 3. For indenters of small semi-angle, however, c is appreciably higher and increases more rapidly as α decreases. However, for any given indenter, since p_m

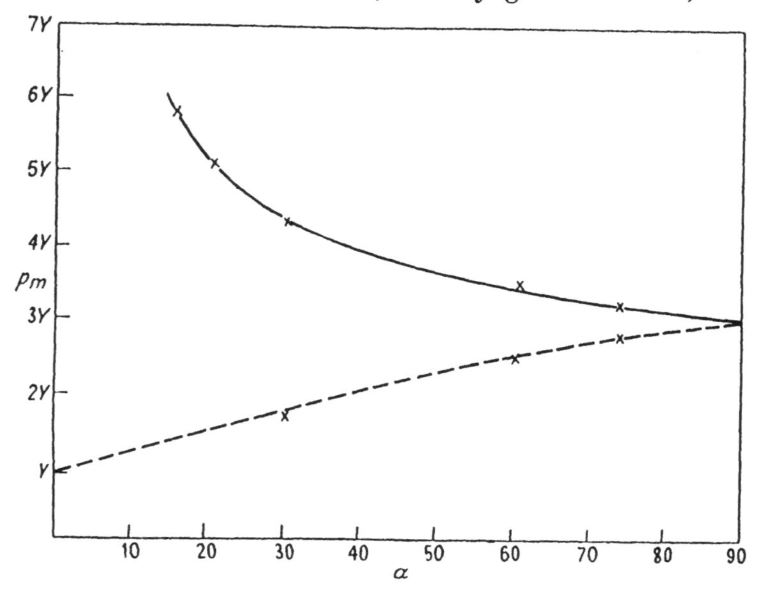

FIG. 8. Yield pressure p_m as a function of semi-angle α of cone. Experimental behaviour of work-hardened copper deformed by steel. Full line: deformation of a flat surface by a hard cone. Broken line: deformation of a cone by a hard flat surface.

remains constant, the area over which plastic flow occurs is directly proportional to the load.

With a metal which work-hardens, the amount of work-hardening produced by the indentation itself will depend on the shape of the indenter. In general a sharp cone will produce more work-hardening than a blunt one, so that it will produce a greater increase in the 'representative' elastic limit of the metal. This factor, together with the increased value of c, will make the yield pressure for a sharp cone appreciably higher than for a blunt one. Nevertheless for an indenter of fixed semi-angle the effective deformation remains the same whatever the size of the indentation produced, since a conical or pyramidal indentation in an infinite flat surface is geometrically similar whatever its size. Consequently for a given indenter, p_m remains independent of the load.† This means that even with a material capable of work-hardening, the area over which plastic flow occurs is directly proportional to the load W.

Similar considerations apply if the pyramid or cone is of a softer metal and its tip is squeezed flat by a harder flat surface. In this case, however, the pressure resisting deformation will decrease with the semi-angle, since a semi-angle of 0° is equivalent to a cylinder for which $p_m = Y$. Thus as α decreases from 90° to 0° p_m will decrease from $2{\cdot}9Y$ to Y. A typical result for work-hardened copper cones is shown dotted in Fig. 8 and confirms this characteristic. It is seen that for blunt cones varying in semi-angle from about 60° to 90° the constant of proportionality connecting p_m with Y does not vary very rapidly and has a mean value of about 2·7. If the material of which the cone or pyramid is made is capable of work-hardening, the arguments used above still apply. In general, a sharp cone will work-harden more than a blunt one. Nevertheless, for a cone of fixed semi-angle the shape of the deformed tip will be geometrically similar whatever its size. Consequently p_m will remain constant and the area over which plastic flow occurs for both annealed and work-hardened metals will be directly proportional to the load.

The area of real contact

It follows from these results that when metal surfaces rest on one another the peaks of their asperities readily deform plastically, the mean pressure being given by $p_m = cY$. Y is some 'representative' measure of the elastic limit of the deformed metal at the tip of the asperities. The factor c depends on the shape and size of the surface irregularities, but for conical and pyramidal asperities of a wide range of angles and for hemispherical asperities, c has a value of about 3. Thus the yield pressure of the surface irregularities is approximately equal to $3Y$.

With conical and pyramidal asperities the yield pressure is independent of the amount of deformation that has occurred so that the area of real contact is directly proportional to the load W. With hemispherical asperities this is only true for metals that are highly worked. For annealed metals A is more nearly proportional to $W^{4/5}$. However,

† Experiments show (Tabor, 1948) that the representative deformation produced by the pyramidal diamond used in the Vickers hardness test is about 8 per cent., whilst the factor of proportionality connecting the yield pressure with the representative value of the elastic limit is about 3·4. This enables the Vickers hardness number to be calculated from the stress–strain characteristics of the metal; there is close agreement between the observed and calculated values for a wide range of materials and degrees of work-hardening.

the surface asperities themselves are usually work-hardened by the very process of preparing the surface. *Consequently we may expect that in most practical cases for all types and shapes of surface irregularities the real area of contact will be very nearly proportional to the load. That is,*

$$A = \frac{1}{p_m} W, \tag{8}$$

where p_m is the mean yield pressure of the asperities.

In contrast, as we saw from equation (1), elastic deformation implies that

$$A = k'' W^{\frac{2}{3}}. \tag{9}$$

Area of real and apparent contact

We have already seen that for bodies of small radius of curvature the condition for plasticity is reached at extremely small loads. The results in Table II, however, show that for bodies of large radius of curvature the load at which the onset of plastic deformation occurs will be very much greater. For example, for an extremely hard spherical surface of radius 1 cm. pressing on highly worked mild steel the initial deformation is elastic up to a load of about 5,000 gm. In practice, however, it is almost impossible to obtain perfectly smooth surfaces. Consequently for loads below 5,000 gm. the fine surface irregularities will be deformed plastically whilst the bulk of the underlying metal will deform elastically. This means that as the load is increased from zero to 5,000 gm. the area of the surface asperities over which plastic flow has occurred will increase proportionally to the load. On the other hand, the apparent area of contact, where the deformation is still essentially elastic, will increase as $W^{\frac{2}{3}}$. When the load exceeds 5,000 gm., plastic deformation of the bulk will commence, and at a load of a few hundred kilograms the main part of the specimen will have reached the condition of full plasticity. There is now plastic deformation on a macroscopic scale, and the mean pressure over the large deformed area will be a measure of the elastic limit of the material in bulk. These are the conditions which apply in most practical 'hardness' measurements.

An extreme case of this effect is shown by some profilometer records recently made by Dr. A. J. W. Moore (1948) on a work-hardened copper surface in which a series of fine parallel grooves had been cut. A hard steel cylinder was placed parallel to the grooves and pressed into the surface. Profilometer records of the resulting indentations formed at various loads are shown in Fig. 9. Fig. 9 (*a*) is for a light load where the tips of the asperities have been deformed plastically, whilst no plastic deformation has occurred in the underlying material. In Fig. 9 (*b*) there is a slight deformation of the bulk material, whilst in Fig. 9 (*c*), for a very heavy load, the bulk deformation has been very severe. Even in this case, however, it is interesting to note that the irregularities retain their identity and are clearly visible even at the bottom of the indentation.

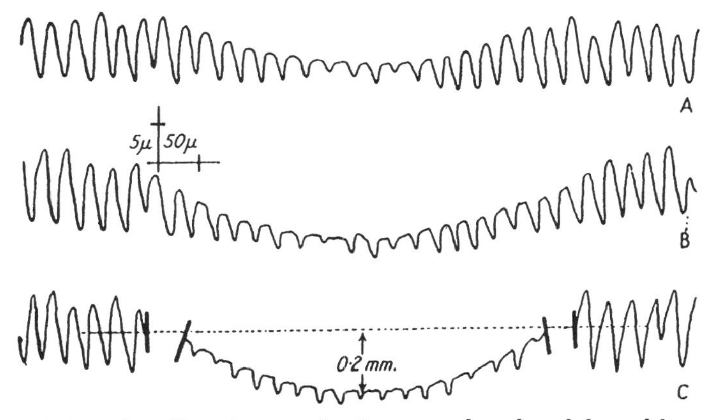

FIG. 9. Talysurf profilometer records of a grooved surface deformed by a hard cylindrical indenter. (*a*) Deformation produced by a small load. The asperities are deformed plastically but the bulk of the metal is still deformed elastically. (*b*) A further stage in the deformation process. (*c*) Under a heavier load both the asperities and the bulk of the metal are deformed plastically, but the asperities still retain their individual identity. The real area of contact (the flattened tips of the asperities) is appreciably less than the area over which macroscopic deformation has occurred.

It is clear from Fig. 9 (*c*) that marked plastic deformation has occurred over two ranges. For the asperities it has occurred at their tips, where the elastic limit has been raised by work-hardening to a high value, say Y_h. Over these asperities the mean yield pressure is given by $p_h = cY_h$. For the bulk material below the asperities, plastic flow has also occurred. Here the elastic limit is relatively low, say Y_l, and the mean pressure over the macroscopic indentation will be given by $p_l = cY_l$. The real area of contact is the sum of the flattened tips of the asperities, whilst the apparent area of content is the region covering the macroscopic indentation. The ratio of the real to the apparent area will be approximately equal to $p_l/p_h = Y_l/Y_h$. In the given example, although the original specimen was highly work-hardened, the ratio is about 1:2.†

This effect is also shown strikingly by an oblique section examination

† It is possible that the friction between the indenter and the tip of the asperity will also serve to increase the effective yield pressure of the asperity.

of similar indentations formed in moderately work-hardened copper. The result for a very deep indentation is shown in Plate II. The figures confirm the remarkable persistence of the asperities even when the deformation of the bulk metal beneath them has been very heavy. The enlarged pictures, Plate II, *c* and *d*, show that the flattening of the asperities is relatively small, although the work-hardening involved appears to have been considerable. Here again the real area of contact is about one-half the apparent area.

The greatest discrepancy between the real and apparent area of contact occurs when flat surfaces (or spherical surfaces of opposite curvature) are placed in contact. The apparent area of contact is the area of the surfaces themselves. The real area of contact is the summed area of all the surface irregularities which are touching and which support the load. Suppose, for example, steel flats of area 20 sq. cm. are placed in contact. The apparent area of contact will be 20 sq. cm. and it will be independent of the load. In fact, however, the surfaces will be supported on their irregularities and these will crush down until their cross-sectional area is large enough to support the load. For a steel for which p_m of the asperities is, say, 100 kg./mm.^{2} the area over which the asperities flow plastically will be proportional to the load and will be equal to 10 sq. mm. when the load is 1,000 kg. Thus when the surfaces are pressed together with a force of a ton the area of real contact will be 1/200th of the apparent area. For a load of 2 kg. the area of intimate contact will be 1/100,000th of the apparent area. The plastic flow of the asperities provides the real area of contact which supports the load. The stresses in the asperities are taken up by the elastic deformation of the underlying metal.

The effect of removing the load

So far we have only considered the effect of applying a given load. We may now ask what happens when the load is removed. It is of course clear that within the range of elastic deformation the process of deformation is completely reversible and a removal of the load enables the surfaces to return to their original configuration.

Let us consider in greater detail the case when plastic deformation occurs. We again take the simple model of a hard sphere penetrating the surface of a softer metal under a load W to form an indentation of diameter d. Suppose we remove the sphere and then replace it with the original load W. The indentation should now be in exactly the same condition as it was when it was originally formed. Thus any change which has occurred in the configuration of the surfaces between removing and replacing the indenter must be reversible, i.e. it must be elastic. This means that the surface of the indentation 'recovers' elastically when the load is removed and deforms elastically when the load is reapplied. Consequently the 'recovered' indentation will have a radius of curvature larger than that of the indenter. This is well established in practical hardness measurements and is referred to as a 'shallowing' effect. Suppose the indenter has a radius of curvature r_1 and the recovered indentation a radius of curvature r_2. When the indenter is replaced in the recovered indentation and the original load W applied, elastic deformation of both surfaces occurs. The radius of curvature of the contacting surfaces reaches a value r, where $r_2 > r > r_1$, and the diameter of the region of elastic deformation reaches a value of d. The whole of this process follows Hertz's classical laws describing the elastic deformation of spherical surfaces of radii of curvature r_1 and r_2, and there is close agreement between the observed value of r_2 and that calculated from the elastic constants of the materials.† At any stage in this process if the load is reduced, there is a relaxation of elastic stresses, and the area over which the surfaces touch is still given by the elastic equation. At the stage where the load reaches the value W the elastic deformation has reached its end. The whole of the surface is in a state of 'incipient' plasticity, and if the load is increased by a small amount, further plastic deformation occurs and the indentation increases in size.

We may now take the results in Fig. 6 one stage farther and describe graphically the deformation of a metal surface by a hard spherical indenter as the load W is *increased and then reduced*. Fig. 10 shows the area of contact A as a function of W for a metal which does not work-harden. Over the range OL the deformation is elastic and A is proportional to $W^{\frac{2}{3}}$, as in Fig. 6. At L the onset of plasticity commences and the value of p_m gradually increases with the load. At M full plasticity is reached, the yield pressure is now almost constant, and the area A increases linearly with W along the straight line MN. If the load is removed at N, the surfaces recover elastically and the area A varies again as $W^{\frac{2}{3}}$. The curve NQO is reversible so long as the load does not exceed W_N. When the load exceeds W_N further *plastic* deformation occurs and the area increases along the line NN'.

† The elastic energy which is stored in the surfaces and which is recoverable when the load is removed may be readily estimated. Simple calculations show that this energy is responsible for the rebound observed in the plastic deformation of surfaces during impact (see Chapter XIII).

Similar considerations apply to any bodies which deform one another plastically. When the load is reduced there is relaxation of elastic stresses in the bodies and the surfaces separate according to the laws of elastic deformation. For clean metals the separation may be hindered if there is strong adhesion between the surfaces when the load is first applied. The metallic junctions so formed may not be broken by the

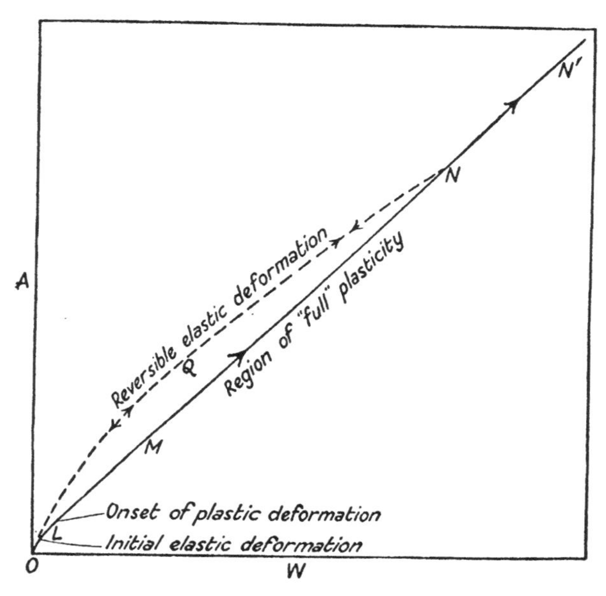

Fig. 10. Area of contact A as a function of the load W for increasing and decreasing load. Once full plasticity has been reached (M) the area A is proportional to W (full line MNN'). For decreasing load the area is determined by the elastic equations and A is proportional to $W^{\frac{2}{3}}$ (broken line NQO).

released elastic stresses when the load is reduced and the area of contact may not decrease appreciably. As we shall see in Chapter XV, this is particularly marked with soft metals. With harder metals or with contaminated surfaces, however, the junctions appear to break readily on reducing the load and the decrease in area of contact follows the elastic equations. This applies both to the surface irregularities and to any macroscopic plastic deformation that may occur. For example in Fig. 9 (c) both the tips of the surface irregularities and the contour of the macroscopic indentation have recovered elastically. The reapplication of the original load will cause both types of surface to deform elastically until they reach the contour they occupied during the original deformation.

Electrical Resistance as a Measure of Area of Real Contact

So far our discussions have been based on the theory of plastic and elastic deformation and on visual examination of the surfaces *before* and *after* they have been placed in contact. We may now consider a method which provides a measure of the area of real contact whilst the surfaces are actually in contact. This consists of determining the electrical

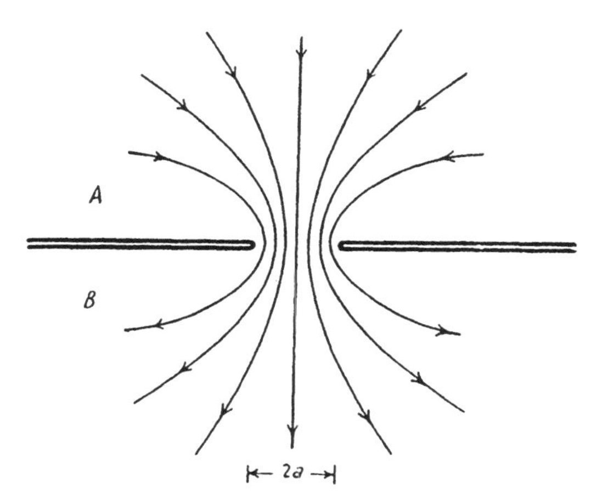

Fig. 11. Flow of current through a circular constriction of radius a in a massive conductor of electrical conductivity λ. The constriction produces an effective 'spreading resistance' R of amount $R = 1/(2a\lambda)$.

resistance between the touching surfaces (Holm, 1946; Bowden and Tabor, 1939).

Let us suppose that two massive metal specimens A and B of specific conductivity λ make contact over a circle of radius a. If current flows from A to B it is concentrated into the constriction and spreads out on either side of it (Fig. 11). This constriction of the current stream introduces a spreading resistance R (*Ausbreitungswiderstand*), the value of which depends on λ and a. Maxwell considered a similar problem in 1873 when he determined 'the correction which must be applied to the length of a cylindrical conductor of radius a when its extremity is placed in metallic contact with a massive electrode'. He showed that to a very close approximation, if a is small compared with the size of the electrode, the additional resistance produced by the spreading of the current is equal to $1/(4a\lambda)$. In the above case the contact between the two massive specimens is equivalent to two such junctions in series, so that the spreading resistance from A to B is given by

$$R = \frac{1}{2a\lambda}. \tag{10}$$

This model assumes that over the circle of contact the surfaces are within range of one another's fields of force so that, from the point of view of electrical conduction, the specimens A and B behave as a continuous single conductor of uniform conductivity λ. Those portions of A and B which are a few Ångströms apart will also be partly conducting on account of the 'tunnel' effect, though the effective conductivity will be very much lower than for a metal. For portions of the surfaces more than about 5 A apart the tunnel effect is negligible. Consequently the contact resistance, if the region of contact is essentially metallic, provides a reasonably close measure of the real area of contact between the surfaces, that is, the area over which one surface is within molecular range of the other.

In practice, however, most surfaces are covered with oxide layers and other contaminant films, the resistance of which may readily dominate the contact resistance. If we assume the surface films to be of constant thickness on each surface and to have a resistance per sq. cm. of σ, the combined resistance of the junction will be approximately

$$R_1 = \frac{1}{2a\lambda}+\frac{2\sigma}{\pi a^2}. \tag{11}$$

It is seen that the 'true' spreading resistance part is proportional to $1/a$; the film resistance to $1/a^2$.

Some measurements were made of the electrical contact resistance R between various metal surfaces under various experimental conditions. In nearly all cases the resistance was measured by a current-potential method, as R is often small compared with the resistance of the leads. Typical arrangements of crossed cylinders and of a sphere on a flat are shown in Fig. 12.

When the metal surfaces were placed together with a light load the resistance was at first variable. If a vibrating tuning-fork was held against the surfaces the resistance decreased slightly and reached a steady state which was comparatively reproducible. The technique was first developed by Meyer (1898). For very heavy loads this procedure was not effective, but a slight relative movement of the surfaces produced the same effect. It is probable that this process breaks through the contaminating films at the regions of contact.

If the geometry of the surfaces is such that the region of contact is a circle and is localized in one region, as in Fig. 12, we may form an

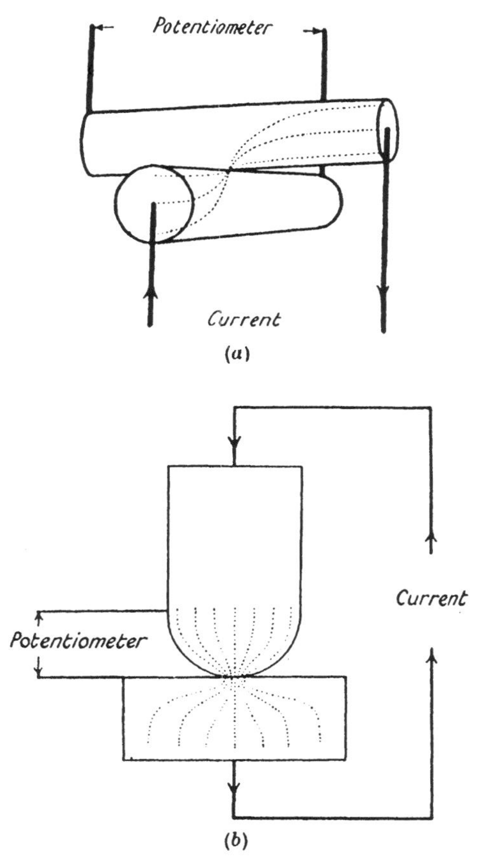

FIG. 12. Arrangement for measuring contact resistance.
(a) Crossed cylinders. (b) Sphere on a flat.

estimate of the actual area of contact between the surfaces in several ways:

i. From the electrical resistance. Assuming the contact to be essentially metallic the area of contact is given by πa^2, where a is obtained from equation (10). In applying this equation we may ignore the effect of pressure and work-hardening on the specific conductivity of the metal, since these are second-order effects.

ii. From microscopic examination of the permanent impressions left in the surfaces.

iii. From the yield pressure p_m of the metals. If W is the load and the specimens are fairly highly worked, $W = Ap_m$. The value of p_m may be found from simple indentation experiments.

Some typical results (Bowden and Tabor, 1939) for silver and steel surfaces are given in Table III.

TABLE III

Surfaces	*Load kg.*	*R ohms*	*Area of contact, A sq. cm.*		
			From R	*Visually*	*From p_m*
Silver					
Crossed cylinders . .	0·5	100×10^{-6}	0·0002	..	0·0002
	5·0	30×10^{-6}	0·002	..	0·002
	50·0	9×10^{-6}	0·018	0·019	0·02
	500·0	$1{\cdot}9\times10^{-6}$	0·15	0·19	0·2
Steel					
Crossed cylinders . .	1	$1{\cdot}0\times10^{-3}$	0·00012	..	0·0001
	5	$4{\cdot}9\times10^{-4}$	0·00061	..	0·0005
	50	$1{\cdot}6\times10^{-4}$	0·0045	0·0045	0·005
	500	$4{\cdot}9\times10^{-5}$	0·042	0·045	0·05
Sphere on flat . . .	5	$4{\cdot}5\times10^{-3}$	0·00065	..	0·0005
	50	$1{\cdot}6\times10^{-3}$	0·0045	..	0·005
	500	$4{\cdot}7\times10^{-5}$	0·045	..	0·05

Three conclusions follow from the results given in this table. First the contact resistance is extremely small even for very light loads. For this reason considerable care must be taken in the electrical measurements. Secondly, there is good agreement between the three methods of determining A. This means that when the contact occurs over a single well-defined region the electrical measurements provide a reasonable measure of A, provided the surfaces have been subjected to vibration in the appropriate way. Thirdly, in such cases the area of contact is not very different for surfaces of different shapes; A is determined essentially by the load and the yield pressure.

Effect of load on contact resistance

If the surfaces are placed together and subjected to vibration the contact resistance reaches a low, relatively reproducible, value. If the load is increased there is very little reduction in resistance (unless the surfaces are again vibrated), presumably on account of the relatively high resistance of the surface layers. If, however, the load is reduced after the surfaces have been vibrated we may expect that the variation in resistance with decreasing load will follow a well-defined relationship, since the separating surfaces are now in a 'reproducible' state. For example, the area of contact as the surfaces separate should follow the elastic equations so that 'a' should decrease as $W^{\frac{1}{3}}$. Consequently the resistance should increase as $W^{-\frac{1}{3}}$, if the resistance is mainly metallic, and as $W^{-\frac{2}{3}}$ if the resistance is mainly due to very thin oxide films. Numerous experiments show that as the load is decreased the resistance increases in a fairly reproducible manner and follows a relation of the type $R = kW^{-n}$, where n has a value ranging from about 1/2·2 to 1/2·6 according to the material of which the surfaces are made. This result suggests that the contact is essentially metallic, but that small amounts of oxide are still present which are sufficient to modify the ideal relation $R = kW^{-\frac{1}{3}}$.

The essentially metallic nature of the contact when the surfaces have been subjected to vibration is confirmed in several other ways. Firstly, the resistance is independent of the current over a very wide range of currents, that is, it is ohmic. This is a well-marked characteristic of metals, and metallic oxides do not as a rule show this behaviour. It is true, however, that small amounts of oxide sufficient to affect the R–W relation would not produce a very noticeable change in the resistance–current behaviour. Secondly, if the surfaces are deliberately covered with a thin film of paraffin oil and then subjected to vibration the resistance is essentially the same as for clean dry surfaces. This suggests that the high pressures and the vibration break up and squeeze out the oil film. Although the oxide film is considerably more rigid, we may expect that there is a similar breaking through of the oxide layer. The ease with which these oxide films may be broken down is demonstrated in Chapter IV. Thirdly, as we saw in Table III, there is close quantitative agreement between the area of contact observed visually and that calculated from the contact resistance, assuming metallic contact. Here again, however, there may still be small quantities of oxide between the surfaces which may play a less significant role in this type of measurement but a more important part in others. Holm has, in fact, shown, by carrying out parallel measurements at liquid-air temperatures where the metallic resistance becomes negligible, that unless the most rigid precautions are taken the oxide resistance may easily be comparable with the metallic resistance.

It follows that for surfaces prepared in the atmosphere, deductions derived from the contact resistance must be viewed with some caution. However, if suitable precautions are taken the results in Table III show that this direct method of determining the electrical resistance can provide a reasonable measure of the real area of contact between metal surfaces.

The contact of flat surfaces

Steel surfaces, finely ground and lapped flat to a few fringes, were placed in contact and the electrical resistance across them measured. One pair of flats had an area of 0·8 sq. cm., the other of 21 sq. cm. The measurements showed two striking results (i) At any given load, the contact resistance of both pairs of flats was almost the same although their apparent area of contact was about 25:1. (ii) At any given load the contact resistance for the flats was of the same order of magnitude as that observed with crossed cylinders. For example, at a load of 5 kg. the contact resistance for the crossed cylinders was about 5×10^{-4} ohms, corresponding to an area of contact of about 5×10^{-4} sq. cm. For the 21 sq. cm. flat which has an apparent area 40,000 times as great, the contact resistance was only one-half this value, i.e. about $2{\cdot}5\times10^{-4}$ ohms. Assuming that for both types of surfaces the contact is metallic to the same degree, it is clear that only a small fraction of the area of the flat surfaces can be in intimate contact; they must touch only at the tips of the highest asperities. This would also explain the observation that the contact resistance is almost the same for the large and small flats.

It is difficult in the case of flat contacts to make an exact estimate of the real area of contact from the conductance measurements alone. The value of the conductance must depend both on the size of the metallic bridges and on their number. Since the spreading resistance of each bridge is inversely proportional to its diameter, and the area of contact is proportional to the square of the diameter, it follows that for a given resistance the area of contact is inversely proportional to the number of bridges. We do not know this number with any certainty, though in the case of flat surfaces it is clear that that number of points of contact on which they are supported cannot be less than three. If we assume the number of bridges is, in fact, three, we may calculate the area of contact from the contact resistance. The results indicate that only a very small fraction of the macroscopic area of the surfaces is in intimate contact.

It is, however, not very satisfactory to assume that the number of legs remains constant at this value. A more satisfactory type of calculation may be carried out if we assume that the yield pressure p_m of the asperities is approximately the same as that of the bulk metal and that the real area of contact A is determined essentially by the ratio $A = W/p_m$. For the steel in these experiments, $p = 100$ kg./sq. mm. If we assume that the surfaces are supported on n equal bridges of radius a, the contact resistance when the bridges are relatively far apart is given by $R = 1/(2an\lambda)$. Combining this equation with the relation

$$A = n\pi a^2 = W/p_m$$

we obtain

$$n = \frac{\pi p_m}{4\lambda^2 R^2 W} = \frac{1{\cdot}39 \times 10^{-6}}{R^2 W}, \tag{12}$$

$$a = \frac{2\lambda W R}{\pi p_m} = 4{\cdot}77 W R. \tag{13}$$

Results for the 21 sq. cm. steel flats are given in Table IV.

TABLE IV

Load *kg.*	$A = W/p$ *sq. cm.*	*Fraction of macroscopic area in contact*	R 10^{-5} *ohms*	n	a 10^{-2} *cm.*
500	0·05	$\frac{1}{400}$	0·9	35	2·1
100	0·01	$\frac{1}{2000}$	2·5	22	1·2
20	0·002	$\frac{1}{10000}$	9	9	0·9
5	0·0005	$\frac{1}{40000}$	25	5	0·6
2	0·0002	$\frac{1}{100000}$	50	3	0·5

The absolute values of a and n in this table must be viewed with some reserve since the presence of a small amount of oxide will have an appreciable effect on both these parameters. For example, if half the resistance is due to an oxide film the value of n will be increased fourfold whilst the value of a will be halved. Nevertheless it is probable that the values given in the table are of the right order of magnitude. The main point brought out is that the effect of increasing the load is to increase the number and average size of the bridges. Further, it should be noted that the number of bridges is not very large even at the heaviest loads, whilst the area of the bridges lies between 10^{-3} and 10^{-4} sq. cm. These results are in complete agreement with the discussion on p. 22 and are consistent with the view that the surfaces are held apart by small irregularities which flow under the applied load until their total cross-section is sufficient to support the load.

It follows from the discussion in this chapter that the real area of contact between metals is small and that at these points of real contact the metals are deformed plastically. As we shall see in a later chapter, welded metal junctions are formed at the points of intimate contact and these junctions play a fundamental part in the mechanism of friction and wear.

REFERENCES

R. F. BISHOP, R. HILL, and N. F. MOTT (1945), *Proc. Phys. Soc.* **57**, 147.
F. P. BOWDEN and D. TABOR (1939), *Proc. Roy. Soc.* **A 169**, 391.
H. HENCKY (1923), *Z. angew. Math. Mech.* **3**, 241.
H. HERTZ (1886), *J. reine angew. Math.* **92**, 156.
R. HOLM (1946), *Electric Contacts*. Almquist and Wiksells, Uppsala. Readers in this field will find this original book of Dr. Holm's of considerable interest.
A. J. ISHLINSKY (1944), *J. Appl. Math. Mech.* (*U.S.S.R.*), **8**, 233. English translation: Ministry of Supply, A.R.D. Theoretical Research Translations, No. 2/47.
MASSACHUSETTS INSTITUTE OF TECHNOLOGY. *Conference on Friction and Surface Finish*, 1940.
C. MAXWELL (1873), *Electricity and Magnetism*, **1**, article 308.
A. MEYER (1898), *Öfvers. VetenskAkad. Förh.*, Stockh., **55**, 199.
A. J. W. MOORE (1948), *Proc. Roy. Soc.* **A 195**, 231.
H. R. NELSON (1940), *Conference on Friction and Surface Finish*, M.I.T. 217.
H. O'NEILL (1934), *The Hardness of Metals and its Measurement.* London.
R. E. REASON, M. R. HOPKINS, and R. I. GARROD (1944), *Report on Measurement of Surface Finish by Stylus Methods.* Taylor-Hobson, Leicester, England.
D. TABOR (1948), *Proc. Roy. Soc.* **A 192**, 247.
S. TIMOSHENKO (1934), *Theory of Elasticity.* New York: McGraw-Hill.
S. TOLANSKY (1945), *J. Sci. Instruments*, **22**, 161.
—— (1948), *Multiple-beam Interferometry of Surfaces and Films.* Oxford Univ. Press.
R. C. WILLIAMS and R. W. J. WYCKHOFF (1944), *J. Appl. Phys.* **15**, 423.

For a number of surface problems discussed in this monograph, the most useful general work is Professor N. K. Adam's admirable book *The Physics and Chemistry of Surfaces* (Clarendon Press, Oxford).

II

SURFACE TEMPERATURE OF RUBBING SOLIDS

The very rapid friction of two thick bodies produces fire.

LEONARDO DA VINCI

WHEN one solid body slides over another most of the work done against the frictional force opposing the motion will be liberated as heat between the surfaces. This heat will be carried away from the surfaces by conduction and radiation, but quite primitive calculations indicate that even under moderate conditions of speed and load the surface temperatures may reach very high values. It is of course difficult to measure these temperatures by ordinary methods. If we embed thermometers or thermocouples in the solids near the rubbing surfaces, we find that the temperature rise is very small. This is partly because of the relatively large heat capacity of the thermometer (or thermocouple), but is mainly due to the fact that the temperature falls off very rapidly as we move away from the actual rubbing interface. We may, however, measure the surface temperatures directly by using the surfaces themselves as a thermometer. If the sliding surfaces are made of two different metals, the thermo-electric potential generated on sliding provides a measure of the surface temperature. It is evident that the thermo-electric effects and the frictional processes occur at the same points of intimate contact between the surfaces. These thermo-electric measurements therefore yield information concerning the temperature of the surface layers of the metals at the points where they are actually rubbing. The first part of this chapter describes an investigation, by thermo-electric measurements of this type, of the surface temperatures developed between sliding metals. The second part of the chapter deals with the surface temperatures developed between non-conducting solids, where the thermo-electric method is inapplicable.

SURFACE TEMPERATURE OF SLIDING METALS

Calculation of Surface Temperature

Consider a cylinder whose face slides over a surface with a velocity of v cm. per second (Fig. 13). If all the frictional work is liberated as heat, then the amount Q of heat developed is given by

$$Q = \frac{\mu W g v}{J} \quad \text{calories per second,} \tag{1}$$

where μ = coefficient of kinetic friction, W = load on the cylinder, g = constant of gravity, J = mechanical equivalent of heat.

This heat will raise the temperature of the sliding bodies. We assume that the temperature across the whole of the rubbing interface is constant and that there is a steady temperature drop along the length of the cylinder. We now consider the flow of heat through any element

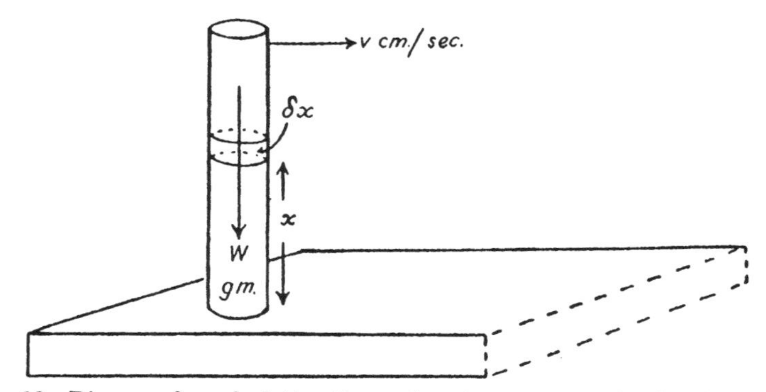

Fig. 13. Diagram for calculating the surface temperatures developed between sliding metals. The narrow cylinder is assumed to make thermal contact over the whole of its lower face.

of the cylinder at a distance x from the surface. The amount of heat gained by conduction (assuming the thermal conductivity k is independent of temperature) will be

$$k\pi r^2 \frac{d^2T}{dx^2}\delta x,$$

where r = radius of cylinder. The heat lost by emission will be

$$\sigma 2\pi r(T-T_0)\,\delta x,$$

where σ = cooling coefficient assuming Newton's law of cooling, T = temperature of emitting surface, T_0 = temperature of the surroundings. In the steady state these must be equal and

$$k\pi r^2 \frac{d^2T}{dx^2} = \sigma 2\pi r(T-T_0). \tag{2}$$

Hence

$$T-T_0 = Ae^{-\sqrt{(2\sigma/kr)}x}, \tag{2 a}$$

where the constant A is still unspecified.

Of the frictional heat Q which is liberated some will go into the upper body and some into the lower. We do not know exactly how it will be divided, but we may consider that a certain fraction α will go into the top cylinder. All the heat which goes into the cylinder must be emitted from it. If it is so long that the top end is sensibly at room temperature we may write

$$\alpha Q = 2\pi r\sigma \int_0^\infty (T-T_0)\,dx. \tag{3}$$

From (2a) and (3) it follows that

$$T-T_0 = \frac{\alpha Q}{\pi r}\frac{1}{\sqrt{(2\sigma kr)}}e^{-(2\sigma/kr)x}, \tag{4}$$

so that at the rubbing interface where $x = 0$ the rise in temperature is given by

$$T-T_0 = \frac{\alpha\mu Wgv}{J\pi r}\sqrt{\left(\frac{1}{2\sigma kr}\right)}. \tag{5}$$

We may apply this calculation to an example which can be realized experimentally. If a constantan cylinder 1 mm. in diameter, loaded with 100 gm. wt., is slid over a mild steel surface with a velocity of 200 cm./sec, the measured value of the kinetic friction is 0·3, k is 0·05 cal. cm.$^{-1}$ sec.$^{-1}$ °C.$^{-1}$ and σ is about 0·001 cal. cm.$^{-2}$ °C.$^{-1}$ Inserting these quantities in equation (5) and assuming that α is $\frac{1}{2}$ (a reasonable assumption for sliding surfaces of comparable conductivities), we find that the rise in temperature of the constantan surface is given by

$$T-T_0 \approx 200°\text{ C.}$$

This calculation is made on the assumption that the area of contact is that of the end of the cylinder. This is certainly not so (see Chap. I). Even with carefully prepared surfaces, the actual rubbing area of contact may be only a very small fraction of the apparent area, so that the temperature rise in the small region of contact should be very much greater. The above calculations make sweeping assumptions and can only be regarded as crude approximations which would give minimum temperatures. Nevertheless it is apparent from them that on simple theoretical considerations we should expect the surface layers to reach very high temperatures.

Measurement of surface temperature

The experimental arrangement (Bowden and Ridler, 1936) and the method of measuring the surface temperature is shown in Fig. 14. One of the metals A is in the form of a flat annular disk, which can be rotated with a uniform velocity about the point O. A wire of the same metal A leads down the axis of rotation and dips into a mercury cup M. This is connected by a copper wire to one terminal of a high-frequency galvanometer or cathode-ray oscilloscope G. The second metal which constitutes the other half of the thermocouple is in the form of a cylinder B, which rests on the disk. The metal B is connected by a copper wire to the other terminal of the galvanometer. All the metal junctions except the sliding one at S are at room temperature. A method which depends on the same principle has been devised by Shore (1925) and Herbert (1926) and applied to the measurement of

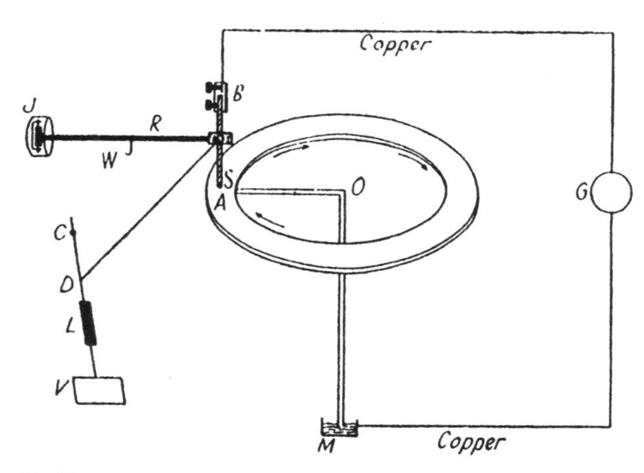

FIG. 14. Experimental arrangement for measuring the thermal p.d. developed between sliding metals.

the temperature of cutting tools. The cylinder B is attached to a rigid arm R which is carried on a gimbal J, so arranged that the cylinder can move freely up and down or to and fro. Any required load can be applied to the cylinder by adding weights to the arm at W. The apparatus is heavily constructed and mounted on massive concrete blocks. The rotating disk is mounted on ball bearings and is very accurately machined. The surface of the disk can be ground flat to a few fringes. The disk can, by a suitable arrangement of belts, be driven at different rates so that the sliding speed can be varied from a fraction of a centimetre per second up to high speeds. The gimbals J are not rigidly fixed but are carried on an arm which is so arranged that a rotation of a screw head causes the cylinder to move across the surface of the disk. The reason for this will be apparent later.

When the metal disk rotates, the frictional force between the surfaces at S will tend to drag the cylinder with it. This movement is prevented by a fine wire which is fixed to the cylinder and comes off at a tangent to the disk. If a measurement of the average friction is required the other end of this wire is attached to a damped pendulum or a damped spring which records the average force acting on the wire. If an analysis of the frictional force is required it is attached to an oscillograph.

The surface temperature is determined from the maximum thermo-electric potential developed between the rubbing surfaces and recorded on the galvanometer. Subsidiary calibrations are carried out with thermocouples of the same combination as the sliding surfaces. There are, however, two possible sources of error in these measurements. Small changes in resistance take place between the surfaces during sliding and this might alter the apparent p.d. The contact resistance is only a few hundredths of an ohm. Since in these experiments the thermocouple is always in series with a high resistance, any changes in the contact resistance between the surfaces can have no appreciable effect. A second possible source of error is connected with the fact that the thermo-electric potential varies with the state of working of the metals which form the couple. Because of this, any work-hardening of the metals during sliding will cause a change in their thermo-electric power. It is clear, however, that this effect is also small. The thermo-electric potential between worked and annealed specimens of a metal varies slightly with the direction of measurement on the specimen, but it is always small. For constantan it is about $0{\cdot}7\times10^{-6}$ V./°C., and for iron and steel it is of the same order (Elam, 1935). The average thermo-electric potential of the constantan–steel couple formed by the surfaces was about 25×10^{-6} V./°C., so that the state of working of the surfaces can only affect the real potential by a few per cent. This variation is within the limits of experimental error. Bridgman (1918) has shown that pressure can also cause a change in the thermo-electric power of a metal, but the magnitude of this effect is even smaller than that discussed above.

Temperature of sliding metals

Experiments with a mild steel cylinder sliding over a mild steel surface gave no potential difference even at the highest loads and speeds. This is to be expected if the potential is a true thermo-electric one. It shows that the electromotive force is not due just to rubbing, nor to the disturbance of oxide or films on the metal surface, nor to any electromagnetic effect. In order to test this further a number of experiments was made with a series of fusible metals sliding on steel. Cylinders of gallium (m.p. 32° C.), Wood's alloy (m.p. 72° C.), lead (m.p. 327° C.), and constantan (m.p. 1,290° C.) were slid on a steel surface and the thermal e.m.f. measured. Some typical results for the maximum temperatures developed are shown in Fig. 15, where the rise of surface temperature calculated directly from the measured potential is plotted against the speed. As will be shown later, the temperature (as well as the friction) fluctuates during sliding and the temperature values plotted are the peaks of the galvanometer record. For each metal the results were similar: the temperature increased as the speed

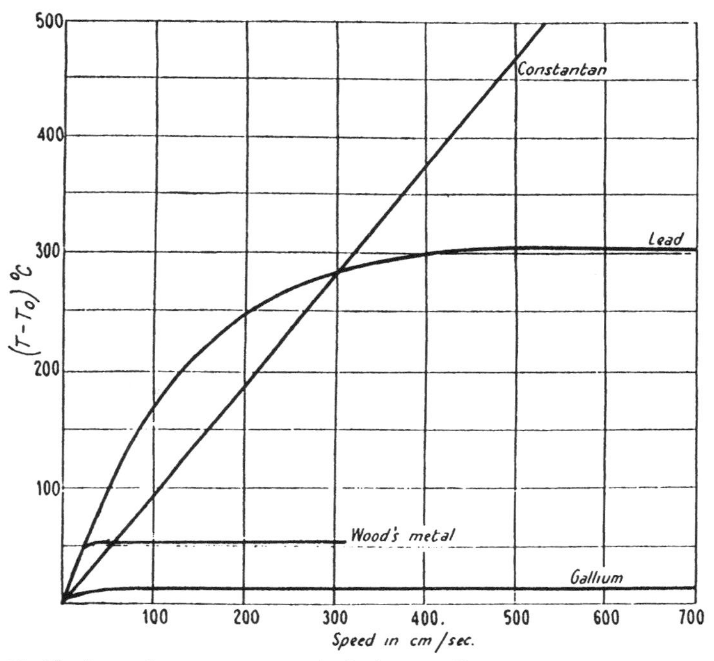

FIG. 15. Maximum temperatures reached when small cylinders of gallium, Wood's metal, lead, and constantan are slid on a steel surface. (Load 100 gm.) The temperature does not exceed the melting-point of the metal.

increased and rose to a maximum value which could not be exceeded. *This maximum corresponded numerically to the melting-point of the metal* as determined from separate calibrations of the thermo-couple. The temperature of the room (T_0) was 17° C., so that with gallium the flattening occurred at 32° C., with Wood's alloy at 72° C., and with lead at 327° C. The melting-point of constantan (m.p. 1,290° C.) was not approached at the loads and speeds employed and no flattening of the curve occurs. It will be seen from Fig. 15, Fig. 17, and Plate III*a* that temperatures of 500 or 1,000° C. are easily reached on the metal surfaces, even under moderate conditions of load and speed. There is no obvious sign of this heating: the mass of the metal appears to be quite cool, and it is clear that the intense heating is confined to a thin layer in the region where rubbing actually occurs.

The fact that the intense heating is confined to a thin layer at the surface of contact is shown in an interesting way by the behaviour during an experiment. If the cylinder of metal B was allowed to run in the same track for a short time, the thermo-electric potential fell off and reached a low value. This was not because of any real fall in surface

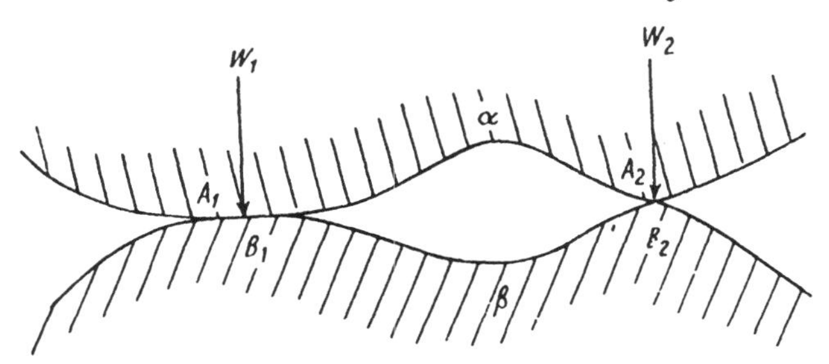

FIG. 16. Conditions of contact between metal surfaces.

temperature, but because a thin layer of metal B was rubbed off on to the steel and the sliding surface was now B/B, and therefore gave no thermo-electric potential. The thickness of the metal layer necessary to cause this falling off was very small. Frequently the layer was too small to be visible. In practice the cylinder was kept moving gradually across the steel surface so as to expose continually a fresh rubbing surface. Under these conditions there was no falling off in the potential.

It is apparent that the surface temperature will not be uniform over the whole surface of the metal. Although the surfaces were carefully prepared they cannot be perfectly flat, and the conditions of contact may be represented diagrammatically in Fig. 16. It is only the areas in contact $A_1 B_1$, $A_2 B_2$, etc., which are heated directly by friction, and it is only these areas which constitute the thermo-junction. The areas α, β are not heated by direct friction, nor can they make any contribution to the thermo-electric potential.

Even the rubbing areas will not necessarily all be at the same temperature. The shape of the surface irregularities and the distribution of load will vary from point to point, so that the surface temperature at $A_1 B_1$ may be different from that at $A_2 B_2$. The electromotive force measured at any instant is that of a number of thermo-junctions connected in parallel, all of which are not at the same temperature. The observed potential will be the integrated effect of all the thermocouples formed by the parts of the surfaces which are in contact. It is apparent that some points at the surface may be at a temperature appreciably higher than that indicated by the observed thermo-electric potential.

Temperature of lubricated surfaces

The experiments were repeated with various lubricants on the polished surfaces. The conditions were those of 'boundary lubrication', that is, the surfaces were not separated by a thick layer of fluid lubricant, but

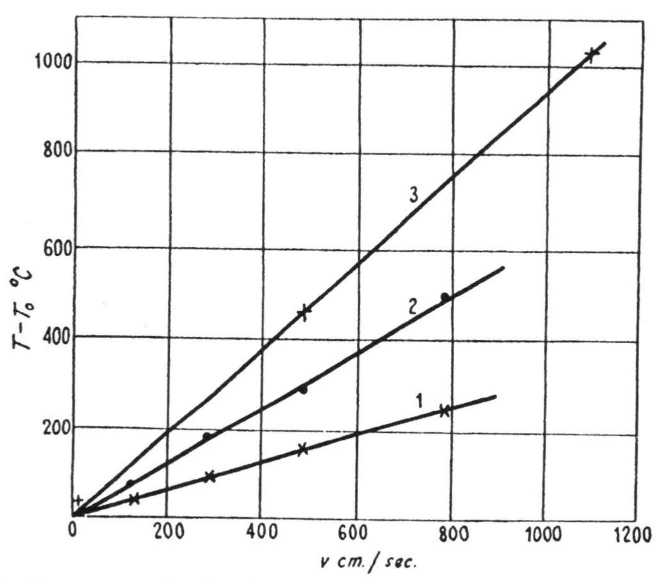

FIG. 17. Temperature developed at the points of rubbing contact between a constantan pin and a steel disk. Load 102 gm. 1, oleic acid; 2, commercial lubricant; 3, no lubricant. It is seen that even in the presence of a lubricant, temperatures of several hundred degrees C. are readily reached at moderate loads and speeds.

by a lubricant film of molecular dimensions (see Chap. IX). As will be shown later, there is strong evidence that this adsorbed film does not remain intact, but is continuously being destroyed and repaired during sliding. The fact that an electromotive force is developed on sliding provides additional evidence that metallic contact occurs through the boundary film.

The surface temperatures reached by steel sliding on constantan lubricated under 'boundary conditions' with various lubricants are shown in Fig. 17.

It is evident from these experiments that even in the presence of lubricant films the surface temperatures reached may be very high, and may exceed several hundred degrees centigrade at relatively small loads and sliding speeds. It should again be emphasized that in these experiments there is no obvious sign of bulk heating: the surfaces slide as though they are well lubricated and the mass of the metal is quite cool.

The fact that these high temperatures are reached even by well-lubricated surfaces obviously has an important bearing on the theory and practice of boundary lubrication. It is true that these high temperatures are localized at the points of contact, but it is just at these points that friction and lubrication are occurring. This will be discussed in more detail in Chapters IX and XII.

Intermittent nature of surface temperature

It has already been pointed out that, in general, there are marked fluctuations in the surface temperature during sliding. These fluctuations are of great interest since they yield information concerning the detailed processes that occur between the sliding surfaces. They may be analysed by means of a sensitive direct current amplifier and a cathode-ray oscillograph. Fluctuations in the thermo-electric potential lasting less than 10^{-4} sec. can be followed by the cathode-ray beam and a continuous record of the temperature obtained on a moving film camera.

Some typical results (Bowden, Stone, and Tudor, 1947) for a constantan cylinder sliding on a steel surface are shown in Plate III. 1 for a load of 500 gm. and a sliding speed of 300 cm./sec. It is seen that the temperature is fluctuating very rapidly, and temperature flashes of 1,000° C. which may last less than 10^{-4} sec. are recorded.

The results obtained when the surfaces are lubricated with a mineral oil are similar except that the peak temperatures are reduced. It is clear that very high temperatures and very rapid fluctuations can occur at the points of real contact through the lubricant film.

Surface temperature and thermal conductivity

If the sliding solid is a poor thermal conductor the frictional heat will not be conducted away so rapidly, and the surface temperature will be correspondingly higher. According to equation (5) the surface temperature should vary inversely as the square root of the thermal conductivity k.

In Fig. 18 the rise in surface temperature of cylinders of copper, nickel, lead, constantan, Wood's metal, and bismuth sliding on polished steel under identical conditions of load and speed is plotted against $1/\sqrt{k}$ for these metals.

These results are not corrected for slight differences of friction and real area of contact, but it is clear that the surface temperature is greater the smaller the thermal conductivity of the metals. For example, the rise in surface temperature of bismuth, which is a poor thermal conductor, is some five times as great as that of copper under the same experimental conditions. These results are of further interest since they

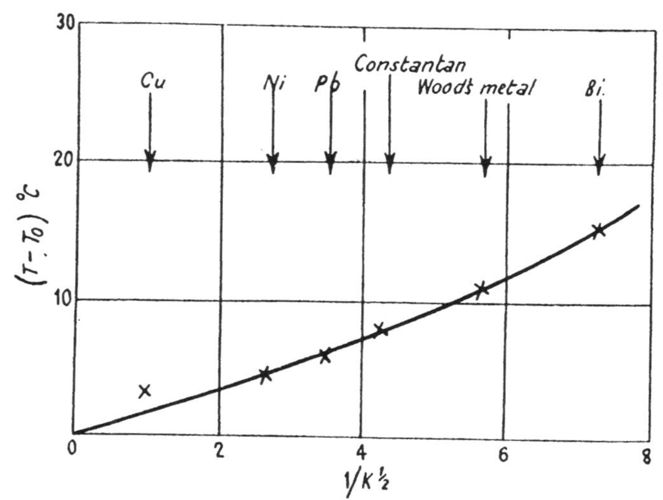

FIG. 18. Temperature rise of rubbing surfaces as a function of their thermal conductivity k (load 32 gm., sliding speed 20 cm./sec.).

enable us to form some idea of the temperature which might be reached by rubbing surfaces of non-conductors such as glass or silk. Under the same conditions of load and speed, we should expect that with surfaces of glass ($k = 0{\cdot}0017$), for example, the temperature rise would be greater by an order of magnitude than that observed on copper ($k = 0{\cdot}92$). With rubbing surfaces of silk ($k = 0{\cdot}0001$) it would be greater still. We should, therefore, expect that with non-conductors local high temperatures at the points of rubbing contact should occur very much more readily than with conducting surfaces.

Surface Temperature of Non-conducting Solids

Obviously the thermo-electric method cannot be used with non-conducting solids such as glass or quartz, but it is possible to show the existence of local high temperatures by visual means. If one or both of the surfaces is made of transparent material, such as glass or quartz, and the apparatus so arranged that a clear image of the rubbing surfaces can be seen, it is found that, when sliding starts, a number of tiny stars of light appear at the interface between the rubbing surfaces. It is clear that they correspond to small hot spots on the surface.

In the experiments described here the lower surface was usually a flat glass disk, and a mirror was mounted beneath it in such a way that the region of contact between the slider and the disk could be clearly seen. In some of the experiments, when a transparent slider was used, the top of the slider itself was optically polished and observations made through it. The lower surface was set in motion at a fixed speed and the load on the slider gradually increased. When the load reached a sufficiently high value a number of small, dull red luminous points was observed on the surface of the slider. The position of these points changed from instant to instant as the points in intimate contact wore away and new regions came into contact. If the speed (or load) was increased the spots became brighter and whiter, corresponding to the higher temperatures reached. (The possibility that these luminous spots are due to some triboluminous effect other than a thermal heating must be considered, but the evidence is against this.) The results in the following sections refer to the conditions under which the first dull red hot spots were observed visually in a completely darkened room. The experiments were carried out for clean surfaces and for surfaces wetted with a mixture of glycerine and water.

TABLE V

Incidence of hot spots with sliders of different melting-points on glass

Composition of alloy	*hardness V.D.H.*†	*m.p. °C.*	*Visual hot spots on clean glass*
80 Au, 20 Sn	230	300	None
80 Au, 20 Pb	108	420	,,
75 Au, 25 Te	120	450	,,
73 Ag, 27 Sn	93	480	,,
70 Ag, 30 Sb	120	480	,,
80 Ag, 20 As	170	500	,,
50 Au, 50 Cd	..	520	,,
92 Au, 8 Al	221	570	Hot spots
Constantan	130	1,200	,,
Nickel	170	1,450	,,
Iron	130	1,500	,,
Tungsten	..	3,000	,,

Temperature at which hot spots become visible

By using sliders of different melting-point it is possible to fix approximately the temperature at which the hot spots become visible. Experiments were carried out with a number of metals and metallic alloys, the

† V.D.H. Vickers Diamond (Pyramid) Hardness in kg./sq. mm.

melting-points of which covered a suitable range. These were selected so that they did not oxidize readily even at elevated temperatures. The results are collected in Table V.

The experiments showed that when metals or alloys melting *below* 520° C. were slid on glass or quartz, no hot spots could be seen even at the highest speeds and loads. With a gold–aluminium alloy melting at 570° C., however, and with all metals melting *above* this, the hot spots were readily seen. This would fix the temperature at which the hot spots first become visible to the eye at between 520 and 570° C.

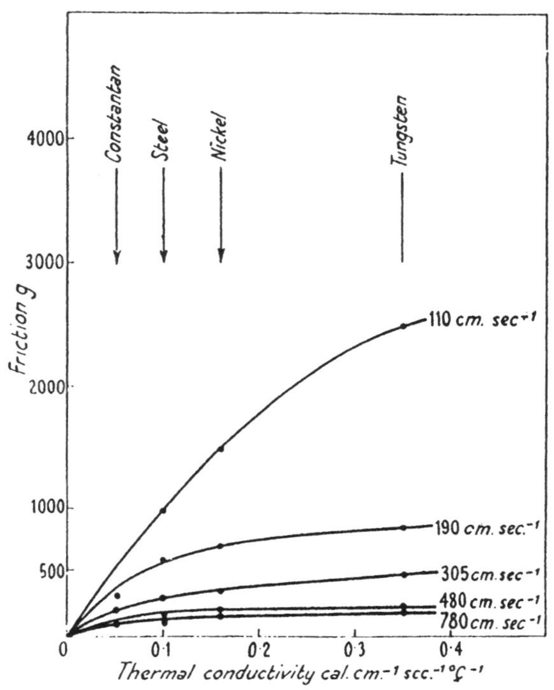

FIG. 19. The generation of visible hot spots between metal pins sliding on a glass surface. The vertical ordinate gives the frictional force at which hot spots appear for various sliding speeds, using sliders of constantan, steel, nickel, and tungsten. The smaller the thermal conductivity of the pin, the more readily the hot spots are produced.

Thermal conductivity and the incidence of hot spots

Four hard metals of widely differing thermal conductivity were selected as sliders. The metals used were constantan ($k = 0{\cdot}05$), steel ($k = 0{\cdot}10$), nickel ($k = 0{\cdot}16$), and tungsten ($k = 0{\cdot}35$ cal cm.$^{-1}$ sec.$^{-1}$ °C.$^{-1}$).

Clean surfaces. The results for clean surfaces are shown in Fig. 19. In this figure the frictional force at which visible hot spots occur (for a number of fixed sliding speeds) is plotted against thermal conductivity. It is seen that, in all cases, hot spots occur more readily the lower the thermal conductivity of the slider. This is particularly marked at the lower sliding speeds. For example, at a surface speed of 110 cm./sec. a tungsten slider gives hot spots when the frictional force is 2,600 gm., whilst with a constantan slider hot spots occur when the frictional force is only 350 gm.

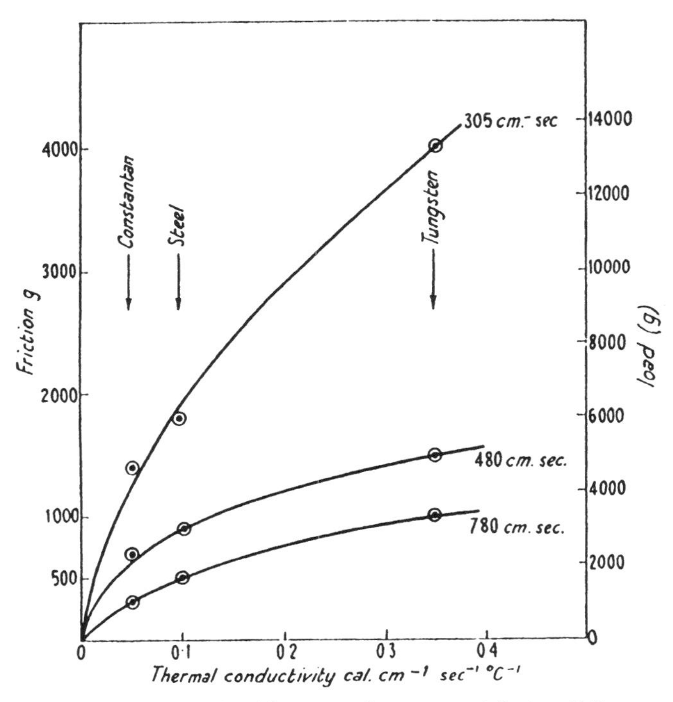

FIG. 20. The generation of visual hot spots between metal pins sliding on a glass surface covered with a mixture of glycerine and water. The results are similar to those in Fig. 19, except that the frictional force necessary to produce visible hot spots is 6 to 7 times higher.

Wet surfaces. The results for surfaces flooded with a mixture of glycerine and water are shown in Fig. 20. It is seen that the curves are of the same general form as those obtained with clean surfaces. The main difference is that higher frictional forces (six- to sevenfold) are required to produce visible hot spots when the surfaces are flooded with liquid. Although this difference is relatively large, it is evident that the presence of the liquid film is not able to prevent the occurrence of extremely high local temperatures as a result of frictional heating.

Photographic recording of hot spots

Although the occurrence of a transient hot spot is readily observable visually, the intensity is too low to affect a photographic plate. If, however, the slider is run over the same track a number of times, the cumulative effect is sufficient to produce a record on a photographic plate. A Super XX plate ($6\frac{1}{2} \times 8\frac{1}{2}$ in.) was held in a frame mounted on a steel turntable with the emulsion side upwards. A glass plate of the

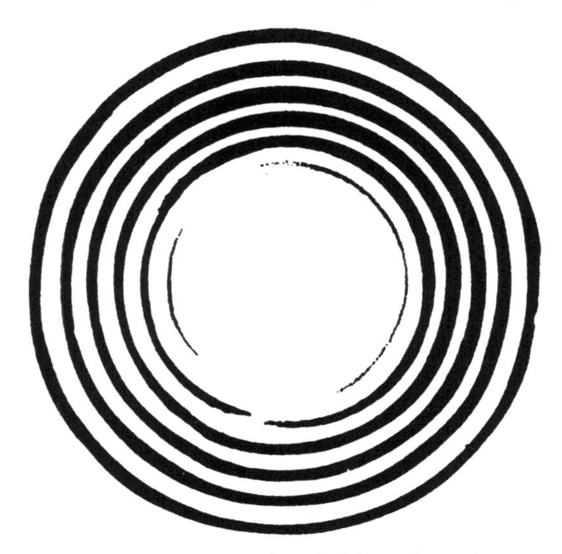

FIG. 21. Photographic hot spot trace of steel slider on lapped glass surface. Load 1,200 gm. Speed of sliding at innermost visible track 70 cm./sec. approximately.

same size was clamped on top, its upper surface being ground to a uniform grease-free finish. The slider rested on the glass surface under a given load, and was allowed to run on the same track for 2 min. It was then moved in 1 cm. and again run for 2 min. The process was repeated and, in this way, a series of concentric tracks was obtained at various radii and therefore at various peripheral speeds. On developing the plate a number of concentric dark rings appeared. The innermost visible ring on the plate gave the lowest speed at which hot spots could be recorded photographically under these conditions. A typical plate is shown in Fig. 21. It is seen that, with a load of 1,200 gm., a sliding speed of approximately 70 cm./sec. is just sufficient to produce a trace on the photographic plate. The results obtained for clean surfaces using four sliders of different thermal conductivities are shown in Fig. 22. In this figure the minimum speed to give a record on the plate is plotted against the thermal conductivity for a number of fixed loads. The results are similar to those obtained by the visual method and the actual values of the loads and speeds agree with those obtained by visual observation. It was also found that the results obtained photographically for surfaces lubricated with glycerine and water were similar to those obtained visually.

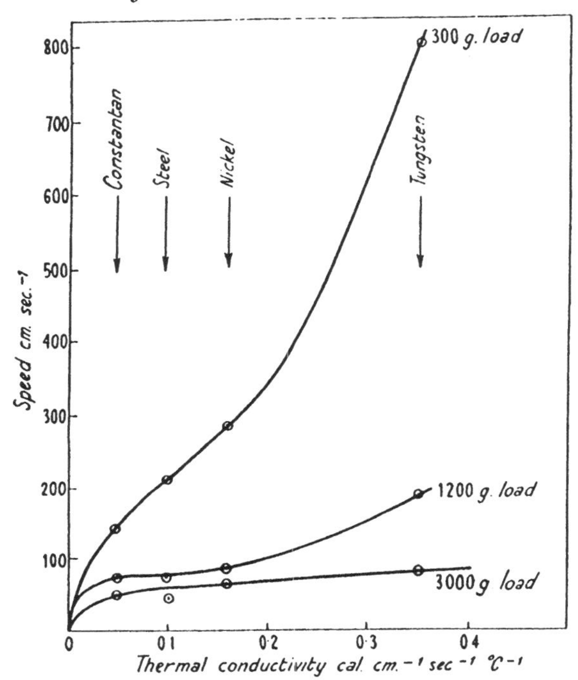

FIG. 22. Conditions necessary to produce photographic records of hot spots for metal sliders rubbing on a clean glass surface. The sliders used were constantan, steel, nickel, and tungsten and the results are similar to those obtained by the visual method as in Fig. 19.

Effect of grit on the incidence of hot spots

Some experiments were carried out to investigate the effect of grit on the incidence of hot spots.. Carborundum of known particle size was used; this was mixed into a thick paste with glycerine and water and added to the glass disk. The disk was set rotating and the load increased until hot spots became visible.

It was found that with hard metal surfaces the presence of carborundum particles had very little effect. With metals such as constantan (V.D.H. 130) and tungsten (V.D.H. 750), the load and speed necessary to produce hot spots was approximately the same as that required with wet surfaces in the absence of abrasive; also the size of the particles (which was varied from 10 to 56 μ) made no appreciable difference to the incidence of hot spots. Apparently the grit does not become embedded in the surface and so has relatively little effect. With soft metals, however, the presence of carborundum particles causes the hot spots to appear at much lower loads and speeds. Also hot spots are readily obtained with low-melting metals such as tin or lead which normally would not give them at all. Apparently the metal acts as a lap, and the sliding really occurs between the embedded abrasive particles and the glass. Particles of 20 μ and upwards are very effective: smaller particles less so. The temperature gradient set up in the abrasive particle during running must be quite high. For example, if we assume that visual hot spots correspond to a temperature of about 550° C. and that the temperature of the metal slider cannot rise above its bulk melting-point, then, in the case of the lead slider, when hot spots are observed there must be a temperature drop of about 200° C. through the abrasive particle, i.e. over a distance of about 10^{-3} cm. This implies a temperature gradient of the order of 10^{5} ° C./cm. in the abrasive particle. These high gradients will, of course, be transient since the metal will rapidly soften around the particle allowing it to sink in, whilst the load will be borne by new abrasive particles coming into contact with the surface.

Influence of size and shape of slider

For any given combination of surfaces the incidence of hot spots is determined primarily by the rate of liberation of frictional heat, that is by the product μWgv, where μ is the coefficient of friction, W the load between the surfaces, and v the speed of sliding. It is not greatly influenced by the size and shape of the slider. If a large flat slider is used the conditions of load and speed necessary are much the same as for a small curved one. The main difference is that with large flat surfaces the hot spots may be thinly distributed over a wide area instead of being concentrated into a smaller one. This is in harmony with the view that contact between the solids occurs only locally at the summit of the surface irregularities, so that the real area of contact is very small and bears little relation to the apparent area of the surfaces. It means that even with light loads the pressure at the points of real contact is high and it is just at these points that the rubbing and the liberation of frictional heat occurs. It is, of course, common knowledge that if surfaces are rubbed hard enough they get hot, but a point brought out by these experiments is that the loads and speeds necessary to give detectable hot spots are very low. For example, with constantan sliding on glass with a load of about 1 kg., visible hot spots (temp. 520–570° C.) can be seen when the sliding speed is as low as 1 or 2 feet per second. If the upper slider of metal is replaced by a poor conductor such as quartz ($k = 0{\cdot}0035$), the hot spots appear even more readily.

Measurement of transient hot spots [*A*]

An additional method for studying the occurrence of transient hot spots is by the use of infra-red sensitive photocells such as the lead-sulphide cell. These cells can be constructed with very small time

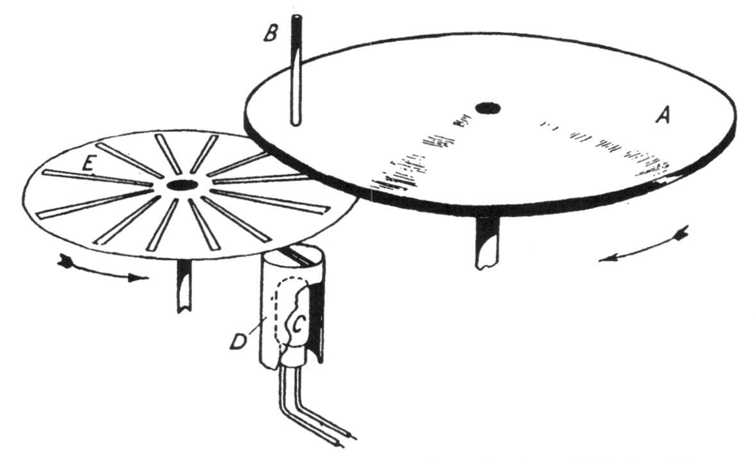

FIG. 23. Experimental arrangement for the investigation of frictional hot spots using a lead sulphide infra-red cell.

constants so that transient hot spots of very short duration can be detected and measured. In addition the cells are capable of responding to radiations of wave-lengths greater than the visible spectrum, so that hot spots of comparatively low temperature can be detected. It is of course clear that in the application of this method one of the sliding bodies must be transparent to infra-red radiation.

Mr. Thomas has recently applied this technique to a more detailed study of the growth and decay of hot spots produced between sliding solids and to a determination of their size and temperature. The general arrangement of the apparatus is shown diagrammatically in Fig. 23. The lower surface *A* is in the form of a glass disk about 0·2 cm. thick and can be rotated by a motor to give linear speeds ranging from 100 to 700 cm./sec. The upper surface *B*, which is usually of metal, is in the form of a cylinder with the rubbing face about 1 mm. in diameter. The loading of the upper slider and the frictional measurements are made in a manner similar to that shown in Fig. 14. The photo-sensitive cell *C* (which has a time constant of about 10^{-4} sec.) is enclosed in a brass holder *D* below the glass disk. In the top of the brass holder is a narrow slit in line with the direction of motion and immediately beneath the centre of the slider. Between the glass disk and the cell a chopper *E* may be inserted so that the radiation may be chopped at about 4,000 cycles/sec. The output from the cell is suitably amplified and observed on a cathode-ray oscillograph.

The cell may be calibrated by placing a platinum filament of known temperature in place of the slider *B* and observing the amount of radiation received by the cell. In deducing surface temperatures from this calibration certain assumptions must be made concerning the size and emissivity of the hot spots since the photocell provides a measure of the total radiant *energy* reaching it rather than the temperature of the radiating source. For this reason more satisfactory results may be obtained by determining the spectral distribution of the radiation from the hot spot. This may be done by comparing the total radiant energy reaching the cell when it is exposed directly to the glass disk with that observed when a filter is placed between the disk and the cell. If the absorption/wave-length characteristics of the filter are known the ratio provides a direct measure of the temperature of the hot spot and is independent of its area and of its emissivity.

The reliability of this method may be demonstrated by measuring the temperature at which hot spots first become visible. The photocell method yields a value of about 600° C. for the highest temperatures observed. As we have seen, the earlier method described gives a value of about 570° C. The agreement is very satisfactory. The method also confirms the earlier observation that the surface temperature is limited by the melting-point of the surfaces. For example, with a slider made of a gold aluminium alloy (m.p. 570° C.) the temperature of the largest group of hot spots reaches an upper value of about 600° C., which is, within the experimental limit, the same as the melting-point of the alloy. That melting takes place is confirmed by the fact that the intermittent smears of the metal are clearly visible on the surface of the glass disk. However, it is interesting to note that there is some evidence for the occurrence of *occasional* hot spots which are considerably hotter. These, however, occur very infrequently.

Finally, the investigation again shows that surface temperatures of several hundred degrees centigrade are developed readily between rubbing solids, although the mass of the bodies remains quite cold.

A modification of the above method has also been used to estimate the size and temperature of the hot spot throughout its duration. For this purpose the chopper was used and alternate segments were covered with a perspex filter. A typical oscillogram for a single hot spot is shown in Plate III. 2. From the ratio of successive peaks the temperature at any instant in the life of the hot spot can be deduced. Similarly, the area of the hot spot may be derived from these temperature values and the absolute values of the radiant energy reaching the cell. An analysis of this type shows that at a given load and speed the hot spots extend over a wide range of temperatures and sizes, and that both temperature and area vary throughout the duration of the hot spot.

In addition, the results bring out two further points which are of some interest and which are consistent with the theoretical treatment described in the next section. The first is that the time taken for the hot spot to reach its maximum temperature depends on its size; the larger the hot spot the longer it takes to reach its maximum temperature. This is shown clearly in Table VI for the steel slider where the time increases with the load. (Since these hot spots usually occur singly, they constitute the main region of contact at any instant, so that their area may be considered as being approximately proportional to the load.)

The second conclusion is that under uniform conditions of load and speed the actual temperature of the hot spot and the rate at which the hot spot reaches its maximum temperature depend on the thermal conductivity of the slider as well as on the area of the hot spot. The higher the thermal conductivity, the lower the maximum temperature reached, but the greater the rate at which the maximum temperature is attained. The effect of the thermal conductivity of the slider on the rate of rise of the hot spot temperature is shown in the last three lines of Table VI for sliders of steel, constantan, and tungsten, for which the thermal conductivities are 0·10, 0·05, and 0·35 cal./sec./cm./°C. respectively (the load and speed are approximately the same). The corresponding times are 2, 3, and $0{\cdot}5 \times 10^{-3}$ sec. respectively.

It is interesting to note that a time of rise of 10^{-3} sec. at a sliding speed of 380 cm./sec. corresponds to a linear displacement of the disk of about 4 mm., that is, about four times the diameter of the tip of the slider. This suggests that the larger hot spots observed are not produced by the simple collision of an asperity on one surface with an asperity

TABLE VI

Slider material	*Thermal conductivity cal./sec./cm./°C.*	*W gm.*	*v cm./sec.*	*Time of rise of temperature of largest hot spot sec.*
Steel	0·10	150	380	$\simeq 0{\cdot}5\times10^{-3}$
,,	0·10	300	380	$1{\cdot}5\times10^{-3}$
,,	0·10	450	380	$2{\cdot}0\times10^{-3}$
Constantan	0·05	450	380	$3{\cdot}0\times10^{-3}$
Tungsten	0·35	350	380	$\simeq 0{\cdot}5\times10^{-3}$

on the other, but by the relatively prolonged rubbing of a slider-asperity on the surface of the disk.

The work shows that the dependence of the hot spot behaviour on load, speed, thermal conductivity, and hardness is in general agreement with the theoretical model described in the next section. In addition, the behaviour is somewhat complicated by the thermal softening of the glass at the higher rubbing temperatures (see above). Nevertheless, over a wide range of experimental conditions for various metal sliders rubbing on a glass disk, the hottest hot spots have an area of the order of 10^{-3} sq. cm. and a duration of about 10^{-4} to 10^{-3} sec. It is interesting to note that this duration is of the same order as that observed by the thermo-electric method in the sliding of metal surfaces.

A More Exact Calculation of the Surface Temperature

The derivation of equation (5) is based on a relatively crude physical picture in which it is assumed that the whole of the interface of the slider is at the uniform temperature T and that the whole of the heat is dissipated by emission from the surface. Although this may be approximately true for cylindrical sliders of very small diameter where the real area of contact may be an appreciable fraction of the cross-section of the slider, it cannot in general be considered valid. In most cases the points at which rubbing occurs are distributed over the rubbing surface at relatively large intervals and the temperature over the rubbing surface will be very far from uniform. Under such conditions the frictional heat is dissipated into the bulk of the solids surrounding the points of real contact and the heat flow does not depend on the emissivity of the surfaces but on their bulk conductivities. A detailed analysis of the heat flow occurring when massive surfaces rub at a single small region has been developed by Blok (1937) and also by Jaeger (1942). Although their treatment involves relatively complicated mathematics, it is not difficult to obtain similar results by a much simpler treatment. This we shall now do.

We assume that two massive surfaces I and II of specific thermal conductivities k_1 and k_2 touch over a small circular region of radius a. As a result of friction at this region a quantity of heat Q is developed per second and this heat flows away into the two metal surfaces. Suppose a portion Q_1 flows into body I and Q_2 into body II, where

$$Q = Q_1 + Q_2.$$

We now assume that a steady state is reached at which the junction attains a steady temperature T whilst the bulk of the bodies remains at a temperature T_0 (approximately room temperature). We may now define the thermal conductance of this junction by the relation

Heat flow per sec. = thermal conductance × temperature drop. (6)

By analogy with the electrical case described in Chapter I the thermal conductance from the junction into body I is $4ak_1$ and that into body II is $4ak_2$. Then

$$\begin{aligned} Q_1 &= 4ak_1(T-T_0) \\ Q_2 &= 4ak_2(T-T_0) \end{aligned} \quad \text{and} \quad Q = Q_1 + Q_2. \tag{7}$$

Hence
$$T - T_0 = \frac{Q}{4a}\frac{1}{k_1+k_2}. \tag{8}$$

Then, if the load is W, the coefficient of friction μ, and the sliding speed v,

$$Q = \frac{\mu W g v}{J}.$$

Hence
$$T - T_0 = \frac{\mu W g v}{4aJ}\frac{1}{k_1+k_2}. \tag{9}$$

The result given by Jaeger for a square junction of side $2l$ at low sliding speeds is

$$T - T_0 = \frac{\mu W g v}{4{\cdot}24lJ}\frac{1}{k_1+k_2}. \tag{10}$$

Except for the constant this is identical with equation (9). As we have seen, these equations have been calculated for the steady thermal state. As we should expect, therefore, the equations are only valid at relatively low sliding speeds. At higher surface-speeds the slider is continuously cooled by the oncoming portions of the rotating surface. As a result the temperature rise is less than that given by equation (9), and in such cases Jaeger gives the following equation for a square junction of side $2l$:

$$T - T_0 = \frac{x_1^{\frac{1}{2}}\mu W g v}{3{\cdot}76lJ\{1{\cdot}125k_2 x_1^{\frac{1}{2}} + k_1\sqrt{(lv)}\}}, \tag{11}$$

where body I is the smooth surface (the disk), body II the surface which carries the junction (the slider), and $x_1 = k_1/\rho_1 c_1$ where ρ_1 is the density and c_1 the specific heat of body I. The main effect is to make the temperature rise $T-T_0$ increase less rapidly than the first power of v.

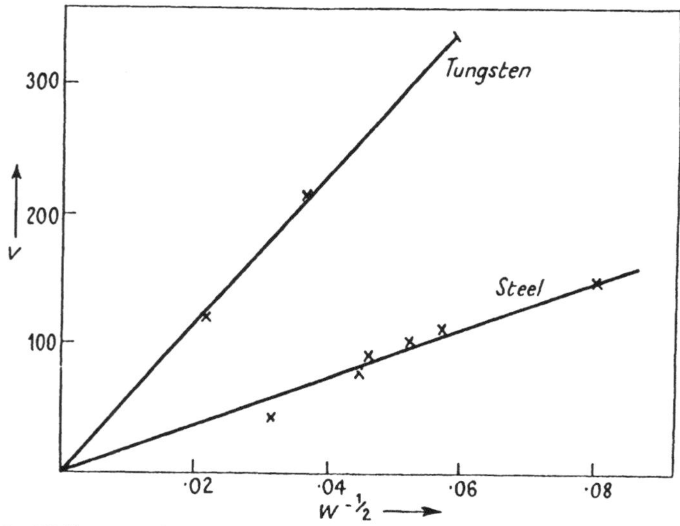

FIG. 24. Sliding speed v necessary to produce hot spots of a fixed temperature as a function of the load W. It is seen that for both tungsten and steel sliders, v is proportional to $W^{-\frac{1}{2}}$ in agreement with equation (12).

It is interesting to investigate the simple relation given by equation (9) in greater detail. We assume that the area of contact between the surfaces is determined by the yield pressure p_m of the softer of the two sliding metals. If the geometry of the surfaces is such that contact occurs over a single circular region of radius a, then as we have seen in Chapter I, $\pi a^2 p_m = W$. Equation (9) then becomes

$$T-T_0 = \frac{\mu v g \sqrt{(W p_m \pi)}}{4J(k_1+k_2)}. \tag{12}$$

We may compare these results with those obtained from the simpler model. Firstly we see that in both cases the temperature rise is proportional to the sliding speed v. Secondly the temperature rise increases with the load. On the simple model where the effect of load on the area of contact is ignored it increases linearly with W. Here, however, where the effect of load on the area of contact is taken into account the temperature rise increases only as $W^{\frac{1}{2}}$. Thus for a fixed pair of surfaces, a constant temperature rise should occur when $vW^{\frac{1}{2}}$ is a constant, that is, v should be proportional to $W^{-\frac{1}{2}}$. In some careful measurements of the incidence of visual hot spots (where $T-T_0$ is a constant) Thomas has recently shown that this relation is indeed approximately obeyed. His results for steel and tungsten cylinders sliding over the surface of a glass disk are plotted in Fig. 24. It is seen that in each case the sliding velocity at which hot spots are just observed is proportional to $W^{-\frac{1}{2}}$.

The third and most important difference between the two models is that the simple model does not indicate the way in which α, the fraction of heat going into the upper slider, depends on the ratio k_1/k_2. In this model α is assumed constant even for the most diverse combinations of materials. Consequently the surface temperature depends only on the conductivity k_2 of the upper slider and is proportional to $k_2^{-\frac{1}{2}}$. In contrast, the present treatment shows that for low sliding speeds $\alpha = k_1/(k_1+k_2)$ and the surface temperature is proportional to $1/(k_1+k_2)$. Thus the conductivity k_1 of the lower surface is as important as that of the upper slider. This is shown in Table VII (taken from Jaeger's paper), which shows the variation of $1/(k_1+k_2)$ and $k_2^{-\frac{1}{2}}$ for various combinations of sliding solids.

TABLE VII

Slider substance 2	*Lower surface substance 1*	k_2	k_1	$\frac{1}{k_1+k_2}$	$k_2^{-\frac{1}{2}}$
Copper . .	Mild steel	0·918	0·144	0·94	1·04
Mild steel .	,,	0·144	0·144	3·47	2·63
Lead . .	,,	0·0827	0·144	4·41	3·48
Bismuth .	,,	0·0194	0·144	6·12	7·18
Copper . .	Copper	0·918	0·918	0·54	1·04
Glass . .	,,	0·0017	0·918	1·09	24·0
Silk . .	,,	0·0001	0·918	1·09	100·0
Silk . .	Glass	0·0001	0·0017	550·0	100·0

It is seen that even according to the Blok–Jaeger model for ordinary values of the thermal conductivity a fairly large variation in the temperature can be obtained by varying the conductivity of *one* of the sliding surfaces. Further, it is seen that this variation is of the same order as that obtained with the $k_2^{-\frac{1}{2}}$ law. However, the effect of low thermal conductivity on the surface temperature will be very marked only if both the surfaces are poor conductors. These conclusions apply to the model derived for low speeds of sliding. At higher sliding speeds, where the cooling effect of the moving surface becomes important, the temperature rise is still determined by the conductivity of both surfaces but, as equation (11) shows, it is less dependent on the conductivity of the moving surface.

In using equation (12) there are two main difficulties which should be borne in mind. The first is that the value of p_m should refer to the yield pressure of those parts of the surfaces at which rubbing actually occurs. If the temperature rise is high, this value may be appreciably less than the values at room temperature. For example, when high-melting metals produce visible hot spots on glass surfaces the temperature rise (approximately 500° C.) must produce a considerable softening of the glass, and it is probable, as Mr. Thomas has recently shown, that in all such cases the area of real contact is determined by the mean yield pressure of the glass at this temperature, rather than by the hardness of the metals themselves. The second difficulty arises if, as is usually the case, the surfaces touch over more than one region. In such circumstances the calculation of the temperature rise is subject to the same uncertainties as those discussed in Chapter I in estimating the area of real contact from electrical resistance measurements.

Nevertheless equation (12) provides a reasonably satisfactory description of the main factors involved in the generation of frictional hot spots when the sliding speed is low and the sliding surfaces are large. When the sliding surfaces are small so that the real area of contact is comparable with the apparent area, equation (5) is more applicable. In most cases, however, it is probable that the real temperature relation lies between the two equations. At higher surface speeds, of course, equation (12) should be replaced by equation (11) and equation (5) by a similar modification derived by Jaeger.

The experiments described in this chapter show that when solid bodies slide on one another extremely high surface temperatures may be developed even under moderate conditions of load and speed. The high temperatures are confined to a very thin layer of the bodies close to the rubbing interface and there may be relatively little bulk heating. The high temperatures are, indeed, localized at a number of very small points or hot spots between the rubbing surfaces where intimate contact occurs and their position changes from instant to instant as the surface asperities wear away and new points come into contact. A more detailed experimental analysis shows, in fact, that both the area of the hot spots and their temperature vary throughout the life of the hot spot. In general the hot spots show very rapid fluctuations and the time taken for the hot spot to reach its maximum temperature depends on the area of the hot spot and on the thermal conductivity of the surfaces. Although, therefore, the duration depends on the experimental conditions, the results show that over a wide range of materials, loads, and speeds the duration of the hot spot is of the order of 10^{-4} to 10^{-3} sec. Similarly the area of the hottest hot spots between metals sliding on glass is of the order of 10^{-3} sq. cm. This is probably considerably larger than the size of hot spots occurring between metal surfaces and is due to thermal softening of the glass.

As the theory indicates, the surface temperature produced by frictional heating increases with the sliding speed and with the load. It is also profoundly influenced by the thermal conductivity of the surfaces. The lower the conductivity of either of the surfaces, the higher the temperatures observed, but the highest surface temperatures are obtained only if both surfaces possess low thermal conductivities. However, even under the most severe conditions of sliding the surface temperature cannot, in general, exceed the melting-point of the bodies.

Even in the presence of lubricant films surface temperatures exceeding several hundred degrees centigrade may be reached at relatively small loads and sliding speeds, although there may be little bulk heating of the bodies and the surfaces may appear to remain well lubricated. These high temperatures occur through the lubricant film at the points of most intimate contact. This observation is of great significance in the study of boundary lubrication since (as we shall see in Chap. XII) the local high temperatures may produce a marked deterioration in the lubricating properties of the lubricant film at the very points where the surfaces are rubbing. In addition, the high temperatures may cause volatilization and decomposition of the lubricant.

The intense local heating produced at the surface of rubbing solids has an important bearing on a number of surface phenomena, such as the abrasion and seizure of metals, the 'frictional welding' of plastics and other materials, and the initiation of chemical reaction and chemical decomposition under friction and impact. The part that surface temperatures play in the surface flow and polish of solids and in the surface melting of ice will be discussed in the following chapter.

REFERENCES

H. Blok (1937), *Inst. Mech. Eng.* **2**, 222, 'General Discussion on Lubrication'.
F. P. Bowden and K. E. W. Ridler (1936), *Proc. Roy. Soc.* **A 151**, 610.
—— M. A. Stone, and G. K. Tudor (1947), ibid. **A 188**, 329.
P. W. Bridgman (1918), *Proc. Amer. Acad. Arts Sci.* **53**, 269.
C. F. Elam (1935), *Distortion of Metal Crystals*, p. 147. Oxford University Press.
E. G. Herbert (1926), *Proc. Inst. Mech. Eng.* **2**, 289.
J. C. Jaeger (1942), *Jour. & Proc. Roy. Soc. N.S.W.* **76**, 203.
H. Shore (1925), *Jour. Wash. Acad. Sci.* **15**, 85.

III

EFFECT OF FRICTIONAL HEATING ON SURFACE FLOW

In this chapter we shall discuss the effect of surface temperatures produced by frictional heating on surface melting and surface flow. The first part deals with the process of polishing; the second with the sliding on ice and snow.

Polishing and Surface Flow of Solids [*A*]

The usual method of polishing surfaces is to rub them together with a fine powder between them. By this process a rough surface having visible irregularities is changed into one where the irregularities are invisible. If the surface gives specular reflection the height of these irregularities will be less than half a wave-length of visible light. The classical work on polishing is that of Sir George Beilby (1921), who showed that the top layer of the polished solid is different in structure from that of the underlying material. It has lost its obvious crystalline properties and has apparently flowed over the surface, bridging the chasms and filling up the irregularities in it. This effect may be shown by polishing a metal specimen with a suitable polishing powder until its surface is covered with the typical polish layer. If part of the surface is lightly etched, the etched portion reveals the original pre-polish scratches in the surface. This is seen in Plate IV. 1 for a polished copper surface, prepared by Mrs. Honeycombe, which was etched below the line *AB*. The original scratches are clearly revealed, and it is apparent that the polish layer which was smeared over the surface has been dissolved away by the etchant.

The mechanism of the polishing process has been a subject of discussion for many years. Newton, Herschel, and Rayleigh considered that polishing was essentially due to abrasion. Beilby's view was that it is a surface tension effect—that the polisher tears off the surface atoms and the layer below this 'retains its mobility for an instant and before solidification is smoothed over by the action of surface tension forces'. The work described in the last chapter, however, and experiments (Bowden and Hughes, 1937) which will be described in this, suggest that the polishing is due to a high temperature softening or melting at the points of rubbing contact.

Some information about the structure of the Beilby layer has been provided by electron diffraction. Experiments with metal surfaces show that the crystals near the surface are broken down in size and reach the crystal size of the bulk metal only at a depth of several hundred Ångströms. The surface layer itself consists either of very fine crystals or of an amorphous layer, but the electron diffraction technique is not capable of distinguishing unambiguously between these two states. The work of Hopkins (1935), Cochrane (1938), Finch (1937), and Glocker (1942) supports the view that the surface layers are melted by the polishing process and as a result of the high thermal conductivity of the underlying metal are frozen rapidly to form an amorphous film about 20 A thick. In the course of time, as Cochrane has shown, the surface film may recrystallize since the amorphous state is not stable. On the other hand, Raether (1947) has recently suggested that the polish layer is always microcrystalline, but the crystal size is so small that the interference pattern obtained approaches that obtained from amorphous material. If the crystal size is sufficiently small the two views are, of course, identical.

With non-metals, such as calcite, the conclusions are somewhat different. There is considerable distortion of the underlying material to a depth of several thousand Ångströms, but the crystals are not appreciably broken down in size. Further, if the crystal is polished along a cleavage plane the surface layers show no amorphous properties. If, for example, calcite is polished along a cleavage face the polish layer consists of large crystal blocks tilted at small angles relative to the underlying material; but the layer still retains its single crystal structure. Hopkins suggests that the surface layers which are melted during polishing solidify relatively slowly on account of the poor thermal conductivity of the underlying material, and during this process the underlying bulk material is able to impose its crystal structure on the solidifying surface layer. If, however, the crystal is polished along a plane making a sharp angle with a cleavage face the polish layer is amorphous, but it becomes crystalline again on heating to a temperature well below its melting-point.

With ionic crystals such as rock-salt, Raether finds that the polish layer consists of very fine crystals and he suggests that the action of polishing very brittle materials of this type is a simple mechanical one of breaking down the crystal size. Very hard substances such as diamond and tourmaline show a single crystal structure, unaffected by polishing, and Raether concludes that the nature of the polished surface of non-metals is governed by the hardness or brittleness of the material. The harder the material, the less the surface is deformed, and the structure of the polish layer covers a complete range from amorphous and micro-crystalline to single crystal. On the other hand, with metals which can undergo plastic flow, however hard they may be, the surface material can be sheared along crystal planes without breaking up its general cohesion so that the surface layer always consists of very small crystals. This simple classification is not accepted by other workers, and Finch in particular has emphasized the surface flow theory for non-metals as well as metals.

With most metals it seems clear that the polish layer, whatever its structure, will consist of a 'fudge' of oxide and metal. Dobinski (1937) has indeed shown that in some cases the polish layer consists predominantly of the oxide, and Raether has suggested that the increased resistance of many polished metals to corrosion is due to the protective oxide film which is spread over the surface. In addition, as Brockway and Karle (1947) have recently shown, the polish layer will often include considerable quantities of polishing material unless special precautions are taken.

Whatever the nature or structure of the polish layer, it is evident from the above discussion and from the conclusions of the previous chapter that the frictional heat generated during polishing may play an important part in the polishing process. The hot spots produced at the points of rubbing contact between the polishing medium and the specimen may readily cause a local softening or melting of the surface layers, and this material will be spread over the surface of the specimens by the polishing action itself and solidify or crystallize to form the polish layer. On this view, then, the polishing process is in general due to surface softening or melting rather than to simple mechanical abrasion.

We may perform a simple experiment to test this hypothesis. If polishing is due primarily to a mechanical abrasion and wearing away of the specimen we may expect the relative *hardness* of the specimen and of the polisher to be of major importance. If, however, it is due to surface melting it is the *relative melting* points which will be the determining factor. If the polisher melts or softens at a *lower* temperature than the specimen it will melt or flow first and will have comparatively little effect on the specimen.

Influence of melting-point

Loss of weight. The first set of experiments (Bowden and Hughes, 1937) were carried out on the loss of weight experienced by various metals when they slide on a polisher of given material. The results in Table VIII are for cylinders of lead, Wood's alloy, and gallium sliding over a polisher of thick filter-paper under conditions of constant load and speed. The loss of weight may be compared with the hardness and the melting-point.

TABLE VIII

Cylinders 0·2 cm. diameter; load 100 gm.; speed 110 cm./sec.

Metal	*Melting-point* °C.	*Vickers' hardness*	*Loss of weight* gm./cm. *of sliding*
Lead . .	327	5	$0{\cdot}6\times10^{-8}$
Wood's alloy .	69	25	$3{\cdot}7\times10^{-8}$
Gallium . .	30	6·6	$53{\cdot}0\times10^{-8}$

There is no doubt as to the general trend. Lead is softer at room temperature than Wood's alloy, yet it loses much *less* weight. It is the low-melting metal which loses the most weight. A microscopic examination of the filter-paper showed small globules of the metal adhering to the fabric of the paper.

The result of rubbing on a block of pure camphor (m.p. 178° C.) is shown in Table IX.

TABLE IX

Nominal pressure 60 gm./cm.^{2}; speed 205 cm./sec.

Metal	*Melting-point* °C.	*Vickers' hardness*	*Loss of weight* gm./cm. *of sliding*
Lead . .	327	5	$< 0{\cdot}1\times10^{-7}$
Wood's alloy .	69	25	$3{\cdot}2\times10^{-7}$
Gallium . .	30	6·6	$165{\cdot}0\times10^{-7}$

Again the loss in weight depends primarily on the melting-point and not on the hardness of the metal as measured at room temperature.

Surface flow. The second series of experiments consisted of a microscopic examination of the surfaces of various materials when they slide over different polishers. Generally the polishing material was in the form of a fine powder which was embedded in a lead or camphor polishing block. During the polishing experiment the surfaces were flooded with water. A fine pin-scratch was made in the surface before each experiment so that the same areas on the surface could be compared in the photomicrographs at various stages. The result of rubbing Wood's metal (m.p. 75°) with a camphor block (m.p. 178° C.) is shown in Plate IV. 2. Although the Wood's metal (V.D.H. *c.* 25) is very much harder than the camphor, it melts at a lower temperature and it will be seen that surface flow and polishing of the alloy occurs. On the other hand, camphor will not polish tin (m.p. 232° C.), although its hardness is only *c.* V.D.H. 4. Similarly camphor will not polish lead, white metal, or zinc, which melt at a higher temperature. A polisher using a powder of oxamide (m.p. 417°) will readily cause flow of all these metals but does not produce any effect on speculum metal (m.p. 745°) (see Plate IV. 3) or copper (m.p. 1,083°) which melt at temperatures well above 417° C. Lead oxide (m.p. 888°) will polish speculum metal and all metals melting below it, but has little effect on nickel and molybdenum, which melt above it. These in turn are readily polished by the high-melting oxides such as chromic oxide and stannic oxide.

Similar results are obtained with glasses, quartz, and some non-metallic crystals; calcite, for example (Plate IV. 4), shows little flow on cuprous oxide which melts slightly below it, but is readily polished by zinc oxide which melts above it.

Surface flow below the melting-point. Although these results point to the view that polishing only takes place at temperatures above the melting-point of the specimen, some experiments show that appreciable flow sometimes occurs at temperatures which must be well below the melting-point. This occurred with metals of very high melting-point such as nickel (m.p. 1,452° C.), palladium (m.p. 1,555° C.), and molybdenum (m.p. 2,470° C.) on polishers which melt above 1,000° C., for example, cuprous oxide (m.p. 1,235° C.). Again gold (m.p. 1,063° C.) showed some surface flow on oxamide (m.p. 417° C.). These results, however, are not inconsistent with the view that polishing is due to surface flow. It is well known that the mechanical strength of many metals and solids falls to a low value at temperatures well below the melting-point. The rounding of sharp metal crystals and the low value of tensile strength, hardness, and other mechanical properties, at these temperatures show that metals may lose their rigidity and resistance to shear at comparatively low temperatures. For such solids, surface flow would be expected to take place at temperatures well below the melting-point and experiment shows that, in certain cases, this can indeed occur. The rate of flow and polish is, however, very much less and may take hours, instead of the few minutes required by a high-melting polisher.

Mechanism of polishing

These experiments provide strong evidence, not only that high local temperatures occur, but that they play a large part in the process of polishing. In many cases the frictional heat will raise the temperature to a sufficiently high value to cause a real melting of the solid at the points of sliding contact. The molten solid will flow or will be smeared on to cooler areas, and will very quickly solidify to form the Beilby layer. Polishing under these conditions is rapid. If the sliding is gentle or the melting-point of the polisher is low, the surface of the solid may not reach the temperature of melting. Polish and surface flow may still occur under these conditions, provided the temperature reaches a point at which the mechanical strength of the solid is sufficiently low for it to yield under the applied stress or under surface tension forces. Polishing under these conditions is usually a slower process.

The relative hardness of solid and polisher as normally measured at room temperature is comparatively unimportant. This is shown clearly in the case of Wood's alloy and tin on camphor, or speculum metal and nickel on lead oxide. The harder metal of low melting-point is polished, while the softer metal of higher melting-point hardly flows at all. Similarly zinc oxide, which is comparatively soft (Mohs' hardness 4), readily polishes quartz (Mohs' hardness 7). The amount of surface flow is governed, not by the properties of the solids at room temperature, but by their relative mechanical properties at the high temperature developed between the sliding surfaces.

It is not suggested that the smoothing or polishing of solid surfaces can occur *only* as a result of surface flow. In the case of diamond, for example, surface melting can hardly occur. Again, with many substances the surface irregularities may be removed by chemical attack: electrolytic polishing of metals is a clear example of this. In the polishing of solids such as diamond it is possible that chemical attack by oxygen and a burning away of the surface irregularities may be important.

The action of a typical polisher

We may now consider briefly the action of a typical polisher. This usually consists of a lead or pitch lap or polishing block in which are embedded fine particles of the polishing powder, for example, alumina or rouge. The polishing powder rubs on the surface of the specimen, usually in the presence of a liquid such as water (Fig. 25). At the points where rubbing occurs high localized temperatures may be developed, and as we saw above, the amount of surface flow will depend on the relative mechanical properties of the solids at these high temperatures.

It is clear that increasing the load will increase the polishing rate partly because it will increase the area over which the surface layers are deformed and partly because it will increase the amount of heat developed and consequently the amount of surface melting or softening. Similarly an increase in speed will lead to an increase in the amount of heat developed and hence in the rate of polishing. It follows that, other conditions remaining constant, the polishing rate should increase with the load and the speed.

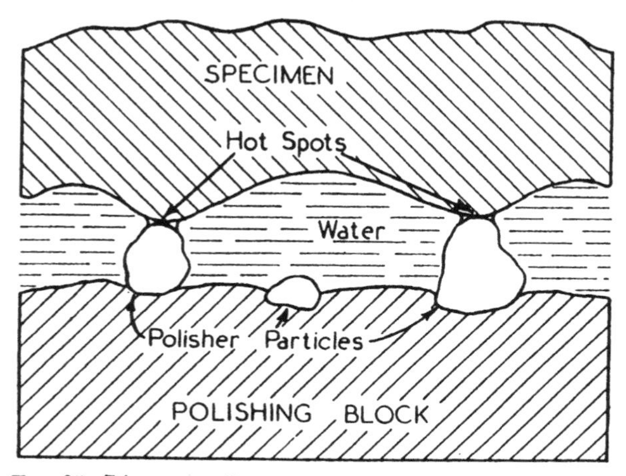

FIG. 25. Diagrammatic representation of a typical polisher.

A third variable in the polishing process is the liquid medium. As we saw in the previous chapter, the presence of the liquid does not prevent the occurrence of localized hot spots. It does, however, appear to localize the heating by preventing it from spreading to a great distance from the hot spot. In this way it prevents large-scale surface melting. If the supply of liquid is insufficient and the load or sliding speed is high, this large-scale surface melting or softening of the specimen may occur and particles of the polishing powder may become embedded in its surface over a considerable area. This effect can be observed, for example, when silver is being polished with rouge. Under certain conditions the silver surface takes on a reddish tinge which is due to embedded rouge. In the same way the presence of the liquid prevents a large-scale melting of the polishing block. In the case of a pitch block, for example, if this is not prevented, the pitch is melted off and adheres to the surface of the specimen.

It is clear that the physical properties of the liquid, its heat capacity, thermal conductivity, latent heat, and boiling-point will all influence the rate of cooling and may all have considerable influence on its action as a polishing medium. In addition its chemical properties may be important since at the high temperatures developed there may be rapid local chemical attack. In certain special cases the liquid may play a major part in the polishing process. For example, Grebenschikov (1931, 1935) has suggested that in the polishing of glass the water used as the liquid medium dissolves the top surface layers and forms a silica gel of appreciable thickness; the polisher then wipes this layer away wherever it projects above the common level. It is clear, however, that with most other materials and especially with metals this mechanism is not possible. The polishing in these cases is essentially due to the surface flow produced by frictional heating.

The Mechanism of Sliding on Ice and Snow [A]

Define to me why one who slides on the ice does not fall.

LEONARDO DA VINCI

Another phenomenon where surface melting may play a part is in the sliding of solids on ice and snow—in skating or ski-ing. It is well known that the friction under these conditions may be remarkably low ($\mu = c.\ 0{\cdot}03$). The suggestion has often been made that in skating or ski-ing the surfaces are lubricated by a layer of water formed by pressure melting (Reynolds, 1901), but few experiments have been made to support or to disprove the suggestion.

Pressure melting

Experience shows that skis slide quite readily on snow at $-20°$ C. If we consider that the ski makes contact with the snow over the whole of its under surface, the pressure for an average man of 75 kg. weight on skis of area 5,000 cm.^{2} is 15 gm./cm.^{2} This pressure can only form a water layer if the snow is at $-0{\cdot}00012°$ C. or a higher temperature. The real area of contact is, of course, less than the apparent area. In the case of metals the real area may be a minute fraction of the apparent area of contact, and the mutual pressure between the two surfaces may be the flow pressure of the metal (see Chap. I). With a powder such as snow, which can pack down and conform to the shape of the ski, we might expect the area of contact to be greater—perhaps not less than 1/1,000th of the area of the ski. If this were the case, then the pressure is still only great enough to melt snow at $-0{\cdot}12°$ C. In order to have sufficient pressure to melt snow at the low temperature of $-20°$ C. the real area of contact must be less than $0{\cdot}031$ cm.^{2}, i.e. about 1/100,000th of its apparent area.

These calculations of the pressure melting are made on the assumption that the pressure is applied equally to the solid and to the liquid phases. Poynting (1881) has pointed out that if the pressure is applied to the solid alone, the lowering of the melting-point is about $11\frac{1}{2}$ times as great. (This effect and its bearing on flow pressures has been discussed by Johnston (1912) and Jeffreys (1935).) We cannot, on theoretical grounds, rule out the pressure-melting theory, but if it is true it means that the real area of contact is extremely small. The pressure itself is, of course, not sufficient to cause melting. Heat must be supplied from some source which is at a higher temperature than that corresponding to the equilibrium melting pressure.

Melting due to frictional heating

Simple calculations show that appreciable surface melting may be produced by the frictional heat liberated at the sliding interfaces. In the case of a ski sliding on ice at $-20°$ C., for example, if the coefficient of friction is $\mu = 0{\cdot}05$ and the weight carried is half that of an average man, the amount of frictional heat liberated when the ski moves forward a distance l of 1 cm. is given by

$$Q = \frac{\mu M g l}{J} = \frac{0{\cdot}05 \times 37{\cdot}5 \times 981 \times 1000}{4{\cdot}8 \times 10^7} = 0{\cdot}044 \text{ calories.}$$

This heat is concentrated at the points of contact of the snow crystals. If the ski were making rubbing contact over its whole area, this heat would be sufficient to raise the temperature from $-20°$ to $0°$ C. and to melt a layer over 6 molecules in thickness over this area. In actual fact, the real area of contact must be very much less, and the heat is concentrated at the points of contact of the snow crystals. Again we do not know the real area of contact, but if the area of contact is about 1/1,000th of the ski, then the thickness of the water layer melted would be 2×10^{-4} cm. or 10^4 molecular layers. It is clear that a great part of the heat must be lost by conduction from the points of contact to the surrounding snow and to the ski. These crude calculations cannot be taken very seriously, but they do show that the frictional heat liberated is considerable, and that the retention of a small fraction of it may be sufficient to cause local melting at the points of contact of the snow or ice crystals.

Formation of a water layer

Some experiments were carried out (Bowden and Hughes, 1939) at the Jungfraujoch Research Station in Switzerland to determine whether a water layer is formed at all, and if so, whether it is due to pressure melting or to frictional heating. A miniature ebonite ski was constructed and two metal electrodes were sealed into the ski about 0·2 cm. apart. They projected at the bottom surface and were ground flat so that they were level with the bottom surface of the ski. When placed on salty ice at $-20°$ C. the resistance between the electrodes was extremely high, of the order of 2×10^6 ohms, since there is virtually no conducting material between them. As sliding commenced there was little change in the resistance and the friction was relatively high. As the ice was allowed to warm up, the friction gradually fell, but little variation in the electrical resistance was observed. In the neighbourhood of the melting-point, however, there was a sudden fall in resistance to a value of about 2×10^4 ohms, showing that a continuous water film had formed between the surfaces. The film was still invisible, but a crude estimate from the electrical resistance suggests that its thickness was of the order of 10^{-2} cm. or less. Under these conditions the friction had the low value of $\mu = 0{\cdot}03$. Thus the high resistance at the lower temperatures of sliding indicates that surface melting occurred only at isolated points of contact. This would also account for the higher values of the coefficient of friction. We are therefore justified in concluding that during sliding on ice a water film is formed between the surfaces and its presence is largely responsible for the low friction observed. This view is supported by the fact that the coefficient of friction is not independent of the load but in general decreases somewhat with increasing load. This departure from Amontons' law (see later chapters) is consistent with the view that the surfaces are separated by a relatively thick liquid film so that the surfaces are sliding under partly hydrodynamic conditions.

Effect of temperature

Similar experiments were carried out on the influence of temperature on the friction of different solids sliding on ice. The results are shown in Fig. 26. It will be seen that the friction increases markedly as the temperature falls and at a temperature of $-80°$ C. it is some five or six times as great as it is at 0° C. The value for the coefficient of friction ($\mu_k = 0{\cdot}1$) at these low temperatures is of the same order of magnitude as that observed on other crystalline solids such as calcite. The large influence of temperature on the friction of ice is in marked contrast to the behaviour of most other solids where temperature has only a small effect. The behaviour is consistent with the view that the low friction is due to a lubricating water layer, and the effect is marked with ice since the underlying material retains its rigid character even though the surface layers are molten. With metals this is not the case and, as we shall see in Chapter V, the friction of metals does not vary in this striking way with the temperature. Further, in agreement with this view, the friction rises at low temperatures since it becomes increasingly difficult for a water layer to be formed.

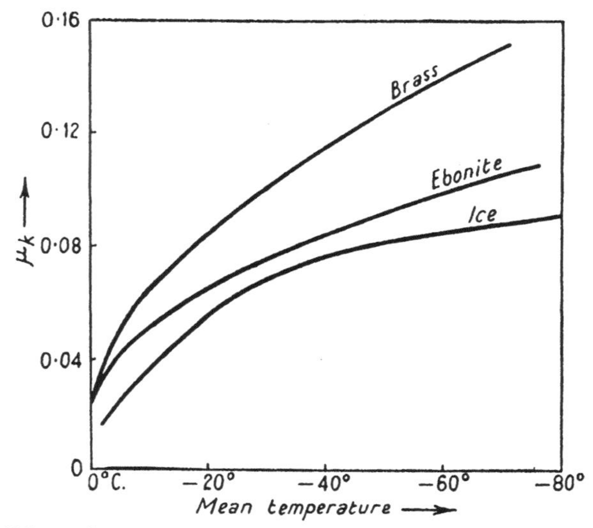

Fig. 26. Effect of temperature on the friction of brass, ebonite, and ice sliding on ice. The friction increases markedly as the temperature falls and is lower for the slider possessing the lower thermal conductivity.

Effect of thermal conductivity

The influence of the thermal conductivity of the ski on the friction at low temperatures sheds further light on the mechanism of water-layer formation. If sufficient pressure is applied to the ice to lower the melting-point to the actual temperature of the ice it is, of course, capable of melting. An appreciable quantity cannot melt, however, unless heat is supplied from some source at a temperature higher than the pressure–melting-point equilibrium. Both the heat capacity of the ice and its thermal conductivity are small, and this heat can most readily be supplied from some outside source. If the temperature of the atmosphere is higher than either of the ice surfaces, it could be supplied by conduction from the air. Under these conditions we should expect that the friction of a good thermal conductor would be less than that of a bad one. The friction of a brass ski on cold ice should be *less* than that of an ebonite one.

If, however, the lubricating film is formed by frictional heating, the converse will be true. The frictional heat is liberated at the interface between the sliding surfaces, and if the ski is a good thermal conductor, the heat will be carried away rapidly and less will be available for surface melting. On this view the friction of a brass ski on cold ice should be *greater* than that of an ebonite one. Fig. 26 shows the results obtained using a miniature ski of brass and of ebonite.

At temperatures near 0° C. the frictions of both skis were the same. At lower temperatures, however, the results showed that the friction of the brass was considerably greater than that of the ebonite. The lower the temperature the more pronounced this difference usually became. These results provide evidence that the frictional heating plays an important part in the formation of the water film.

Static and kinetic friction and the influence of speed

Some simple experiments were carried out on the sliding of ice on ice at various temperatures and speeds. The main results comparing the friction to *start* sliding μ_s with the friction during sliding itself μ_k are incorporated in Fig. 27. (The measurements were not carried out at exactly the same load so that, as Amontons's law is not obeyed, the values are not exactly comparable. The effect of this factor is, however, small.) It is seen that at all temperatures the static friction is much higher than the kinetic friction, though at 0° C. the difference is less marked.

These results support the view that the water layer is produced by frictional heating. If the pressure-melting view is correct, there is no particular reason why the static friction should be so much higher than the kinetic friction. A water film should be present at the points of contact, even though the surfaces are stationary. If, however, the water is melted by frictional heating, we should expect the static friction to be very much higher than the kinetic friction. This view is again substantiated by the effect of speed of sliding. Provided the speed of sliding is sufficient to produce a continuous layer of water, it should have little effect on the kinetic friction. This is observed. If, however, the sliding speed is too low to produce melting at the points of real contact, the friction should rise. This again is observed. At very low sliding speeds there is an appreciable increase in the coefficient of kinetic friction.

Some additional work which supports the view that frictional heating can produce local surface melting and therefore a low value of friction has been obtained by Mr. R. Hutchison (unpublished), who measured the friction of solids sliding on benzophenone, dinitrobenzene, and sodium hyposulphite. With these substances, which contract on melting, the application of a uniform pressure would lower the melting-point and so hinder melting. Experiments showed that at very low speeds of sliding, the coefficient of friction was relatively high ($\mu = 0{\cdot}2$). At

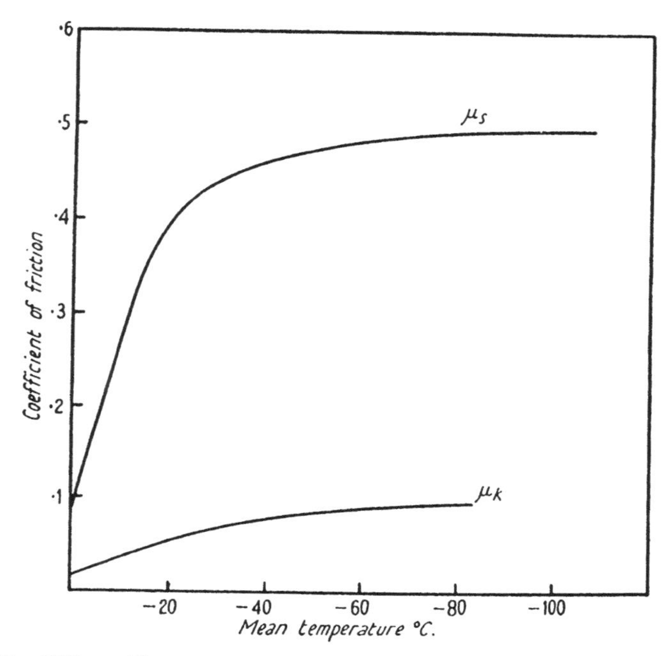

Fig. 27. Sliding of ice on ice. Influence of temperature on the static friction μ_s and the kinetic friction μ_k.

higher speeds of sliding, however, where the frictional heating was appreciable so that local surface melting could occur, the kinetic friction fell to a very low value ($\mu = 0{\cdot}03$) which is comparable with that observed on ice.

These observations have an interesting bearing on sledging and ski-ing. Few quantitative measurements of the friction of sledges seem to have been published, but there is general agreement that the friction increases at low temperatures. Many Arctic explorers (Wright, 1924, p. 44; J. M. Scott, 1933, p. 273; Cherry-Garrard, 1922, pp. 456–7) have recorded that at very low temperatures, −30° to −40° C., the friction between the snow and the runners became so great that the sensation was that of pulling a sledge over sand. Wright, summarizing the conclusions of the Scott Polar Expedition of 1911–13, says:

> 'Quite apart from any question of the hardness of the snow, however, the surface temperature has an important influence. Our opinion was that the friction decreased steadily as the temperature rose above zero Fahrenheit (−18° C.), the presence of brilliant sunlight having an effect, which was more than a psychological one, on the speed of advance. Below zero Fahrenheit (−18° C.) the friction seemed to increase progressively as the temperature fell, as if a greater and greater proportion of the friction were due to relative movement between the snow grains and less to sliding friction between the runner and snow.'

This steady increase in friction as the temperature of the ice or snow falls is clearly shown in Fig. 26. The effect is less marked at very low temperatures, and it is probable that below −40° C. very little surface melting occurs under these conditions.

The influence of the thermal conductivity of the sliding body on the friction, shown in Fig. 26, is also borne out by practical experience. It will be seen from this figure that at low temperatures the friction of a good thermal conductor is considerably greater than that of a poor one. Nansen (1898, pp. 445–6) compared two sledges, one having nickel-plated runners and the other maple runners. The temperature was low—the actual value is not given, but the mean temperature during that month was −36·8° C. (−34·2° F.). He found that the friction of the metal was higher: 'The difference was so great that it was at least half as hard again to draw a sledge on the nickel runners as on the tarred maple runners.'

The thermal conductivity is also important in ski-ing. Many skis are fitted with brass or steel edges, although sometimes vulcanite or composition edges are used. The friction measurements show that the latter should be faster at low temperatures. If metal must be used, one of low thermal conductivity such as German silver or constantan should be better. Apart altogether from the thermal conductivity of the surfaces their hydrophobic properties are also important. Skis coated with a wax or lacquer, or skis made of plastic, in addition to being poor thermal conductors are hydrophobic, so that water films will not adhere to the surface. However, it seems in general that the formation of a water layer between the ski and the ice or snow is the main condition for facilitating sliding. As these experiments show, it is the thermal conditions favouring surface melting by frictional heating which are the most important, particularly at temperatures considerably below 0° C. This view has received direct confirmation in an extensive series of practical tests recently carried out by Klein (1947) on the sliding characteristics of aircraft skis on ice and snow.

REFERENCES

SIR GEORGE BEILBY (1921), *Aggregation and flow of solids*, 1st ed. London, Macmillan & Co.

F. P. BOWDEN and T. P. HUGHES (1937), *Nature*, **139**, 152.

—— —— (1937), *Proc. Roy. Soc.* **A 160**, 575.

—— —— (1939), ibid. **A 172**, 280.

L. O. BROCKWAY and J. KARLE (1947), *J. Coll. Sci.* **2**, 277.

A. CHERRY-GARRARD (1922), *The Worst Journey in the World*, **2**. London, Constable & Co.

W. COCHRANE (1938), *Proc. Roy. Soc.* **A 166**, 228.

S. DOBINSKI (1937), *Phil. Mag.* **23**, 397.

G. I. FINCH (1937), *Trans. Far. Soc.* **33**, 425.

R. GLOCKER (1942), *Schriften d. dtsch. Akad. d. Luftfahrtforschung.*

I. V. GREBENSCHIKOV (1931), *Keramika i Steklo*, **7**, 36.

—— (1935), *Sotsialisticheskaya Reconstruktsiya i Nauka*, **2**, 22.

H. G. HOPKINS (1935), *Trans. Far. Soc.* **31**, 1095.

—— (1936), *Phil. Mag.* **21**, 820.

H. JEFFREYS (1935), *Phil. Mag.* **19**, 840.

J. JOHNSTON (1912), *Journ. Amer. Chem. Soc.* **34**, 788.

G. J. KLEIN (1947), Nat. Res. Coun. Canada Aeronautical Report AR–2, *The Snow Characteristics of Aircraft Skis.*

F. NANSEN (1898), *Farthest North*, **1**. London, George Newnes.

J. H. POYNTING (1881), *Phil. Mag.* (5), **12**, 32.

H. RAETHER (1947), *Met. et Corr.* **22**, 2.

O. REYNOLDS (1901), *Papers on Mechanical and Physical Subjects*, **2**. Cambridge University Press.

J. M. SCOTT (1933), *The Land that God gave Cain.* London, Chatto & Windus.

C. S. WRIGHT (1924), *Miscellaneous Data* (*Scott Polar Expedition of 1911–1913*), compiled by Colonel H. G. Lyons. London, Harrison and Sons.

IV

FRICTION AND SURFACE DAMAGE OF SLIDING METALS

That thing which is entirely consumed by the long movement of its friction will have part of it consumed at the beginning of this movement.

LEONARDO DA VINCI.

RETURNING again to the sliding of metallic surfaces, there are, as we have seen, two major experimental observations. First that the area of contact between them is very small so that the pressure at the local regions of contact is very high and is sufficient to cause plastic flow of the metal, and secondly that at sliding speeds frequently used in practice the surface temperatures may rise to very high values. The third point we need to investigate is the type of interaction between the moving surfaces and the physical changes which occur in them during sliding.

THE MEASUREMENT OF FRICTION

The methods used for measuring friction vary very widely. Obviously any method which will give, at the same time, a measure of the normal load between the surfaces and of the tangential force necessary to cause sliding can be used to determine the coefficient of friction. The frictional results described in this book were obtained by a variety of experimental methods which were usually designed to suit the particular conditions of the experiment. These methods include the use of a simple weight hanging on a pulley, the tilting of an inclined plane, the deflexion of a pendulum or of a spring, the measurement of the rate of deceleration of the moving solid, and the use of piezo-electric crystals or resistance strain gauges or electrical capacity methods to measure the normal and tangential forces between the moving surfaces. In general, all these methods give similar results *provided the conditions between the rubbing surfaces are the same*. In many of the experimental methods which are in common use, the moving solid and the measuring system possess a high inertia, so that they only measure the average forces and are not capable of recording any rapid changes or fluctuations which may occur in the frictional force during the sliding process. It is often desirable to analyse these forces in more detail, and for this purpose the electrical methods mentioned above used in conjunction with a cathode-ray oscillograph are suitable. If mechanical systems are used they must possess a low inertia and a rapid response.

One form of mechanical apparatus which has been found useful for a wide range of experimental conditions is shown in Plate V (Bowden and Leben, 1939). The friction recording device has a high natural frequency and the system which applies the load has a natural frequency which is as high or higher. The details of this are shown in Plate V. 2 and V. 3. The friction is measured between a flat sliding lower surface (*A*) and a stationary upper surface which is usually in the form of a small curved slider (*B*). The lower surface is mounted on a carrier (*C*) which runs on rails (*D*) and is driven steadily by water pressure with a force much greater than the opposing friction. This ensures steady uniform motion free from vibration. The speed of the lower surface is adjustable and ranges from about 0·001 cm./sec. to a few cm./sec.

The upper surface is attached to a light rigid duralumin arm (*E*). This is carried on a bifilar suspension of tightly stretched piano wire supported on a rigid frame (*K*). The load is applied between the surfaces by means of a flat spring (*F*) bent into a circle. The extent of the compression of this spring which is adjusted by means of the screw (*G*) determines the load between the surfaces. A flat piece of spring steel (*H*) prevents lateral distortion of the spring when sliding. The normal load acting between the surfaces can be measured by applying an equal and opposite load to the upper surface by means of weights and the pulleys (*L*). In this manner a load of up to 8 kg. can be applied by a device weighing only a few grammes.

Movement of the lower surface thus sets up frictional forces between the two surfaces. The torsion arm is deflected about a vertical axis and the amount of deflexion is proportional to the frictional force exerted between the two surfaces. The magnitude of the deflexion is measured by a light beam reflected from the mirror (*M*) on to a horizontal slit in a moving film camera.

The mounting of the lower surface contains a heating element by which the surface can be heated rapidly during an experiment or held at any desired temperature. A copper/constantan thermocouple (*N*), pressing in contact with the surface, gives a measure of the surface temperature and simultaneously records the temperature on the camera by means of the galvanometer (*P*). The distance travelled by the lower surface is also recorded on the camera. This can be used in conjunction with a small perforated disk throwing time marks on the edge of the film, to record the actual velocity of sliding of the surface.

This apparatus, which operates in general at high loads and at low sliding speeds, is also suitable for measuring friction under conditions of boundary lubrication. Moreover, the low inertia and high frequency (1,000 vibrations/sec.) of the measuring device make it very sensitive to changes in friction. Provision is made for damping the measuring device either with pads of piano felt or with air damping, but in general we have found the apparatus more revealing and sensitive when undamped.

Since the lower surface is electrically insulated, a measure of the electrical conductance between the surfaces, or the changes in it during sliding, may be made and recorded on the camera simultaneously with the friction. Alternatively, if the surfaces are of different metals, the thermo-electric p.d. between them may be measured so that a record of the surface temperatures developed during sliding may also be obtained.

The measurement of the temperature of stationary surfaces

One interesting point which is of some practical importance arises when we attempt to measure the temperature of a surface by placing a thermocouple against it. A very convenient method of doing this is to press a wire of copper and of constantan against the surface and to measure the thermal p.d. generated. This is a measure of the temperature of the surface between the tips of the wires since the surface itself constitutes a third 'neutral' junction. The thermocouple may be calibrated by observing the p.d. at which materials of known melting-point melt when heated on the surface.

There are, however, two sources of error in this method which may, in some cases, be serious. If the thermocouple is calibrated on a surface which is a poor thermal conductor, such as stainless steel, the calibration will provide misleading results on a good thermal conductor such as copper. For example Dr. Greenhill has recently shown that errors as large as 10 per cent. may occur in such cases. This effect is due to the temperature drop at the tips of the wires due to the conduction of heat from the surface. Apart from the influence of conductivity the temperature drop at a metal junction, as we saw in Chapter II, also depends on the size of the junction. Consequently we should expect that the temperature drop at the tip of the thermocouple wires will depend on the force with which the wires are pressed on to the surface. This is observed and may often lead to variable results. These difficulties may be overcome by using thermocouple wires of very fine gauge and making the area of contact relatively large, so that it does not vary critically with the load. One way of doing this is shown in Fig. 28 (*a*);

another which was developed by Dr. Greenhill is shown in Fig. 28 (*b*). This thermocouple gave results that were almost independent of the load and were accurate to within 5 per cent. on copper surfaces after calibrating on a stainless steel surface.

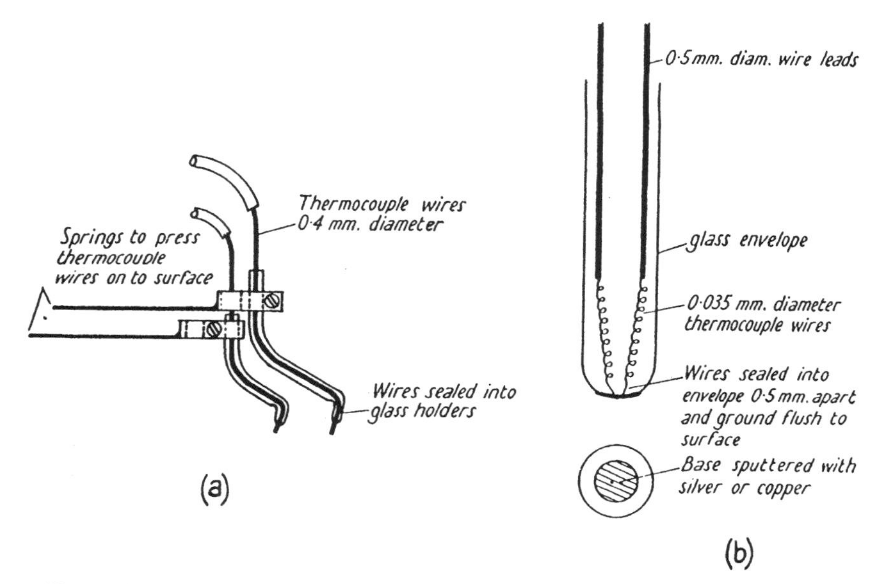

FIG. 28 (*a*) and (*b*). Two types of thermocouples used to measure the temperature of solid surfaces. They are designed to produce only a small temperature drop when pressed against the hot surface.

Although this thermocouple provides a reliable means of measuring surface temperature, another complication arises in investigating friction as a function of surface temperature. This investigation is usually carried out on the apparatus described above by recording the friction and the temperature of the lower surface simultaneously on the camera as the lower surface is heated up. The sliding speeds are so small that the temperature rise due to frictional heating may be neglected. On the other hand, as the cool upper slider passes over the lower hot surface an appreciable temperature drop may occur between the surfaces at the region of contact. Consequently the temperature where the surfaces actually rub may be appreciably less than the temperature of the lower surface as recorded by the thermocouple. This effect is most marked when the lower surface is a poor conductor and the upper slider a good conductor. For example, for copper on constantan the temperature drop is of the order of 35 per cent. The effect is least for a bad conductor sliding on a good conductor: for a constantan slider on copper it is about 5 per cent. For similar metals the temperature drop is intermediate in value and is small for good conductors. Thus when the lower surface is a good conductor a temperature drop of the order of 10° C. in 200° C. may be expected and in general this may be ignored. For a good conductor sliding on a poor conductor, however, the effect is serious. This may be overcome by making the slider in the form of a small metal tip, supported by alumina cement in a steel holder. For copper on constantan the error is reduced from 35 per cent. to less than 10 per cent., as may be seen in Table X.

TABLE X

Surfaces	*Temperature of lower surface*	*Drop in temperature at region of contact*
Copper slider on constantan surface	150	57
	200	73
Constantan slider on copper surface	150	8
	200	11
Copper on copper	155	5
Steel on steel	155	~10
Insulated copper slider on constantan surface	150	15
	200	15

By taking the precautions mentioned above it is possible, when investigating the friction between sliding surfaces as a function of the surface temperature, to measure the temperature to within about 5 per cent. Measurements of this type for lubricated surfaces will be discussed in Chapters IX and X.

Preparation of surfaces

In order to obtain reproducible measurements of sliding friction, great care must be taken in cleaning and preparing the metal surfaces. Since boundary lubrication is a surface phenomenon, a trace of contaminating grease often has a very great effect on the friction. These contaminants may be only 1 molecule thick and may arise from fatty films migrating from the fingers even if the cleaned side is not handled. A trace of abrasive material left in a soft metal will also give results which are not characteristic of the metal. It must be recognized, of course, that however carefully they are 'cleaned' and prepared, all metallic surfaces will, if they are exposed to the air, be covered with a layer of oxide, and probably a physically adsorbed gas and water vapour layer as well. It is only by taking elaborate precautions and working in a high vacuum that these films can be eliminated and measurements made on naked metals (see Chap. VII). Under most conditions, therefore, both of laboratory experiment and of practice, surface films of this nature are normally present on the metals.

Experiment has shown that the frictional behaviour is not greatly influenced by the actual degree of surface finish provided the surfaces are finely ground and flat. For highly polished surfaces, on which a Beilby layer has been formed, the results, though similar in character, are less reproducible, probably because of the altered nature of the Beilby layer. In general, the surfaces were prepared in a very finely ground condition (similar to that shown in Plate I. 3 (*d*), Chap. I) using standardized techniques, and were carefully degreased.

THE FRICTION OF METALS

Measurements made with this apparatus show that under many conditions the sliding between metal surfaces may not be a continuous process but may proceed in a series of intermittent jerks. The friction builds up to a maximum during the 'stick' and falls rapidly during the 'slip', and there are corresponding changes in the area of contact and the surface temperature. The nature of the motion observed depends on the physical and mechanical properties of the friction-measuring system as well as on the intrinsic frictional properties of the surfaces themselves; these will be discussed in greater detail in Chapter V. In what follows we shall deal mainly with the nature of the surface damage produced between sliding metal surfaces (Bowden, Moore, and Tabor, 1943; Bowden, 1945; Moore, 1948).

The most striking conclusion drawn from an extensive study of the friction of metals is that the magnitude of the frictional force and the extent and type of surface damage caused by sliding are determined primarily by the *relative physical properties* of the two sliding surfaces. In particular the behaviour is very dependent upon the relative hardness of the two surfaces and, if the sliding speeds are high, upon their relative softening-points or melting-points. We may for convenience divide them into three main types: (i) a hard metal sliding on a soft one, (ii) a soft metal sliding on a hard one, (iii) similar metals sliding.

Careful examination of the metals shows that some surface damage always occurs even with lightly loaded, well-lubricated surfaces. Plate VI. 1 shows the tracks produced on a flat metal surface when a small curved slider passes over it once. In these experiments, a heavy load (4,000 gm.) was used and the surfaces were unlubricated, so that the effect can be seen more clearly. As we shall see later, the surface damage with loads of only a few milligrams is essentially of the same nature although it is, of course, on a greatly reduced scale. It should be emphasized that in these experiments and in the ones below, the speed of sliding is very small so that the temperature rise due to frictional heating is inappreciable. Plate VI. 1 (*a*) is a typical result obtained when the upper curved slider is a hard metal and the lower flat surface is a softer one. In this case the curved slider was mild steel (V.D.H. 120) and the flat surface was copper (V.D.H. 60). It will be seen that a shallow groove has been ploughed out in the softer copper surface. The coefficient of friction under these conditions is about $\mu = 0{\cdot}9$. Plate IV. 1 (*b*) is characteristic of type (ii) where a curved slider of a softer metal (copper) slides over a harder flat surface (steel). The harder metal is comparatively little damaged, but the softer metal has welded on to it and fragments of it remain adhering to the surface. Again, the coefficient of friction with dissimilar metals is of the order of $\mu = 0{\cdot}7$. With similar metals sliding together, type (iii), the damage is much more profound (see Plate VI. 1 (*c*) for nickel on nickel) and the coefficient of friction is considerably greater ($\mu = 1{\cdot}2$–$1{\cdot}5$).

Highly magnified taper sections which represent a cross-section of these three types of track, cut at right angles to the direction of sliding, are shown in Plates VI and VII. Plate VI. 2 shows a section of the track produced on copper by a curved steel slider. It will be seen that the groove is ploughed out of the copper to a depth of *c.* 1–2 microns, and that the copper is built up at the edges to about the same height. Just beneath the surface the metal is very severely deformed, and the deformation and work-hardening of the copper extend to a great distance (15 microns or more) beneath the surface.

If the upper surface is of a softer metal, this ploughing out is not observed. Instead, small fragments of the softer metal are left welded on the harder one. Characteristic welded junctions of this type, formed when copper is slid on mild steel at a low speed, are shown in Plate VI. 3. The taper section was again cut at right angles to the direction of sliding, and the conditions were similar to those used on the steel surface. The coefficient of friction in this case is about $\mu = 1$. The horizontal magnification is 120. The greater part of the steel surface is undamaged, and isolated fragments of copper *A* are left adhering to the steel and distributed over it. For these junctions the shearing has occurred in the copper itself. In other places, for example the hollows and pits *C* shown in Plate VI. 3, the copper has plucked out small fragments of the steel. Photomicrographs of the copper–steel junctions at higher magnifications are shown in Plate VII. 1 and 2. The white portion F of the steel in Plate VII. 1 is ferrite, the dark portion P is pearlite, and C is a copper fragment *c.* 10^{-3} cm. wide and *c.* 10^{-4} cm. high. It is clear that the copper is welded on to the steel and that the junction has sheared in the copper itself. A similar copper fragment is shown in Plate VII. 2. It is seen that the forces which sheared through the copper have also caused appreciable deformation of the underlying steel, and have actually raised portions of the steel above the general level of the steel surface. Over a large portion of the surface, however, the break has occurred between the copper and the steel, so that there is no apparent change in the surface. The wear of a hard metal by a soft one which is brought about in this way by a welding and plucking action is a much slower process and is very different in appearance from the more normal abrasive wear produced by rubbing with a harder surface (see Chap. XIV).

A taper section of the track formed when similar metals slide (copper on copper) is shown in Plate VII. 3. In this case the friction is higher ($\mu = 1\text{–}1{\cdot}5$) and the damage is great. The ploughing and tearing is very evident over a large area of the track; the metal may be torn out to a depth of 20 microns or more beneath the surface. Beneath this still farther, the metal is deformed and work-hardened to an additional depth which may be greater than 50 microns. It is clear that when similar metals are in contact the local high pressures must cause an equal flow of both surfaces so that they both contribute equally to the formation of welded junctions and when sliding takes place, both surfaces will be distorted and torn. Moreover, since the metal will be work-hardened at the interface the junction will be stronger than the underlying metal, so that the break will occur in the bulk of the metal. This welding and interchange of metal is illustrated in Plate VII. 4, which shows the track on the lower copper surface after the copper slider has passed over it eleven times (instead of once as in Plate VII. 3). The damage is very great, and it is probable that the darker mass A of heavily worked copper has come from the upper slider.

An interesting confirmation that welding takes place between the sliding metals was obtained from some experiments with platinum sliding on silver. At the beginning of the run the behaviour was, as expected, characteristic of dissimilar metals; the coefficient of friction was $\mu = 0{\cdot}5$ (Plate VIII. 1) and the track was a smooth groove in the silver (see Plate VIII. 4). After the slider had moved a short distance the behaviour changed to that characteristic of similar metals. The friction rose to

$\mu = 1{\cdot}3$ (Plate VIII. 2) and the torn track now resembled that found in such cases (Plate VIII. 5). A close examination showed that a small amount of silver had become welded on to the platinum slider (Plate VIII. 6) so that the sliding was in reality silver on silver.

The sliding of steel on steel is of some interest. At first thought we might expect that the behaviour would be characteristic of similar metals. Experiment shows, however, that with some steels this is not the case—it resembles much more closely that of dissimilar metals. The coefficient of friction is about $\mu = 0{\cdot}5$ and the damage is correspondingly smaller. A characteristic track formed on a mild steel surface when a slightly worn slider of the same metal passes over it once is shown in Plate IX. 1. The localized nature of the damage is at once apparent. A considerable area of surface within the track is unchanged. At a number of points, however, penetration, ploughing, and tearing of the metal have occurred. The depth of this localized damage varies from 1 to 10 microns. At other points on the surface the metal is raised above its former level: some of this metal comes from the lower surface and some is probably left behind by the slider. The reason why the friction is more characteristic of dissimilar metals is probably that the steel itself is not a homogeneous alloy. It is probable that the small localized points of contact between the upper and lower surfaces will not, in general, be of identical composition, nor will they possess identical physical properties. An extreme case of this is cast iron, which is very heterogeneous. It gives a low coefficient of friction ($\mu = 0{\cdot}3$) and comparatively little damage when it slides upon itself. It is found that if a very uniform steel is used, such as an austenitic stainless steel, or if pure iron is used, the behaviour is immediately characteristic of similar metals: the friction is high ($\mu = 1\text{–}1{\cdot}5$) and the typical torn track is produced. These results, that the friction and interaction are very dependent upon the *relative* physical properties of the two metals, are quite general, and in the appendix at the end of the book the values of the friction for a number of similar and dissimilar metals are collected. It is apparent that, over a wide range, the friction of dissimilar metals is always considerably less than that of similar metals, and this is accompanied by a corresponding diminution in the surface damage.

In these experiments the speed of sliding was very slow, so that the rise in the surface temperature was inappreciable; at high loads and speeds this will no longer be so, and the nature of the interaction between the metals will be determined by the relative mechanical properties of the metals *at high temperatures*. The relative softening- or melting-points of the metals become an important factor. These observations have an obvious bearing on engineering practice and on the wear and friction of metals sliding under actual working conditions (see Chap. XII).

CHEMICAL AND RADIOACTIVE DETECTION OF METAL TRANSFER [*A*]

When the surfaces are lubricated, the metallic interchange between the surfaces is very greatly reduced, so that its detection even by the taper section method is very difficult. It may, however, be detected by a sensitive chemical method which has been applied by Dr. Moore. A gelatine-coated paper which has been immersed in a suitable electrolyte is placed on the surface of the metal to be examined and a current passed so that the foreign metal is electro-dissolved into the gelatine. If an appropriate reagent is present in the gelatine, very small quantities of metal may be detected. For example dithio-oxamide may be used to detect copper, and Plate IX. 2 shows the patterns obtained by allowing the copper slider to pass once over polished steel (*a*) when clean and (*b*) when lubricated with 1 per cent. lauric acid in paraffin. The black areas indicate copper, and it will be seen that the adhesion of copper to the lubricated steel surface is quite considerable. The track of the slider is about 1 mm. wide, and adhesion has occurred at a number of small points distributed over it. There is a tendency for these points to lie on straight lines in the direction of sliding, and this probably corresponds to high spots on the copper slider. Plate IX. 2 (*c*) and 2 (*d*) show corresponding results when the experiments are carried out on a steel surface which has been finely lapped. The lapping scratches run at right angles to the track, and the pick-up of the copper is concentrated in the regions which correspond to the ridges of the lapping scratches (Bowden and Moore, 1945).

Similar experiments of a more quantitative nature were carried out with copper on platinum, both clean and lubricated with solid potassium stearate. Any copper adhering to the flat platinum surface after the copper slider had passed over it once was removed electrolytically from an appropriate length of track and micro-estimations made. With the clean metals the surface density of copper adhering to the platinum was 2×10^{-5} gm. per mm.^{2} of track. With lubricated metals the amount of copper adhering to the platinum was $1 \cdot 7 \times 10^{-7}$ gm./mm.^{2} If this copper were spread evenly over the platinum in the path of the slider, it would correspond to a layer about 150 A thick; but as the photographs show, it is distributed irregularly over the surface of the track in a number of small discrete particles of varying size.

A still more sensitive method is to use a slider of radio-active metal and to detect any pick-up on the other surface either by a Geiger counter (Sakmann, Burwell, and Irvine, 1944) or by a photographic method (Gregory, 1946). Plate IX. 3 shows the results obtained by Mr. Gregory when a radioactive slider of lead was slid once over a flat steel surface. The steel was then placed in contact with a photographic plate and the presence of any lead detected by a fogging of the plate. Plate IX. 3*a* is for unlubricated surfaces (coefficient of friction $\mu = 0{\cdot}4$) and Plate IX. 3*b* for surfaces lubricated with 1 per cent. stearic acid in paraffin ($\mu = 0{\cdot}1$). By comparison methods it is possible to estimate approximately the amount of lead present.

It will be seen that the photographs bear a general resemblance to those obtained by the chemical method. An estimate of the amount of metallic pick-up shows that for unlubricated surfaces the amount of lead which adheres to the steel is of the order of 4×10^{-7} gm./mm.^{2} of track. For the lubricated surfaces it is about 2×10^{-8} gm./mm.^{2} of track. If this were spread uniformly over the track it would correspond to a film thickness of *c.* 20 A, but again it is concentrated mainly at the high spots on the lapping scratches. It is clear that the reduction in friction produced by a lubricant is accompanied by a marked reduction in the amount of metallic pick-up, but even under the best conditions of boundary lubrication, some localized breakdown of the film and localized adhesion is observed. The bearing of this on the theory of lubrication will be discussed more fully later.

Friction and Surface Damage at Light Loads [*A*]

The experiments so far described were carried out at loads of the order of kilograms. It may be thought that plastic flow is inevitable at these loads and that the conclusions so far drawn are consequently only valid under a restricted range of conditions. It is therefore of considerable interest to examine the friction and surface damage between sliding surfaces at very light loads. An investigation of this kind has recently been carried out by Dr. Whitehead using loads of the order of milligrams. The surface damage was examined with an electron microscope and the friction determined by means of the apparatus shown in Fig. 29 (Whitehead, 1950).

The lower surface is in the form of a small block of metal *A* mounted at the edge of a brass turntable B. The turntable is driven by a suitably geared electric clock-motor and the sliding speed of the surface B is of the order of 0·01 cm./sec. The upper surface or slider C, is in the form of a small hemisphere of diameter about 0·8 mm. and is mounted at one end of a steel wire D. The other end of the wire is suitably clamped at the end E of a lever arm F. The load is applied to the slider by raising the remote end G of the lever arm and so flexing the wire in a vertical

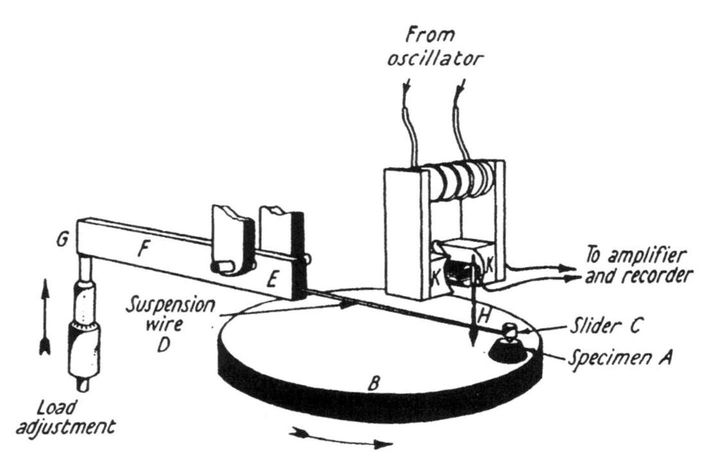

FIG. 29. Apparatus used to measure the friction between surfaces at very small loads. By using suspension wires of various thicknesses the load range may be varied from a few milligrams to 100 gm.

plane. By using wires of various thicknesses a practicable range of loads between 5 mg. and 20 gm. may be obtained. When the lower surface is set in motion it drags the slider with it and flexes the wire D in a horizontal direction until the restoring force is equal to the friction. The deflexion of D in the horizontal direction is thus a measure of the frictional force. This is recorded electrically in the following way. The movement of D deflects the needle H which rotates a moving coil supported between a pair of soft iron pole-pieces (K). The magnetic coils around the pole pieces are energized by an alternating current at 100 cycles/sec. Deflexion of the needle alters the induced voltage in the moving coil and this is connected via an amplifier to a suitable pen recorder. The resultant frequency of the system is limited by the frequency of the pen recorder rather than by that of the electrical circuit and is of the order of 10 cycles/sec. Thus the apparatus will respond to fluctuations occurring in 1/10th sec. or longer, but will not give a reliable record of more rapid fluctuations. A typical friction trace for steel sliding on electrolytically polished aluminium is shown in Fig. 30. It is seen that the motion is irregular and that the coefficient of friction at a load of about 0·12 gm. is of the order of $\mu \approx 1{\cdot}2$. This is essentially the same as that observed at loads of several kilograms.

In examining the surface damage by electron microscopy it is necessary to make a thin film replica of the damaged surface. The simplest method is to pour a 2 per cent. solution of Formvar (polyvinyl formal)

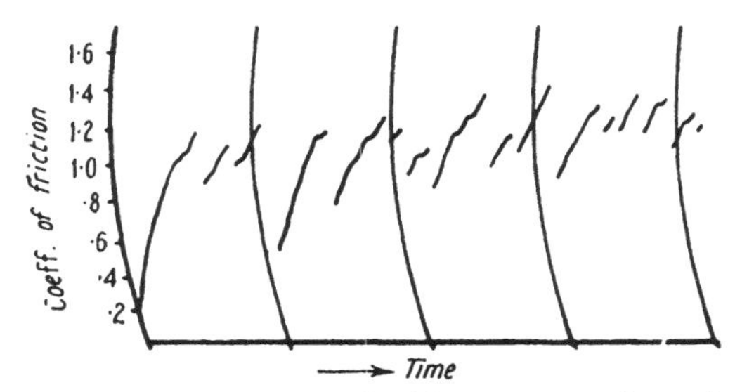

FIG. 30. Typical friction trace obtained with apparatus shown in Fig. 29. The record is for steel sliding on aluminium at a load of 0·12 gm. The coefficient of friction at the 'sticks' lies between $\mu = 1{\cdot}2$ and $\mu = 1{\cdot}4$.

in dioxane over the surface and allow it to drain off and dry. The resulting film of Formvar, which is then stripped off for examination, is about 500 to 1,000 A thick. A typical electron micrograph taken by this method is shown in Plate X. 1. It shows the surface damage produced when a copper slider passes over an electrolytically polished copper surface under a load of 0·64 gm. The coefficient of friction was about $\mu = 0{\cdot}5$, and although the detail is not particularly good it is seen that there is considerable flow and displacement of the copper from the path of the slider. It is also evident that the slider makes contact with the surface at only a small number of points producing a series of parallel tracks. It is also interesting to note that a 'kink' may be produced in one of these tracks whilst the other tracks remain essentially straight and continuous. This effect arises from the fact that the points on the upper surface which rub on the lower surface are far apart in the direction of sliding, so that when the slider is displaced sideways (perhaps as the result of a 'slip') the kinks in each track occur at different positions along the main track. The micrograph covers a small field and only includes one of these kinks. Another micrograph obtained with a Formvar replica is shown in Plate X. 2 for sapphire sliding on sapphire at a load of 0·64 gm. The friction was about $\mu = 0{\cdot}25$ and the surface damage so slight that it was not detectable by optical means. The micrograph, however, shows clearly that the surface of the sapphire has been damaged by the sliding process.

If the damaged surface is of aluminium (or an aluminium alloy) a more satisfactory replica is obtained by oxidizing the surface and stripping off the oxide, which retains the shape of the surface deformation. The specimen is oxidized anodically in a solution of acid sodium phosphate to a suitable thickness and the oxide removed in a solution of mercuric chloride. Free mercury is liberated, the oxide becomes detached from the surface and floats to the surface of the solution. The oxide replica is picked up on a suitable grid, washed, dried, and transferred to the microscope. Typical micrographs obtained in this way are shown in Plate X. 3 and 4 for steel sliding on electrolytically polished aluminium. Plate X. 3 is for clean surfaces at a load of 0·34 gm. when the coefficient of friction was about $\mu = 1$. It shows a typical 'stick' produced in the course of intermittent motion and it is apparent that considerable deformation of the surface has occurred. Plate X. 4 shows a similar track observed on an aluminium surface contaminated with grease. The load was 0·34 gm. and the coefficient of friction about $\mu = 0{\cdot}3$. It is seen that the surface damage is less severe and consists of a series of fine parallel scratches.

The micrographs obtained from aluminium oxide replicas may be further improved by shadowing the replica with an opaque material such as gold and palladium. The shadowing increases the contrast and provides a fairly reliable method of estimating the height of the surface irregularities. Typical results obtained in this way from replicas, shadowed at an angle of 45° with gold and palladium by evaporation *in vacuo*, are shown in Plate XI. 1 and 2. These micrographs show the damage produced on an electrolytically polished aluminium surface by a steel slider. The load was about 0·34 gm. and the coefficient of friction about $\mu = 1$. Plate XI. 1 shows the ploughing and tearing of the aluminium surface and the presence of minute particles of steel buried in the surface. The edge of the track, shown in greater detail in Plate XI. 2, reveals the deformation and flow of the aluminium on a submicroscopic scale. The 'waves' set up at the edge of the track are only about 100 A apart. It is interesting to compare the damage observed on aluminium at these very small loads with the damage produced at much higher loads when the track can be examined with an ordinary optical microscope. Plate XI. 3 shows the track formed at a load of 20 gm. for steel sliding on aluminium. The tearing of the surface and the slip lines at the edge of the track are clearly visible.

Plate XI. 4 shows the edge of a track formed at a load of 2,000 gm. The slip lines are very marked indeed, and it is evident that the deformation produced by sliding extends well beyond the edge of the track itself. There is a marked similarity between this micrograph taken at a magnification of 500 and the electron micrograph in Plate XI. 2 (where the load is 6,000 times smaller) taken at a magnification of 25,000.

These micrographs provide direct evidence for the occurrence of plastic flow, tearing, and plucking of metals when sliding occurs. Although the loads are 10,000 times smaller or less than those described in the previous experiments, the value of the friction, the nature of the motion, and the appearance of the surface damage are essentially the same.

Early Theories of Metallic Friction

In 1699 Amontons published the result of his experimental investigation of the friction of unlubricated solids. He found that the frictional force was independent of the area of the surfaces and was directly proportional to the normal load. In fact he concluded that the frictional force was always equal to one-third of the normal load. To explain this result he assumed that irregularities on the surfaces of the two bodies interlocked, and the relative motion necessitated lifting the load from one position of interlocking to another. This caused a loss of energy which was manifested as a frictional force. Amontons's conclusions were tested by many investigators, notably de la Hire (1732) and Euler (1750). The latter agreed with Amontons in giving to all surfaces a frictional coefficient of one-third. The most systematic work was done by Coulomb (1785), who examined the influence of a large number of variables on the friction. He agreed that the friction was proportional to the load and made the new observation that it was independent of the velocity of sliding. Further work was done by Rennie (1829), Morin (1835), and Hirn (1854). Hirn distinguished carefully between lubricated and unlubricated solids, and he observed that the effect of velocity, surface area, and load differed in the two cases. Few precautions were taken by any of these experimenters to obtain clean and reproducible surfaces. For example, Rennie's surfaces were almost certainly contaminated, since the value he obtained for the coefficient of friction was very low (*c.* 0·1 or less). All these earlier workers agreed with Amontons in assuming that friction was caused by the interlocking of asperities. Rennie pointed out that a more general theory should take into account the bending and fracture of these asperities. The action of lubricants was explained by assuming that the lubricants filled these irregularities and at the same time, by some means, made them more 'slippery'.

These earlier theories were questioned when Lord Rayleigh suggested that the difference between a polished surface and the surface of a fluid might not be very great. This view was confirmed by Beilby, who showed that polishing and grinding were two essentially different processes. The more recent view that friction had its origin in surface forces and was due to molecular cohesion between the solids was then introduced. Ewing (1892) considered that friction was due to the reaction of molecular forces following the molecular displacement. This theory has also been put forward and has received strong support from the extensive work of Sir William Hardy (1936) on static friction. He considered that the friction could be explained in terms of the surface fields of the solids. The results he obtained with lubricants were explained by assuming that the lubricant caused, in a very definite and quantitative way, the reduction of the molecular field of force at the surface of the solid.

An interesting attempt to correlate the interaction during sliding between the surface molecules at the surface of two unlubricated solids has been made by Tomlinson (1929). He considered that the friction was due to the energy dissipated when the molecules were forced into each other's atomic fields and were then separated. He made the further assumption that the molecular fields of force were approximately the same for all substances, and the area of molecular contact could be calculated from Hertz's equation for elastic deformation. He found approximate agreement of experiment with theory, but many of his experimental values for friction were very different from those found by other workers.

These theories differ somewhat, but they all make the assumption that the resistance to sliding is due to the surface fields of force only. It is clear from the experiments which have just been described that the frictional effects are not confined to the surface of the solid but cause distortion and deformation to a very great depth beneath it. Obviously the physical processes that occur during sliding are too complex to yield easily to a simple mathematical treatment, but the experiments show that, under the intense pressure which acts at the summits of the surface irregularities, a localized adhesion and welding together of the metal surfaces occurs. When sliding takes place, work is required to shear these welded junctions and also to plough out the metal. We may inquire how far the friction of metals can be explained in terms of this process.

REFERENCES

AMONTONS (1699), *Histoire de l'Académie Royale des Sciences avec les Mémoires de Mathématique et de Physique*, p. 206.

F. P. BOWDEN (1945), *Proc. Roy. Soc. N.S.W.* **78**, 187 (Liversidge Lectures).

—— and L. LEBEN (1939), *Proc. Roy. Soc.* A **169**, 371.

—— and A. J. W. MOORE (1945), *Nature*, **155**, 451.

—— A. J. W. MOORE, and D. TABOR (1943), *J. Appl. Phys.* **14**, 80.

C. A. COULOMB (1785), *Mémoires de Mathématique et de Physique de l'Académie Royale des Sciences*, p. 161.

DE LA HIRE (1732), *Histoire de l'Académie des Sciences*, p. 104.

EULER (1750), *Histoire de l'Académie Royale des Sciences et Belles-Lettres*, 1748. Berlin, 1750, p. 122.

SIR A. EWING (1892), *Not. Proc. Roy. Instn.* **13**, 387.

J. N. GREGORY (1946), *Nature*, **157**, 443.

SIR W. B. HARDY (1936), *Collected Works*. Cambridge University Press.

G. A. HIRN (1854), *Bull. Soc. Indust. Mulhouse*, **26**, 188.

A. J. W. MOORE (1948), *Proc. Roy. Soc.* A **195**, 231.

A. J. MORIN (1835), *Nouvelles expériences sur le frottement, faites à Metz en 1833*. Paris.

G. RENNIE (1829), *Phil. Trans.* **34**, 143.

B. W. SAKMANN, J. T. BURWELL, and J. W. IRVINE (1944), *J. Appl. Phys.* **15**, 459.

G. A. TOMLINSON (1929), *Phil. Mag.* **7**, 905.

J. R. WHITEHEAD (1950), *Proc. Roy. Soc.* A **201**, 109.

V

MECHANISM OF METALLIC FRICTION [A]

THE ROLE OF SHEARING AND PLOUGHING

THE previous chapter has shown that the friction between metal surfaces cannot be regarded purely as a surface effect. It is clear that during sliding, metallic junctions, large compared with molecular dimensions, are formed and sheared. In addition, if one surface is harder than the other the harder surface asperities plough out the softer material to an appreciable depth below the surface. We may therefore expect that the frictional force will be greatly influenced by the bulk properties of the metals. It is evident that the physical processes which occur during sliding are very complex and do not yield readily to a quantitative treatment. We may, however, express the frictional resistance as the sum of two terms, one of which represents the shearing and the other the ploughing process. In this way it is possible to obtain an approximate expression for the friction in terms of the bulk physical properties of the metals. For a similar analysis see Ernst and Merchant (1940).

In order to separate the forces of shearing and ploughing it is convenient to consider the frictional force between hard sliders of various geometric shapes and a plane surface of a softer metal. Suppose a hard hemispherical slider rests on the softer metal. If the load is W, the hemisphere sinks into the softer metal until the area of contact is sufficient to support the applied load (see Figs. 31 (*a*) and 31 (*b*)). Then if A is the projected area of contact and p is the yield pressure of the softer metal,

$$A = W/p. \tag{1}$$

For simplicity we assume that the surfaces are perfectly smooth so that A, the real area of contact, is the same as the geometrical area. If the slider is rough we may consider any one asperity to be approximately hemispherical, so that the present treatment will apply, on a greatly reduced scale, to each of the asperities.

The force F required to move the slider in a direction parallel to the surface is now considered to consist of two terms. The first is the force S required to shear the metallic junctions formed at the points of intimate contact between the two metals. This may be written as

$$S = As, \tag{2}$$

where s is the force per unit area which, acting in a direction tangential to the interface, is required to shear the junctions.

The second term P is the force required to displace the softer metal from the front of the slider. It will be equal to the cross-section of the

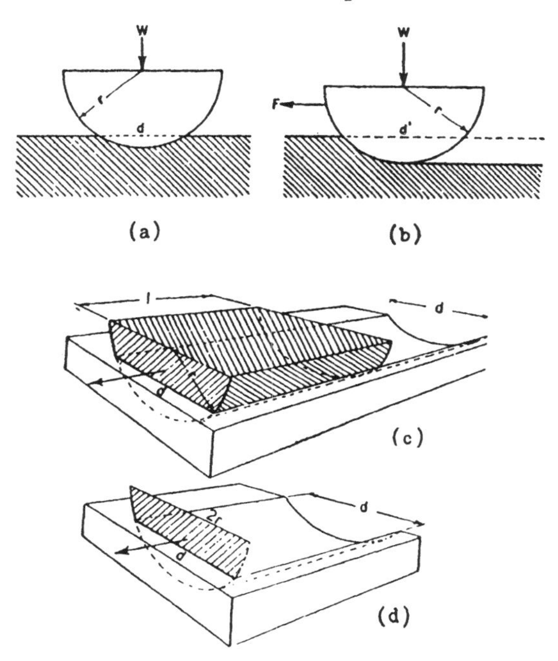

FIG. 31. Deformation of a soft metal by a hard curved surface. (*a*) Hemisphere, stationary. (*b*) Hemisphere, moving. (*c*) Horizontal cylinder normal to direction of motion. (*d*) Flat spade normal to direction of motion. For (*a*), (*b*), and (*c*) the frictional resistance involves a ploughing force P and a shearing force S. For (*d*) the friction involves only the ploughing force P.

grooved track A' multiplied by the mean pressure p' required to displace the metal in the surface. Then

$$P = A'p', \tag{3}$$

where we may expect that p' will be of the same order as p. If d is the track width and r the radius of curvature of the slider, A' is approximately equal to $\frac{1}{12}\frac{d^3}{r}$.

Hence
$$P = \frac{1}{12}\frac{d^3}{r}p'. \tag{4}$$

The total frictional force is then given by

$$F = S+P = As+A'p'. \tag{5}$$

In the above case this becomes

$$F = As+\frac{d^3}{12r}p'. \tag{6}$$

As another type of slider we may consider a cylinder of radius r and length l (Fig. 31 (c)). It is clear that the shearing term S will be equal to the area of the curved surface, that is submerged, multiplied by s. In general this will not be very different from the projected area, ld, multiplied by s. The ploughing term P will be equal to the area of the submerged segment of the front of the cylinder multiplied by the pressure p', so that to a first approximation

$$F = sld+\frac{1}{12}\frac{d^3}{r}p'. \tag{7}$$

We may eliminate the shearing term completely by considering a cylinder of zero length, i.e. a slider in the form of a hemispherical spade (Fig. 31 (d)). In this case the shearing term is zero and the force resisting motion is entirely due to the ploughing term, so that

$$F = \frac{1}{12}\frac{d^3}{r}p'. \tag{8}$$

This is the same as the ploughing term in equations (6) and (7). By simple subtraction we may therefore calculate the shearing term for both the hemispherical and the cylindrical slider. This provides a simple means of determining the separate contributions of the shearing and ploughing terms to the total frictional force.

Experiments were carried out to test the validity of these equations using the apparatus already described. The upper slider, in the form of a spade, a cylinder, and a hemisphere, was made of steel and the lower plane surface of indium, which is a very soft metal. The speed of sliding was very low, of the order of 0·01 cm./sec., so that the rise in temperature due to frictional heating was negligibly small.

The ploughing term P

The theory shows that the ploughing force should, in the case of the spade, be proportional to the cube of the track width. Experimental measurements with steel spades on indium are shown in Fig. 32, and it is seen that this is approximately true. Further the ploughing term should not be influenced by the presence or absence of a lubricant film. This again is directly confirmed by experiment, as may be seen in Fig. 32. These values were obtained when the slider first begins to move. During further sliding the displaced material accumulates in front of the slider and the force resisting motion gradually increases. At this stage a lubricant may have an appreciable effect on the ploughing force.

Using equation (8) the pressure p' resisting the movement of the spade may be calculated. The result for indium is 1,500 gm./mm.[2] compared with the static yield pressure obtained from indentation measurements of 1,000 gm./mm.[2] The higher value of p' may be due

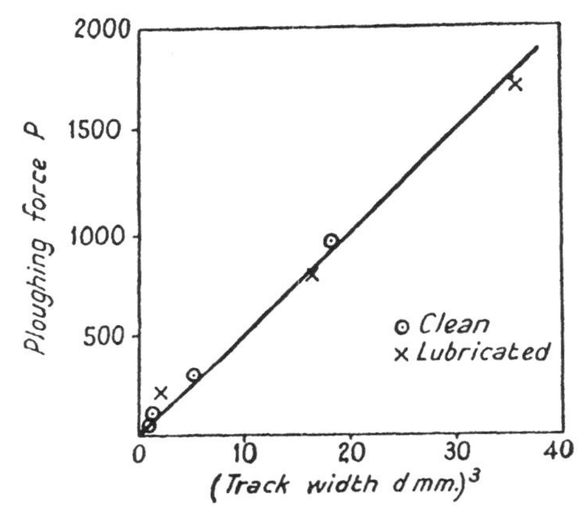

FIG. 32. Force P (in gm.) required to plough through indium with a flat steel spade, as function of the track width d. It is seen that P is essentially the same for clean and for lubricated surfaces, and that it is proportional to d^3.

to the fact that the ploughing experiment involves the flow of the metal at an appreciable rate, whereas the indentation tests are carried out under static conditions.

In spite of the uncertainty as to the exact physical meaning of the pressure p', these results show that the ploughing term may be calculated in terms of the geometry of the moving surfaces and a factor p' that is of the same order of magnitude as the yield pressure p of the softer metal. Since the depth of the track is itself dependent on the yield pressure p of the metal, it should be possible to express the ploughing term as a general function of p. A simple analysis (in which we assume that p' is simply proportional to p) shows in fact that for hemispherical sliders on metals of different hardness the ploughing term should be proportional to $1/\sqrt{p}$, so that it will be less important for hard metals. For conical sliders, on the other hand, the analysis shows that, for a fixed load, the ploughing force is independent of p but increases the more pointed the cone. Finally, for flat surfaces making contact at a number of hemispherical asperities the ploughing term, for a fixed load, will be less the larger the number of the points of contact. In general the analysis suggests that for harder metals the ploughing term is small. For softer metals, however, it may often constitute an appreciable fraction of the total frictional force.

The shearing term S

In Fig. 33 the tangential force F is plotted against the track width d for a steel sphere, a steel cylinder, and a steel spade, all having the same

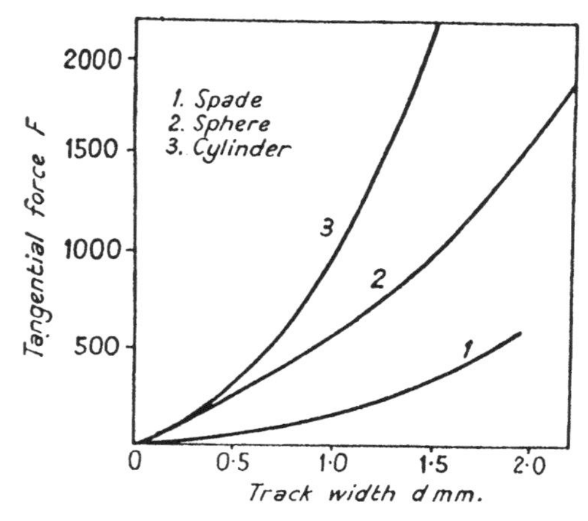

Fig. 33. Frictional force F (in gm.) of curved steel surfaces on indium as a function of the track width d. 1. Spade. 2. Sphere. 3. Cylinder. The difference between curves 1 and 3 gives the shearing term S for the cylinder.

radius of curvature. It is seen that for a given track width the friction is least for the spade, as is to be expected, since the shearing term is clearly absent. Further, as we might expect, the friction for the sphere is intermediate in value between the spade and the cylinder.

The difference between curves 1 and 3 gives the shearing term S for the cylinder, and from equation (7) we see that it should be proportional to the first power of the track width d. In Fig. 34 the experimental value of S thus obtained is plotted against d. It will be seen that experiment is again in good agreement with theory, and from the slope of the curve we obtain a value of $s = 225$ gm./mm.^{2}

The difference between curves 1 and 2 gives, in a similar way, the shearing term for the sphere, and this yields a value of $s = 350$ gm./mm.^{2}

Experiments were also carried out with a curved indium slider on a flat steel surface. The motion proceeded in very marked stick-slips and at each stick the indium left a well-defined 'patch' of indium on the surface, marking clearly the area over which the indium and steel surfaces had welded together. Measurements showed that this area was proportional to the frictional force at the stick, and the shear strength of the steel–indium junctions so formed gave a value of $s = 325$ gm./mm.²

A subsidiary test was carried out to determine the force required to shear solid indium. Cylinders of indium of various diameters were sheared between two flat steel plates sliding over one another in close contact. It was found that the shearing force was proportional to the

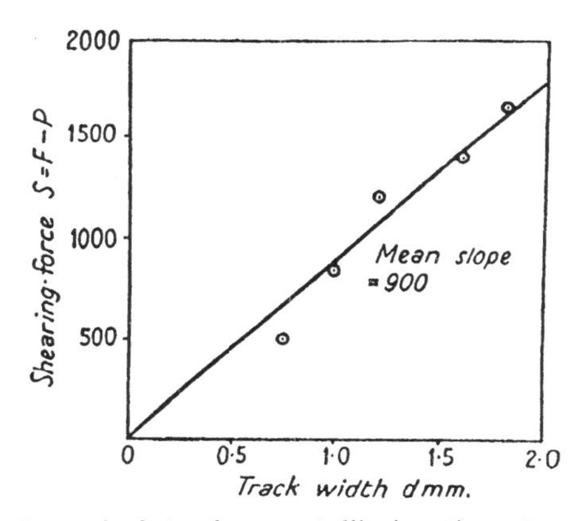

FIG. 34. Force S required to shear metallic junctions formed between steel cylinder and indium. (Difference between curves 1 and 3 in Fig. 33.) The mean slope is 900 gm./mm. Since the length of the cylinder is 4 mm. the shear strength $s = \frac{900}{4} = 225$ gm./mm.²

area of cross-section of the cylinders, and for the same speed as that used in the above friction experiments a value of s was obtained, $s = 220$ gm./mm.² Increasing the speed of shear by a factor of 150 caused s to increase by a factor of only 2.

The results obtained from the friction measurements for a variety of experiments in which steel sliders of different shapes and sizes were used are shown in Table XI.

TABLE XI

Shear strength of steel–indium junctions calculated from friction. (Sliding speed = 0·005 cm./sec.)

Slider	*s in gm./mm.²*
Small steel cylinder (r = 0·25 cm.)	225±30
Large steel cylinder (r = 2·5 cm.)	250±50
Small steel hemisphere (r = 0·25 cm.)	300±50
Large steel hemisphere (r = 2·5 cm.)	270±70
Curved indium slider on flat steel surface	325
Shear strength of pure indium	220

It will be seen that, although the experimental methods were very different in each case, the values obtained for s are in reasonable agreement. Moreover, the value of s (i.e. the shear strength of the steel–indium junction) is very nearly equal to the shear strength of pure indium. We may reasonably assume, therefore, that the shearing has actually occurred within the bulk of the indium itself. This is confirmed

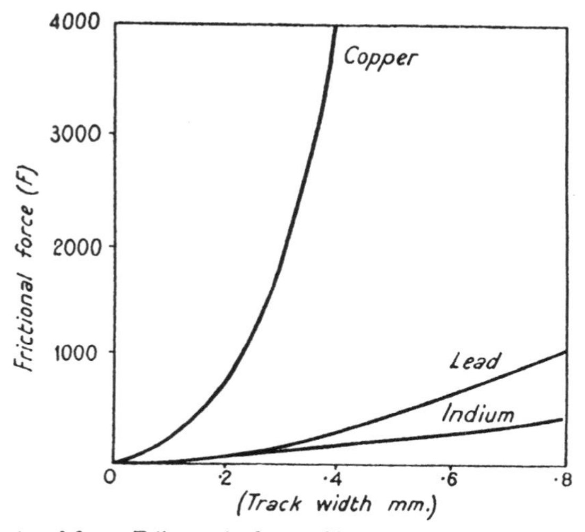

FIG. 35. Frictional force F (in gm.) of a steel hemispherical slider on indium, lead, and copper, as a function of track width. It is seen that for any given track width the friction on copper > lead > indium.

by a micro-examination of the steel slider which shows that the surface is covered with numerous fragments of indium where the softer metal has welded on to the steel.

Similar experiments were carried out with steel sliders on lead and copper surfaces. With lead the ploughing term P is appreciably smaller than with indium. If we neglect it and assume that the frictional force is due entirely to the shearing term S, we obtain a value of 1,800 gm./mm.^{2} for the shear strength s of the lead–steel junctions. This may be compared with the shear strength of pure lead of 750 gm./mm.^{2} obtained by the direct experimental method of shearing small cylinders of the pure metal. With copper surfaces, the ploughing term P is negligibly small and the shear strength s of the copper–steel junctions obtained from the friction measurements is about 28,000 gm./mm.^{2} The shear strength of pure copper is about 16,000 gm./mm.^{2}

The relative frictional properties of a steel hemisphere sliding on indium, lead, and copper are shown in Fig. 35. It is seen that for any given track width the friction of copper > lead > indium and that the values of the friction for a fixed track width are roughly proportional to the shear strength of these metals.

The shear strength of the metal junctions

The shear strength s of the junctions formed between a steel slider and surfaces of indium, lead, and copper are summarized in the second column of Table XII, whilst the shear strengths of the pure metals are, for comparison, given in the third column. As a matter of interest the values obtained for a hard steel ball sliding on hard steel are included at the bottom of the table.

TABLE XII

	Shear strength gm./mm.² calculated from friction measurements	*From shearing of pure metals*
Indium	325	220
Lead	1,600	750
Copper	28,000	16,000
Steel ball	140,000	90,000

For metals such as indium and lead, where the surface asperities flow into one another these values of s are the true values of the shear strength of the junctions formed between the slider and the metal. For metals such as copper and steel, however, where contact occurs at the tips of surface asperities, the real area of contact will be less than that given by equation (1), so that the effective values of s, deduced from the friction experiments, will be somewhat higher than those given in this table. Nevertheless, if we think of the shearing at the tip of any asperity as being a process essentially similar, though greatly reduced in scale, to the processes described above for indium and lead, it is clear that the general picture of the frictional process still remains valid.

The results summarized in Table XII show that the shear strength s calculated from the friction measurements is of the same order as the shear strength of the pure metal, but is somewhat greater. There is also, as we have seen, marked pick-up of the softer metal on the harder slider at the points where metallic junctions have been formed and then sheared. This suggests, therefore, that at those regions where the soft metal flows plastically under the applied load, metallic junctions are formed and shearing takes place within the bulk of the softer metal itself. It should be noted that at the slow sliding speeds used in these experiments the temperature rise due to frictional heating is only of the order of a few degrees. Consequently the metallic junctions must be formed by a 'cold-welding' process. At higher speeds of sliding, however, where appreciable frictional heating occurs, the temperature rise may play a significant role in facilitating the flow and welding of the metallic junctions.

Amontons's Law

We are now in a position to consider the reason for Amontons's observations that the frictional force is independent of the apparent area of the sliding bodies and directly proportional to the load.

As we have just seen, the real area of contact between two metals is given by $A = W/p$, and metallic junctions are formed at these regions of intimate contact. If s is the mean tangential stress necessary to shear these junctions, the total force to shear all the junctions is given by As, so that if the ploughing term is negligible, the frictional force is given by

$$F = As. \tag{9}$$

It follows from the discussion in Chapter I that the real area of contact A depends only on W and p and is almost independent of the apparent area of the surfaces. Consequently F should be independent of the apparent area of the surfaces. This is Amontons's first 'law'.

If now we substitute for A, we obtain

$$F = Ws/p. \tag{10}$$

Thus the frictional force is directly proportional to the load, so that the coefficient of friction is virtually independent of the load. This is Amontons's second 'law', and it holds over an extremely wide range of experimental conditions. Dr. Whitehead, for example, has recently carried out some friction experiments at very small loads using the apparatus shown in Fig. 29 (see Chap. IV). Table XIII gives values of the coefficient of friction observed for loads of a few milligrams and these may be compared with values obtained with the identical surfaces on the large friction apparatus where the load is of the order of kilograms. Similar observations for steel sliding on aluminium are plotted in Fig. 36.

TABLE XIII

Surfaces	*Load gm.*	*Coefficient of friction* μ
Aluminium slider	0·037	1·5
on electrolytically	0·075	1·25
polished aluminium	0·150	1·5
surface	4,000	1·35

It is seen that over a load range of 100,000 or more the coefficient of friction remains sensibly independent of the load, and the same type of agreement is found for surfaces lubricated under boundary conditions (see Chap. IX). It is, however, important to note that Amontons's law holds only when the area of intimate contact increases with the applied load. If by some artificial device, such as the use of thin metallic films,

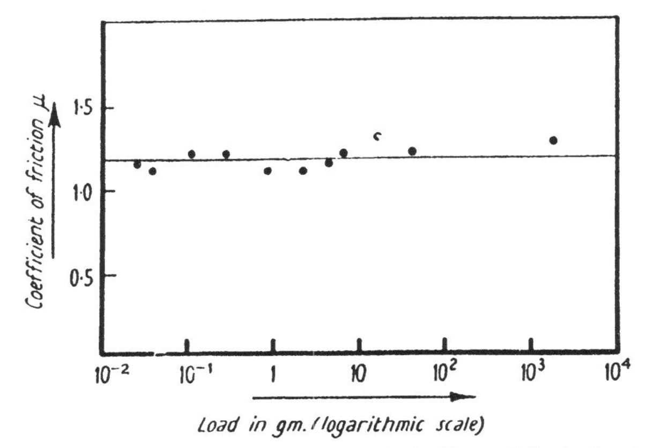

FIG. 36. Friction of steel sliding on electrolytically polished aluminium as a function of load. The coefficient of friction is almost constant for loads ranging from about 10 mg. to 10 kg., i.e. over a load range of 10^6.

the area of contact is prevented from increasing proportionally with the load, this law is no longer valid. The breakdown of Amontons's law for thin metallic films will be discussed in greater detail in the second half of this chapter. A similar effect with oxide films will be discussed in Chapter VII.

There is another case where Amontons's law is no longer applicable. This occurs with soft metals such as indium or lead sliding on a hard metal; it is found that the friction is very variable and depends on the previous history. If the load is first applied and then decreased or removed altogether, the surfaces may continue to 'stick' together and the tangential force necessary to cause sliding may remain high, although there is no normal load between the surfaces. This is not observed with harder, more elastic metals. For such metals the friction decreases when the load is decreased. This is because there is a release of elastic stresses around the deformed region and the small movements that result serve to break the metallic junctions already formed (see Chap. XV). Although on the part of the load–area curve when the load decreases the real area of contact is not proportional to the load (see Chap. I, Fig. 10) the deviation is not very marked and the friction decreases steadily with the load. In the case of soft plastic metals like indium or lead, however, in which the released elastic stresses are very small, this will not always occur and once the clean surfaces are pressed together they may continue to adhere even when the load is reduced or removed completely. Similar effects have been observed on strongly outgassed gold surfaces (see Chap. VII). For this reason measurements of the 'coefficient of friction' of an indium or lead slider on a flat steel plate may have little meaning. If, however, the friction is measured as a function of the real area of contact this difficulty does not arise and consistent results are obtained. For example, frictional measurements of indium sliding on steel show that Amontons's law is not usually obeyed but that the frictional force is directly proportional to the *area* over which the indium and steel surfaces weld together. The shear strength of these steel–indium junctions is about 325 gm./mm.[2] in reasonable agreement with results obtained from other frictional measurements and from direct shearing experiments. This adhesion between soft plastic metals will be discussed more fully in Chapter XIV.

We may now rewrite equation (10) so that the coefficient of friction ($\mu = F/W$) may be expressed in terms of the bulk properties of the materials. From equation (10),

$$\mu = \frac{s}{p} = \frac{\text{shear strength of junctions}}{\text{yield pressure of softer metal}}. \tag{11}$$

As we have already seen, the shearing usually occurs within the softer metal itself, so that s is approximately equal to the shear strength of the softer metal. Consequently μ becomes a function of the physical properties of the softer of the two sliding bodies and may be expressed as

$$\mu = \frac{\text{shear strength of softer metal}}{\text{yield pressure of softer metal}}. \tag{12}$$

Two conclusions at once follow from this result. The first is that s and p, being strength properties of the same metal, vary together and their ratio is roughly the same for the most diverse metals. This explains the observation that the coefficient of friction of a very wide range of metals does not vary by a very large factor: generally it lies between $\mu = 0{\cdot}6$ and $\mu = 1{\cdot}2$. This will be discussed again later in this chapter. The second conclusion is that temperature should not have a very marked effect on the friction of unlubricated metals since s and p should vary together as the temperature is raised or lowered. This is generally found to be the case. Unless the heating is sufficient to affect the nature of the surface films the friction does not depend markedly on the temperature. On the other hand, if the temperature rise is due to frictional heating it will be localized in the surface layers. Consequently sliding at higher speeds or loads may produce a more marked change in s than in p, so that there may be some reduction in the coefficient of friction. This effect will be very marked when the decrease in s is restricted to the surface layers, and when the underlying bulk material retains its high value of p. In general this does not happen with metals to any marked degree. With solids such as ice, however, which retain their hardness at temperatures very close to their melting-point, a very large reduction in friction may occur (see Chap. III).

THE INTERDEPENDENCE OF p AND s

There are two assumptions in the above theory which we may now consider in greater detail. The first is concerned with the use of the shear strength s. We have assumed that if a junction has an area A the force to break it by shearing is sA, and we have compared s with the shear strength of cylinders of the metal constituting the junction. This is valid if the junction is, in fact, sheared as a whole, so that at the end of the shearing process its area is zero. This is approximately true at a 'stick' when the motion is intermittent (see later). If, however, the motion is smooth the junction is never sheared across completely but is continuously being formed and sheared. In such cases the shearing force is more accurately determined by the critical yield stress in shear, that is, the shear stress which just produces plastic flow. It is probable, however, that this parameter will be of the same order of magnitude as the shear strength s, so that the simple treatment given above remains approximately valid.

Another assumption connected with this, but of greater importance, is the interdependence of p and s. So far we have treated the yield pressure and the shear stress as independent variables. According to the theory of plasticity, however, both these stresses must play a part in producing plastic flow of the metals. A simple case will make the position clearer. Supposing a block of soft metal is pressed with one of its flat faces on to a hard flat surface with a normal pressure σ. We consider the block to be a long prism of uniform cross-section so that we may, in effect, treat it as a two-dimensional problem in plasticity. Suppose now the block is subjected to a tangential pull equivalent to a shear stress s (Fig. 37 (a)). We now assume that σ and s reach such values that the block forms a junction with the harder metal across the whole of the interface AB and that shearing takes place in the softer metal near AB, in the slice $ABCD$ (Fig. 37 (b)). Strictly speaking the shear along CD should fall to zero at D and C since there can be no shear stress in the free surfaces AD, CB. For simplicity, however, we

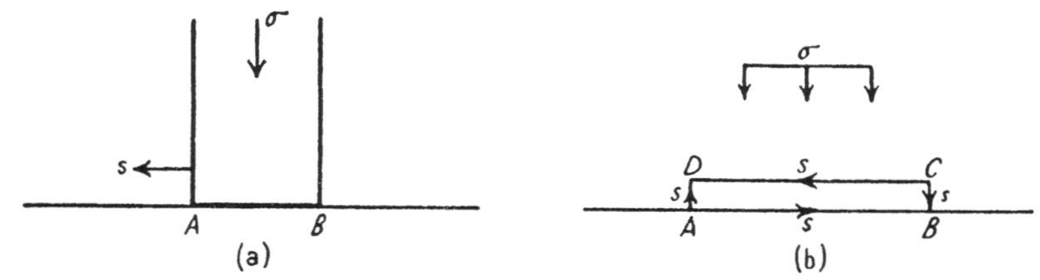

FIG. 37. The shearing of a block of soft metal making metallic contact over the whole of its lower face.

assume a constant average value for s (and also for σ) with shears along AD, CB to give equilibrium at the edges. Then von Mises's criterion for plastic yielding to occur in the slice $ABCD$ is

$$\sigma^2+3s^2 = Y^2, \tag{13}$$

where Y is the yield stress of the material $ABCD$ in extension or frictionless compression. Clearly an infinite number of solutions is possible provided $\sigma < Y$ and $s < (1/\sqrt{3})Y$. Let us consider two extreme cases:

(i) The block rests on the surface under a static pressure $\sigma = Y$. This is the simple condition of compression at the plastic yield stress. According to equation (13) the stress necessary to cause shear is given by $s = 0$. Thus the coefficient of friction at the initial instant of sliding will be zero. In practice, of course, as soon as plastic yielding begins, the cylinder flows and increases the area of contact so that σ decreases and s correspondingly increases.

(ii) The block is pressed on the surface so as to form a junction over the whole interface AB and the normal stress is removed. Suppose the recoverable elastic stresses are not sufficient to break the junction. Then $\sigma = 0$, and according to equation (13) the section $ABCD$ will flow plastically under a shear stress s when $s = (1/\sqrt{3})Y$. This corresponds to a simple shearing process where there is a finite shearing stress and zero normal stress. In this case the coefficient of friction will be infinite. In practice, of course, the recoverable elastic stresses will impose complex normal stresses on the section $ABCD$ and this will probably reduce the value of s. Further, the slightest tangential movement of the surfaces will, in the absence of all normal pressures, tend to separate them completely. However, in the presence of small but finite pressures we may expect very high values of the frictional force if the formation of the junctions across the interface AB can take place. This occurs with very soft metals such as indium (see Chap. XV). It also occurs with other metals when the surfaces are thoroughly outgassed in a high vacuum (see Chap. VII).

Under normal laboratory conditions the coefficient of friction of most 'clean' metals has a value of the order of unity. On the model considered above, this means that $\sigma = s$. Consequently, from equation (13) $\sigma = \frac{1}{2}Y$, that is to say, the normal pressure corresponding to plastic flow in the slice $ABCD$ will be one-half of that necessary to produce plastic flow under pure compression. This means that for a given load the area of contact when sliding commences will be twice that observed under pure compression, i.e. when the surfaces are stationary. An increase of this order is often observed when soft metals such as indium or lead slide on steel.

The model so far considered is, of course, very far indeed from most practical cases. In practice, contact usually occurs between spherical or conical or pyramidal asperities on the surface. Under normal compression the stress σ at each point of real contact is no longer equal to Y but is equal to the yield pressure p where, as we saw in Chapter I, $p \approx 3Y$. When shearing of the junctions formed at these points occurs the relation between σ, s, and Y becomes extremely complex, and has not yet been solved. We may expect that it will be of the form

$$\alpha\sigma^2+\beta s^2 = \gamma Y^2, \tag{14}$$

where α, β, and γ are numerical coefficients so that the solutions will be of the same type as those discussed for the simpler model. Thus we may expect that if the surfaces are loaded normally to produce plastic flow at the real points of contact an extremely small tangential force will initiate shear. A small displacement will, however, deform the softer metal and so produce an increase in the area of contact. This will lead to a decrease in the normal stress and a corresponding increase in the tangential force. This build-up of the frictional force before macroscopic sliding commences has been observed (McFarlane and Tabor). At the other extreme, if the surfaces are very soft or are so clean that welding can occur very readily under very small normal pressures, the tangential stress will be large so that the coefficient of friction will be exceedingly high.

There are thus two established observations which confirm the general validity of equations (13) and (14). The first is the high friction for outgassed metals and for very soft metals. The second is the tendency for the area of real contact to increase when sliding takes place. There is, however, no *a priori* reason for the observed result that under normal laboratory conditions σ is of the same order of magnitude as s. Probably the explanation for this result is that the oxide films on the metal surfaces are not broken down by shear alone; a *normal* pressure of the order of the static yield pressure is needed for the breakdown of the surface films and for the formation of metal–metal junctions. It is only when these surface films are absent that the junctions can form under the smallest normal forces and so give large coefficients of friction.

We may therefore conclude that the shear stress s and the normal stress σ play a combined role in the formation and shearing of the metallic junctions. Under normal laboratory conditions, however, junctions are only formed when the normal stress approaches the yield stress of the material. Consequently the simple picture which considers s and σ (or p) as independent variables gives results which are reliable to within a factor of about 2.

Intimacy of Contact and the Influence of Surface Films

Although the area of contact is, under any given conditions, determined primarily by the mechanical properties of the solids, it is clear that the *intimacy* of the contact and the strength of adhesion at the points of contact will be greatly influenced by the presence of surface films. Under most experimental conditions metal surfaces are covered with a thin oxide layer and other contaminating films, and during sliding these oxide and surface films will be torn and some metallic contact will occur. We should expect that the adhesion and shear strength of these junctions, which consist of metal plus torn fragments of oxide and other contaminants, would be less than that of the pure metal. Experiments described in Chapter VII on the friction between outgassed metals show that this is indeed the case. Similarly experiments show that the deliberate addition of lubricant films reduces the intimacy of contact and the mean strength of adhesion of the junctions.

If over the whole area of real contact A the contact is truly metallic the junction is usually stronger than the softer metal and shearing occurs within the softer metal itself. Consequently the force to shear the junction, as we have already seen, may be written as $s_1 A$, where s_1 is the shear strength of the softer metal. In practice, however, with surfaces prepared under normal laboratory conditions intimate metallic contact does not occur over the whole of the area of contact but only over a fraction α, so that the force required to shear the metallic part of the junctions will be $\alpha s_1 A$. Over the rest of the area of contact, $A(1-\alpha)$, where the contaminant films have not been broken the adhesion will be less intimate and the junctions will be appreciably weaker. The shear strength of the regions where the contaminant films are unbroken will not necessarily be constant. It may vary from a value which approaches s_1 down to very small values. If s_2 is the average value, the force to shear these weaker junctions will be $A(1-\alpha)s_2$ and the total frictional force will be given by

$$F = A[\alpha s_1 + (1-\alpha)s_2]. \tag{15}$$

From the fact that the shear strength of indium–steel junctions as calculated from the friction experiments is almost equal to the shear strength of pure indium, it would seem that for clean steel sliding on indium α is very nearly equal to unity, s_1 to the shear strength of pure indium. On the other hand, the taper section in Chapter IV, Plate VI. 3, for copper on steel suggests that the adhesion is not sufficient to cause shearing within the copper over the whole area of contact, and frequently the breakdown occurs at the steel–copper interface. In this case α must be appreciably less than unity, and s_2 will be less than s_1. As we shall see in Chapter X, the main effect of adding a lubricant to clean metal surfaces is to reduce both α and s_2.

Intermittent Motion

Since the friction between metal surfaces depends on the intimacy of contact α, the yield pressure p, and the shear strength of the junctions, s_1 and s_2, it is clear that if any experimental variations arise which alter one or more of these factors, there will be a corresponding alteration in the frictional force. To a certain extent, therefore, the friction will depend on the experimental conditions under which it is measured. For example, at extremely slow speeds of sliding these factors vary in such a way that the resultant strength of the metallic junctions is often greater than that occurring at higher speeds of sliding. This means, as many workers have shown, that the static friction is often higher than the kinetic friction. If, therefore, one of the sliding surfaces has a certain degree of elastic freedom, the motion may not be continuous, but may be intermittent and proceed by a process of 'stick-slip'. The 'stick' is due to the higher static friction between the surfaces, and the 'slip' to the lower kinetic friction during the slip itself.

This is well exemplified by the friction traces obtained with the Bowden–Leben apparatus described in Chapter IV. A typical record of mild steel sliding on steel (unlubricated) is shown in Fig. 38 (Bowden and Leben, 1939). It is seen that the motion is not steady but consists of fluctuations in which the pull on the slider increases steadily (AB) and then falls very rapidly (BC). A subsidiary calibration shows that

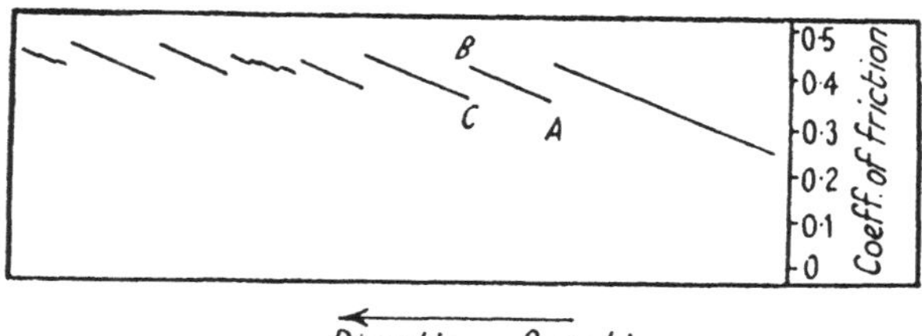

FIG. 38. Typical friction-trace obtained with apparatus described on Plate V for steel sliding on steel. During the interval AB the slider sticks to the lower surface and travels along with it. At B the slider breaks free and slips rapidly to the position corresponding to point C.

the slope of the line AB is such that the upper slider must be moving forward with a velocity identical with that of the lower surface. Thus the surfaces are 'sticking' together for the whole of the time interval between A and C, which amounts to several seconds. Consequently the friction at the maximum B is essentially the static friction between the surface (μ_s). At B a rapid 'slip' of the upper surface occurs and when the force has fallen to C the surfaces 'stick' together again.

We may show by a very simple analysis that intermittent motion of this type is a direct consequence of the fact that the slider has a certain degree of freedom and that the static friction is higher than the kinetic. We consider the fairly general case of two sliding surfaces, where one is driven at a steady uniform speed v and the other (the 'free' surface) is supported by an elastic system so that its deflexion during sliding is a measure of the frictional force. Suppose the inertial mass of this free surface and its supporting parts is m and the force constant is k, so that a deflexion by a distance x corresponds to a force $F = -kx$, where the negative sign indicates that the force is in the opposite direction of increasing x. The equation of motion of the surface in free oscillation, assuming there is *no damping*, is simply

$$m\frac{d^2x}{dt^2} = -kx. \tag{16}$$

Thus the motion is simple harmonic of frequency $n = \frac{1}{2\pi}\sqrt{\frac{k}{m}}$.

Suppose now the surfaces press together with a load W, i.e. a force Wg (point A, Fig. 39). We assume that the surfaces travel together without relative motion until the force on the free surface is equal to the static friction $F_s = \mu_s Wg$. The deflexion of the surface is thus linear with time and the deflexion $BX = (\mu_s Wg)/k$. At the point B slip will occur and we may assume, as a first approximation, that the kinetic friction

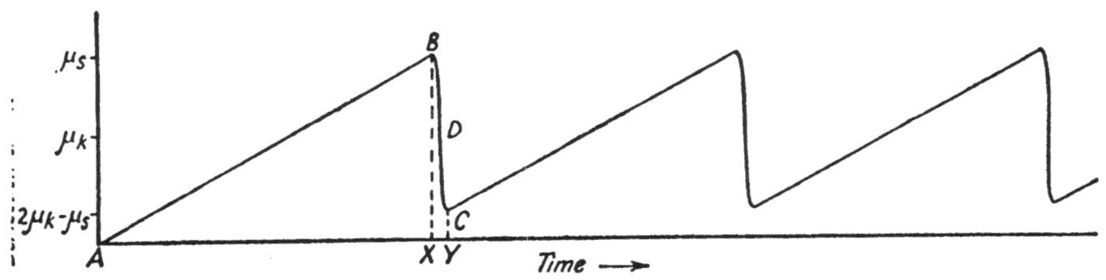

FIG. 39. Analysis of 'stick-slip' or intermittent motion.

during the slip has a constant value $\mu_k Wg$. The equation of motion of the surface is now modified from equation (16) to

$$m\frac{d^2x}{dt^2} - \mu_k Wg = -kx, \tag{16 a}$$

where at the initial instant, say $t = 0$ (point B), the deflexion

$$x = (\mu_s Wg)/k$$

and the forward velocity dx/dt is equal to v. The solution is

$$x = \frac{Wg}{k}\{(\mu_s - \mu_k)\cos\omega t + \mu_k\} + \frac{v}{\omega}\sin\omega t, \tag{17}$$

where $\omega^2 = k/m$. If the velocity v of the lower surface is small compared with the mean velocity of the slip, the last term may be neglected, so that we may write

$$x = \frac{Wg}{k}\{(\mu_s - \mu_k)\cos\omega t + \mu_k\}. \tag{17 a}$$

Thus the motion of the free surface has the same natural frequency as before and it comes to rest relative to the lower surface, that is, it sticks again when $dx/dt = v$. If v can be neglected we may consider that sticking recurs when $dx/dt = 0$, i.e. when $\omega t = \pi$. This is half the natural period of the system and by inserting this value of ωt in equation (17 a) we find that at this instant $x = CY = (2\mu_k - \mu_s)Wg/k$. Thus the size of the slip $BC = BX - CY = (2\mu_s - 2\mu_k)Wg/k$. We see from this relation that the larger μ_k the smaller the slip. In the limit when $\mu_k = \mu_s$ the slip becomes zero, so that the deflexion remains steady at B:

this corresponds to the case of smooth sliding. As μ_k decreases the slips increase in size. In the limit when $\mu_k = 0$ the surface swings beyond the zero position to a deflexion equal and opposite to BX; this corresponds to the most marked stick-slip motion that could ever be obtained, though since μ_k can never be zero the intermittent motion could never, in practice, be as pronounced.

The distance of the mid-point D of the slip from the zero is simply equal to $\frac{1}{2}(BX+CY) = \mu_k(Wg/k)$. Thus, in units of Wg/k, the deflexion at the end of the stick (point B) corresponds to the static friction μ_s, whilst the mid-point of the slip (point D) corresponds to the kinetic friction μ_k.

This analysis is, of course, greatly simplified. In general there will be some damping of the moving parts and under critical conditions this may completely eliminate the occurrence of intermittent motion even though there is a finite difference between μ_s and μ_k. In addition μ_k may be a function of speed so that it will vary during the course of the slip itself. Consequently the detailed behaviour may be very complex indeed, as Morgan, Musket, and Reed (1941) have shown. Nevertheless, the above treatment brings out the main features of the intermittent motion observed between sliding surfaces.

It is interesting to compare the fluctuating frictional force with the electrical conductance (Bowden and Tabor, 1939). Some typical traces for various metal combinations are reproduced in Plate XII. The black traces refer to the friction measurements, the white traces to conductance measurements recorded by an Einthoven string galvanometer. It is seen that with constantan on steel, the conductance and hence the area of contact increases gradually during the stick as the upper surface is pulled into more intimate contact with the lower surface by the increasing tangential force. At the slip there is a sudden fall in the conductance corresponding to a rapid decrease in the area of contact and the friction. The process is then repeated. Similar results are obtained for steel on tin, constantan on zinc, and in general this behaviour is characteristic of harder metals sliding on softer ones.

With tin on steel there is again a sudden decrease in the conductance at the slip. With these combinations, however, the conductance decreases gradually during the later part of the stick as the welded junctions of the upper surface are drawn out thinner and thinner until finally they break at the 'slip'. Similar results are obtained for lead on steel, and in general this behaviour is characteristic of softer metals sliding on harder ones.

With similar metals the friction is very much higher and there are large fluctuations, but no rapid slips are observed. Similarly the conductance is high but shows no rapid variations. In this case both surfaces flow equally and contribute in comparable measure to the formation and the shearing of the metallic junctions. Consequently the changes are less marked and less abrupt than with combinations of dissimilar metals.

When intermittent motion occurs there is also a corresponding fluctuation in the surface temperature. This is clearly shown in Plate XIII, which records simultaneously the frictional force and the thermal e.m.f. generated when constantan slides on steel. With the apparatus described in these experiments the speed of the slip varies from zero to about 5 cm./sec. and the temperature rise at the fastest part of the slip is not greater than 10° or 20° C. Similar values have been obtained by Morgan, Musket, and Reed, using a cathode-ray oscillograph instead of an Einthoven galvanometer to record the thermal p.d. If, however, the sliding system has a smaller inertia and large fluctuations occur in the frictional force, the speed of sliding during the slip may be very much greater. In such cases, as we saw in Chapter II, the surface temperature may be appreciably higher. These temperatures may play an important part in the formation of metallic junctions at the end of the 'slip'. Thus intermittent motion may produce significant fluctuations in the surface temperature and hence in the frictional process as a whole, even though the *mean speed of sliding* may be relatively low. This effect becomes very much more marked with poor thermal conductors and, with materials such as plastics, thermal softening during the 'slip' may play an important part in the sliding process (see Chap. VIII).

Experiments show that with a given apparatus the intermittent motion does not depend on the shape of the surfaces or on the load, i.e. $\Delta\mu$ remains roughly constant for a given combination of metals. The behaviour is, however, profoundly affected by the speed of sliding. As the speed increases $\Delta\mu$ decreases in size until at a critical speed it disappears and the motion is relatively smooth. This is what we should expect. As the surface speed is increased the time during which the surfaces can remain together at the 'stick' is reduced, μ_s becomes smaller, and the difference between the processes of 'stick' and 'slip' becomes less and less marked. Consequently the two coefficients of friction at the stick and during the slip become more nearly equal, with a corresponding decrease in the magnitude of the fluctuations.

The incidence of intermittent motion depends markedly on the relation between μ_s and μ_k and on the general way in which the friction varies with the speed of sliding. Those metal combinations which show a rapidly falling friction–speed characteristic, that is a marked disparity between μ_s and μ_k, will tend to give intermittent motion under a very

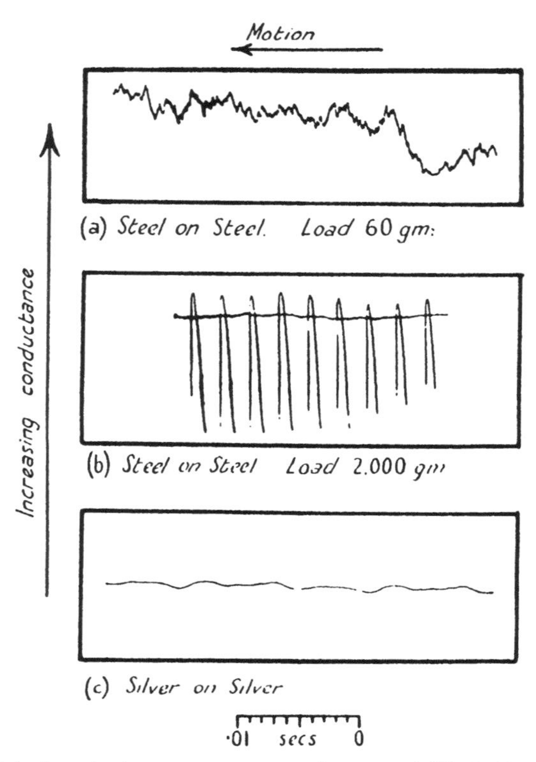

FIG. 40. Electrical conductance measurements between rigidly held metal surfaces. For steel on steel the motion is intermittent and at light loads (*a*) the fluctuations last for 10^{-3} to 10^{-4} sec. At heavy loads (*b*) a resonant frequency in the supporting system is set up and the conductance fluctuations are very marked. The surfaces emit a loud squeak at the resonant frequency. With silver surfaces (*c*) the motion is smooth and the conductance shows no rapid fluctuations at any load.

wide range of experimental conditions. This is shown by some experiments in which the upper surface was held by a rigid steel bar. The friction was not measured, but the electrical conductance was recorded on a cathode-ray oscillograph. A typical result for steel sliding on steel at a speed of 0·025 cm./sec. is shown in Fig. 40 (*a*). It is seen that extremely rapid fluctuations occur in the electrical conductance, each fluctuation corresponding to a relative motion of the surfaces of the order of 10^{-6} cm. At heavier loads the surfaces emitted a loud squeak corresponding to the natural frequency of the bar supporting the upper surface and the conductance showed a similar periodic fluctuation (Fig. 40 (*b*)). Similar results were obtained for Wood's metal sliding on steel. For silver sliding on silver, however, where there is no such marked disparity between μ_s and μ_k, these rapid fluctuations were not observed (Fig. 40 (*c*)) and at heavier loads the squeaking at the resonant frequency of the bar did not occur.

It follows from the above discussion that the incidence of intermittent motion depends on the metal combination considered, on its friction–speed characteristic, and on the velocity of the main forward motion (Blok, 1940; Bristow, 1942). It will also depend on the mechanical properties of the system such as the natural frequency, the inertia, and the damping of the moving parts (Bowden, Leben, and Tabor, 1939). In addition, as Morgan, Musket, and Reed (1941) have shown in a detailed investigation of the slip, the frictional behaviour will depend on the sequence of events during the slip itself. This means that the detailed mechanism of sliding is relatively complex and generalizations as to the type of motion must be viewed with some reserve. Nevertheless, since many moving systems possess an appreciable degree of elastic freedom and since the friction–speed characteristic for many metal combinations is a falling one, we may expect that intermittent motion of the type discussed above will be of frequent occurrence in practice. Even when the moving parts are rigid, the surface irregularities may be capable of microscopic elastic deformations of the order of 10^{-5} cm., as Khaikin, Lissovsky, and Solomonovitch (1940) have shown, using quartz crystals to measure the minute displacements involved. In such cases the elasticity of the surface irregularities themselves may, in the limit, be sufficient to set up vibrations or intermittent motion in the moving parts.

Even in the presence of lubricant films, intermittent motion of a similar nature may occur. This motion is superimposed on the general forward motion of the sliding surfaces and may play an important part in the friction and wear processes involved. As we shall see in a later chapter, an analysis of the motion can give valuable information about the mechanism of wear and the nature and properties of the lubricant films.

Lubricating Properties of Thin Metallic Films

We have seen that of the two terms which are responsible for metallic friction it is the shearing term which is the more important. If we neglect the ploughing term, the frictional force $F = As$. It is apparent, therefore, that if we wish to reduce the friction between clean metals we must make both A and s as small as possible. With most metals, however, this is not possible.

If we choose a metal with a low shear strength it is usually soft, so that for a given load the area of contact A becomes larger (Fig. 41 (*a*)). An obvious exception to this is an anisotropic solid with a plate-like

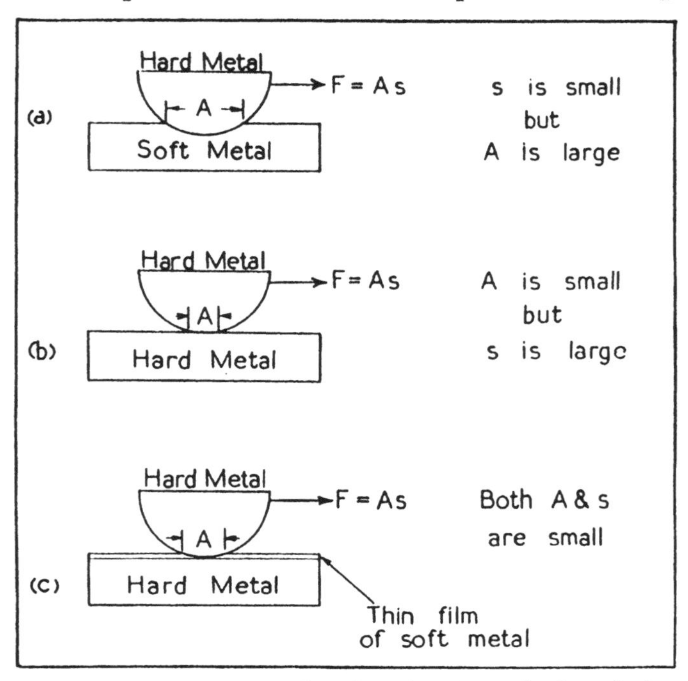

FIG. 41. The friction between metal surfaces is not greatly dependent on their hardness. A low friction may be obtained by depositing a thin film of a soft metal on a hard metal substrate.

structure which is able to withstand a pressure normal to the plates but will shear readily when a tangential force is applied. It is possible that this is the cause of the low friction observed with solids such as talc or graphite (see, however, Chap. VIII). It is difficult to achieve this condition for metals since their physical properties are far less anisotropic. If, on the other hand, we choose a hard metal the area of contact A will be small, but s will be correspondingly great (Fig. 41 (*b*)). For this reason the coefficient of friction of most metals is of the same order of magnitude and lies between about $\mu = 0{\cdot}6$ and $\mu = 1{\cdot}2$.

We may, however, achieve this condition by depositing a very thin film of a soft metal on the surface of a hard one (Fig. 41 (*c*)). Provided the metallic film does not break down, the shear strength s will be that of the soft metal. At the same time A will remain small even for heavy loads, since the load is borne by the hard substrate and there will therefore be little deformation. As a result the friction $F = As$ will be small. It is interesting to examine the general frictional properties of thin metallic films and some study has been made of thin films of indium, lead, and copper deposited on various substrates.

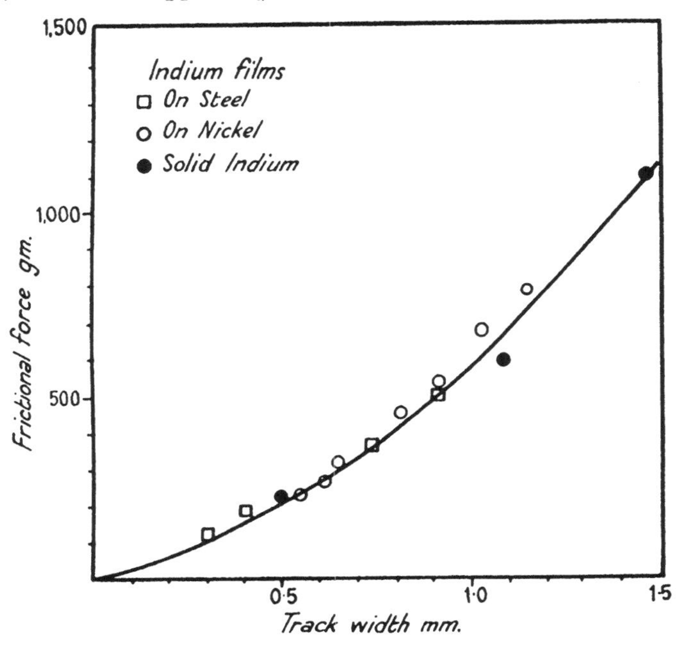

Fig. 42. Friction of hemispherical steel slider on indium films deposited on steel and nickel substrates, as a function of track width. The friction is determined primarily by the track width, whether the friction is measured on the surface film or on the bulk indium.

Friction as a function of track width

It is clear from the previous paragraph that when a hard hemispherical surface slides on a thin metallic film the friction should depend only on the shear strength of the film and on the track width, since this determines the area of contact. It should not depend on the nature of the substrate nor on the applied load. Simple frictional experiments with a steel hemisphere sliding on thin films of indium and lead deposited on various substrates show that this is the case, as is shown in Figs. 42 and 43. In these experiments the track width was altered by varying the thickness of the deposited layer and the radius of curvature of the slider. The main effect of increasing the load was to cause a slight increase in the deformation of the underlying metal and so cause a slight increase in the track width. With a soft substrate such as silver this effect should be more marked than with a hard substrate such as steel. This was observed. In all cases the frictional resistance was determined primarily by the track width whether the friction was measured on the

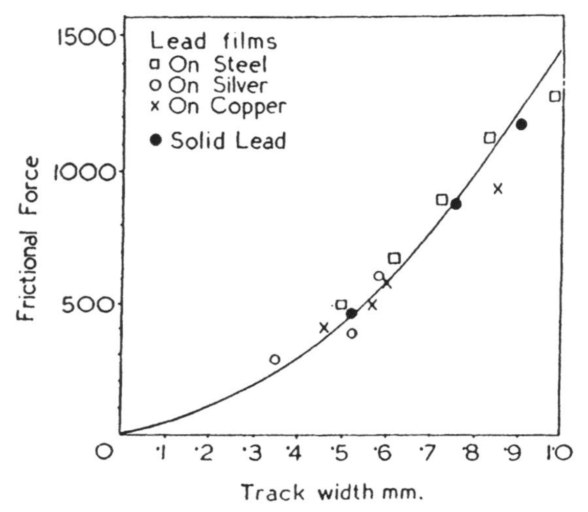

FIG. 43. Friction of hemispherical steel slider on lead films deposited on steel, silver, and copper substrates. The friction is determined primarily by the track width, whether the friction is measured on the surface film or on the bulk lead.

surface film or on the bulk metal. This shows that in these experiments the shearing occurred in the metallic film.

The limiting film thickness

This picture of the action of thin metallic films is valid only when there is no appreciable metallic contact through the film. We may therefore expect that the behaviour will change when the metallic film is less than a certain minimum thickness. This is shown in Fig. 44 for an indium film on tool steel. The film thickness was calculated from the quantity of electricity used to deposit it. The upper curved slider was of hard steel (radius 0·3 cm.) and the load was 4,000 gm. As one might expect, the friction decreases as thinner films are used because the track width and the area of contact A becomes smaller. There is, however, a limit to this and a minimum friction is reached when the thickness is of the order of 10^{-5} cm. With thicknesses less than this, e.g. 10^{-6} cm. (or 30 atomic layers), the film ceases to be effective.

The underlying surface of the tool steel was not highly polished nor was it homogeneous. These factors might easily cause the indium to be deposited preferentially in certain regions, leaving patches of the surface uncovered or only very thinly covered. It is possible that if highly

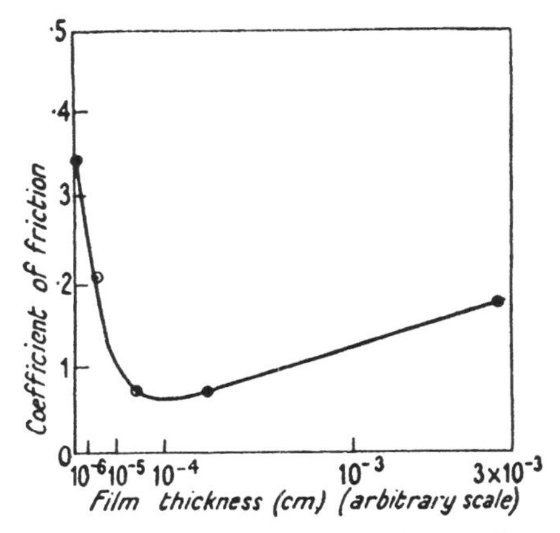

FIG. 44. The effect of film thickness on the frictional properties of thin films of indium deposited on tool steel. A minimum friction is obtained for a film approximately 10^{-4} to 10^{-5} cm. thick.

polished surfaces of a uniform metal were used as a substrate, very much thinner layers, even perhaps one or two molecular layers, might prove effective.

Breakdown of the film

In a similar way, if the load is too high, we may expect breakdown of the metallic film with increased metallic contact between the slider and substrate and a corresponding increase in the frictional force. Typical results which illustrate this effect are shown in Fig. 45 for an indium film 3×10^{-3} cm. thick deposited on a steel substrate. The slider was a steel hemisphere of 0·3 cm. radius of curvature, and the load was increased at the point A from 500 gm. to 8,000 gm. It is seen that with the heavier loads the friction has increased markedly, even though the track width has remained essentially unaltered. The frictional behaviour suggests that the indium film has broken down and that contact is occurring between the exposed steel surfaces. A microscopic examination of the track after the slider has passed over it only once shows that there is, indeed, a progressive breakdown of the indium film as heavier loads are used.

Similar results were obtained with indium films on silver and with lead films on steel and copper. In all cases the load at which the breakdown occurs depends upon the thickness of the film, its strength of adhesion to the underlying surface, and the hardness of the underlying surface. It is also influenced by the shape of the slider and takes place more readily the smaller its radius of curvature. As we shall see in

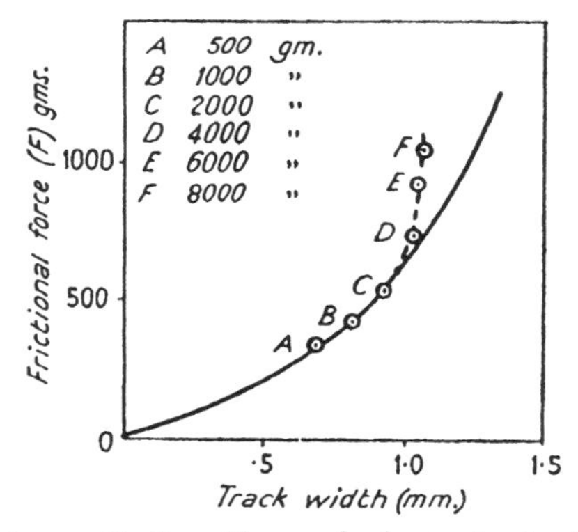

FIG. 45. Breakdown of indium films under heavy loads. If no breakdown occurred the friction would follow the continuous line. Because of breakdown there is metallic interaction between the underlying surfaces and the friction is appreciably higher.

Chapter VII, the breakdown of the oxide film normally present on metal surfaces may, under certain circumstances, produce a similar frictional behaviour.

Wear of films

Apart from the effect of load, the metallic films may also be worn away if the slider traverses the same track a sufficient number of times. Some results obtained with an indium film 4×10^{-4} cm. thick deposited on tool steel (hardness 800 V.D.H.) are shown in Plate XIV. The frictional behaviour is shown in Plate XIV. 1, and it is seen that initially the motion was smooth and the coefficient of friction was about $\mu = 0{\cdot}08$. With successive slidings over the same track the friction gradually rose and after the seventh run small stick-slips set in. The friction and the size of the stick-slips increased and, after 20 runs the friction had risen to $\mu = 0{\cdot}4$. The corresponding tracks are shown in Plate XIV. 2 *a*, *b*, *c*, and *d*. It will be seen that after the seventh run the steel surface is partially exposed and the original lapping marks can be seen. After the twentieth run there is considerable breakdown of the film. Similar results were obtained with indium films deposited on silver surfaces.

The rise in friction with wear of indium films is shown in Fig. 46, and it is seen that the thicker film is more resistant to wear. This behaviour may be compared with the dotted curve, which represents the rate at which nine molecular layers of stearic acid are worn off a steel surface. It is clear that the metallic films resist wear quite well, but they are

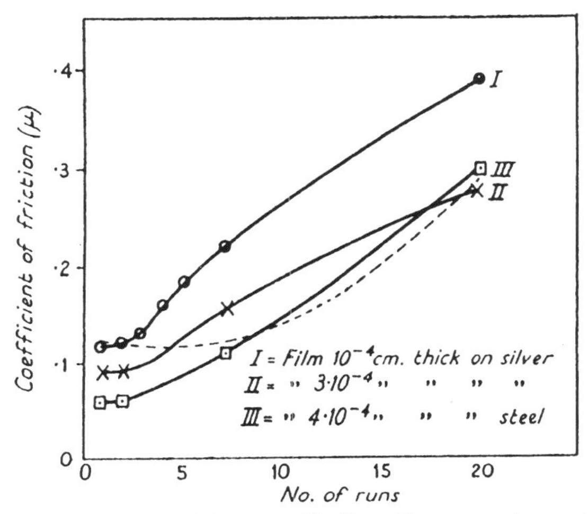

FIG. 46. Rise in friction with wear of indium films on various substrates. For comparison, the rate of wear of nine molecular layers of stearic acid on steel is shown in the dotted curve.

more easily worn off the surface than a much thinner film of the fatty acid (see Chap. IX). One factor that may be of importance here is the surface mobility of the film. The fatty acid molecules have a certain mobility and are able to move over the surface and repair the damaged film. Many metals also possess this ability to wander over solid surfaces, but it is probable that the rate at which this process occurs is less rapid for solid metallic films than it is for the fatty acids.

Effect of temperature

The effect of temperature on the friction of an indium film is shown in Figs. 47 (*a*) and (*b*). The friction decreases steadily as the temperature rises, and reaches a minimum at the melting-point of the indium. When the melting is complete there is a sharp rise in friction (see Fig. 47 (*a*)). If the film is allowed to solidify, the friction again falls to a low value (see Fig. 47 (*b*)).

Similar results were observed with a lead film 3×10^{-3} cm. thick deposited on to tool steel. The friction fell to about half its value over the temperature range 20° to 270° C. At the melting-point there was a further slight fall followed by a sudden rise as the film became completely liquid.

The decrease of friction with increasing temperature is of interest, and should be contrasted with the effect of temperature on the friction of pure metals. As we have already seen, the effect of temperature on pure metals is relatively small, since with increasing temperature the decrease

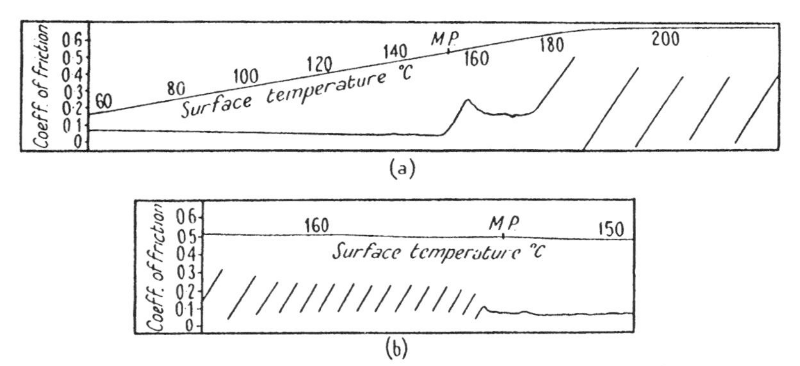

Fig. 47. Friction of steel sliding on indium film as a function of temperature. (*a*) On heating the friction falls slowly but rises rapidly at the melting-point of the indium, 155° C. (*b*) On cooling the friction falls sharply when the film solidifies.

in shear strength is compensated by the increase in area of contact A. In the experiments with the thin films, however, where the substrate has a very high softening temperature, the real area of contact A remains almost constant when the temperature is raised, since the load is taken primarily by the hard underlying metal. We are therefore able to evaluate separately the effects of A and s and to observe the decrease in F as s decreases.

With films of both lead and indium, F falls to about half its value as the temperature is raised from room temperature to the melting-point. The decrease in F is almost linear with temperature. If we consider the case of the lead film and call the frictional force at 20° F_{20} and at 300° F_{300}, we may define γ the temperature coefficient of F by the equation

$$F_{300} = F_{20}(1+\gamma t), \quad \text{where} \quad t = 300-20,$$

then

$$\gamma = -0{\cdot}0015.$$

With the indium films over the temperature range 20° to 150° C. $\gamma = -0{\cdot}0025$. Very close to the melting-point when incipient fusion begins the friction falls still farther, and it is in this region that F reaches its lowest value. Once melting is complete there is a sharp rise and F increases to a high value. For indium deposited on tool steel the value reached is almost as high as that of uncoated tool steel, and it is clear that the molten film affords little protection to the surface. Since the molten indium does not readily wet the tool steel surface, this is not surprising. With the lead film on the silver surface a similar abrupt rise in friction was observed, but the extent of the rise was very much less. The friction of the steel slider on the silver surface in the presence of molten lead was only about one-quarter of that characteristic of steel on clean silver. Molten lead readily wets the silver, and it is clear that it is still able to afford some protection to the surface and to act as a lubricant.

Some subsidiary experiments carried out on mercury films are of some interest in this connexion. It was found that minute quantities of mercury on silver produced a very marked reduction in friction. The mercury readily wets the silver and forms an amalgam with it, and this apparently enables it to act as a fairly effective lubricant. This ability of one metal to wet and spread over the surface of the underlying metal may play an important part in the action of bearing metals and in the friction and lubrication of other combinations of metals.

Metallic Films as Lubricants

Some experiments were carried out to determine the extent to which the coefficient of friction deviates from Amontons's law. The results for tool steel surfaces are shown in Fig. 48 for loads varying from 400 to 8,000 gm. Curve I is for the unlubricated steel, curve II for steel lubricated with a film of mineral oil, and curve III for steel lubricated with a thin film of indium approximately 4×10^{-4} cm. thick. The comparative behaviour is striking. The coefficient of friction for the unlubricated steel and for the steel lubricated with the mineral oil obeys Amontons's law. With the indium film, however, μ decreases markedly as the load increases.

The reason for this difference has already been indicated. With unlubricated steel (or with steel lubricated by the mineral oil) the area over which metallic contact occurs is directly proportional to the applied load. When a thin film of a soft metal is used as the lubricant, however, the increased deformation of the underlying steel resulting from an increase in load produces only a small increase in the area of real contact. Consequently there is only a slight increase in the frictional force as the load is increased. This leads to a corresponding decrease in μ, and as Fig. 48 shows, this effect can be quite marked. The coefficient of friction falls from $\mu = 0{\cdot}2$ at the light load to $\mu = 0{\cdot}04$ when the heaviest load is used. If it were possible to lubricate the surfaces with a metallic film of molecular dimensions, we should expect a closer agreement with Amontons's law.

In some ways the behaviour of the metallic films closely resembles that of ordinary lubricant films. They produce a substantial reduction in the friction, they can cause smooth sliding, and they protect the underlying metal surfaces. In addition, metallic films are worn off the surface by successive sliding over the same track in a manner similar to that of a lubricant film, except that the metallic films are worn away at a greater rate than long-chain paraffins or fatty acid films. A further point of striking similarity is the effect of melting. The transition from smooth sliding to stick-slips when the metallic films are melted is closely analogous to the change observed when solid hydrocarbon films are heated through their melting-point.

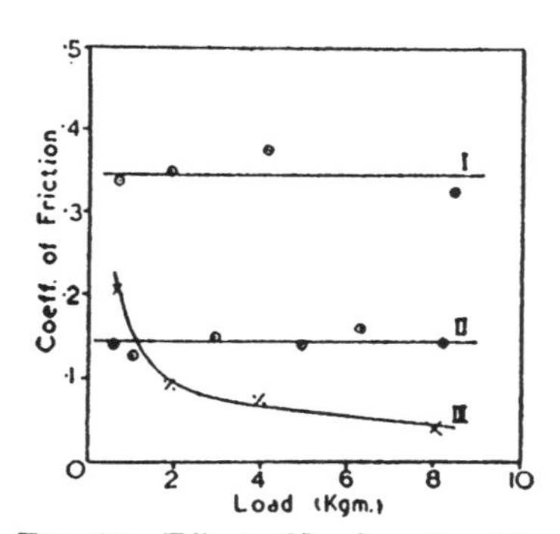

FIG. 48. Effect of load on the frictional behaviour of: I, unlubricated tool steel; II, steel lubricated with mineral oil; III, steel lubricated with indium film of thickness $4 . 10^{-4}$ cm. The coefficient of friction is independent of load for curves I and II, but rapidly decreases with load for curve III.

There are, however, several marked differences between metallic films and lubricant films. As we shall see in Chapter X, even on rough surfaces a lubricant film need only be one or two molecules thick to be effective as a boundary lubricant. A metallic film must be appreciably thicker, of the order of 10^{-5} cm., if it is to be effective. A further striking and fundamental difference is that lubricant films obey Amontons's law and metallic films do not. With metallic films the coefficient of friction decreases as the load is increased, and at high loads the coefficient of friction may be extremely low. With the indium films described in this paper the value of μ under heavy load ($\mu = 0{\cdot}04$) is as low as that observed with the best boundary lubricants. This value is similar to that obtained on ice surfaces (see Chap. III).

These thin metallic films may have important applications in many practical problems, particularly where the loads are very heavy and also when the more conventional lubricants cannot be employed. For example, very thin lead films are effective in deep drawing operations: a film of lead 0·001 in. thick deposited on steel enables the drawing operation to be carried through with dry unlubricated dies and produces an excellent finish on the surfaces (Moore and Tabor, 1943). Similarly Attlee, Wilson, and Filmer (1940) have used films of barium for lubricating the ball races in rotating anode X-ray tubes. In the same way the smearing of a soft metallic film over a harder metal matrix provides a very effective type of bearing alloy. This will be discussed in greater detail in the next chapter.

REFERENCES

Z. J. ATTLEE, J. T. WILSON, and J. C. FILMER (1940), *J. Appl. Phys.* **11**, 611.
H. BLOK (1940), *J. Soc. Aut. Eng.* **46**, 54.
F. P. BOWDEN (1945), *Proc. Roy. Soc. N.S.W.* **78**, 187 (Liversidge Lecture).
—— and L. LEBEN (1939), *Proc. Roy. Soc.* **A 169**, 371.
—— —— and D. TABOR (1939), *Engineer* (London), **168**, 214.
—— A. J. W. MOORE, and D. TABOR (1943), *J. Appl. Phys.* **14**, 80.
—— and D. TABOR (1939), *Proc. Roy. Soc.* **A 169**, 391.
—— —— (1942), *Nature*, **150**, 197.
—— —— (1943), *J. Appl. Phys.* **14**, 141.
J. R. BRISTOW (1942), *Nature*, **149**, 169.
H. ERNST and M. E. MERCHANT (1940), *Conf. Friction and Surface Finish*, M.I.T. 76.
S. KHAIKIN, L. LISSOVSKY, and A. SOLOMONOVITCH (1940), *J. Physics (U.S.S.R.)*, **2**, 253.
J. S. MCFARLANE and D. TABOR (unpublished).
A. J. W. MOORE and D. TABOR (1943), C.S.I.R. (Australia) Tribophysics Division Report A 96.
F. MORGAN, M. MUSKAT, and D. W. REED (1941), *J. Appl. Phys.* **12**, 743.

VI

ACTION OF BEARING ALLOYS [A]

BEARINGS are normally designed to run with a finite film of oil separating them from the rotating journal. The frictional resistance will then depend on the clearance, the load, the speed, and the viscosity of the oil film. Provided the oil film is maintained, the metal surfaces of the journal and the bearing play no part, and there is no wear of either journal or bearing. In practice, however, particularly in starting and stopping, or if impulsive loading or excessive vibration occur, the finite oil film may break down and only thin, boundary films of lubricant may remain. In severe cases, considerable metallic contact may occur. It is during this incidence of boundary lubrication and of metallic contact that the wear and seizure of bearings takes place.

The general mechanism of boundary lubrication will be discussed in Chapters VIII and IX, but even in the presence of boundary films the basic frictional properties of the metals themselves are important. In this chapter we shall discuss the basic frictional behaviour of a number of typical bearing alloys.

It has been pointed out (Bowden and Tabor, 1943) that one factor of great importance is the temperature of the metal surfaces. When the bearing is running, the frictional heat may cause a local softening or melting at the regions where the moving metals are momentarily in contact. It is therefore desirable that one of the constituents of a bearing metal should have a relatively low melting-point, so that if excessive heating occurs the metal may melt and flow to another region and so prevent seizure. For this reason the great majority of bearing alloys contains lead (m.p. 327° C.) or tin (m.p. 232° C.) or some other low-melting constituent.

A number of theories have been proposed to explain the action of typical bearing alloys, but the mechanism was by no means clear. It has been maintained in the past, for example, that an essential feature of a bearing alloy is that it should possess a duplex structure consisting of hard crystals embedded in a relatively soft matrix. It is suggested (see, for example, Bassett, 1937) that the function of the hard crystals is to resist wear and that of the softer constituent to permit a more uniform distribution of the load by allowing any of the hard crystals which are heavily loaded to sink so that the load is spread over a greater area. It is also suggested that the hollows worn in the softer material serve as reservoirs for the lubricating oil. It is certainly true that many successful bearing alloys do possess a structure of this type, as may be seen, for example, in the typical lead-base alloy in Plate XV. 3 and the tin-base alloy in Plate XVII. 3, both of which consist of a soft matrix in which are dispersed a number of hard particles. Nevertheless, other workers have suggested that this theory is not satisfactory and that, in fact, the hard particles play little part in the frictional properties of the alloys.

In any event, the theory of duplex bearing alloys described above cannot be of general validity. There are many cases in which it cannot be true. For example, with many modern bearing alloys the surface layer does not possess a duplex structure at all but consists of a single-phase alloy. Another very wide class of bearing 'alloys' consists of a continuous matrix of the harder metal with a small amount of the softer metal finely dispersed through it. Some of the copper–lead alloys provide a good instance of this (see, for example, the copper–lead alloys shown in Plate XV. 1 and 2). In this case the hard copper forms the continuous phase, so that it is not possible for it to 'sink' into the lead. It is evident that these alloys must function by a different mechanism.

In this chapter we shall discuss the theory of action of two main types of duplex bearing alloy. The first is the copper–lead type of alloy, in which the matrix is the harder of the two phases; the second is the 'white-metal' type of alloy, in which the continuous matrix consists of the softer phase.

Copper–Lead Bearing Alloys

Two typical copper–lead bearing alloys were selected for investigation. The first (see Plate XV. 1), which contained 27 per cent. lead, was non-dendritic; the copper formed the continuous phase, and small droplets of lead were distributed through it. The second (see Plate XV. 2) contained 20 per cent. lead and was dendritic; the lead formed a continuous network between the copper dendrites.

Microhardness measurements show that in both these alloys the copper phase has a hardness of about 60 (V.D.H.), whilst the lead phase has a hardness of about 8 (V.D.H.). There is, however, an appreciable difference in the overall or macroscopic hardness of the alloys. In the non-dendritic alloy, where the copper forms the continuous hard phase, the macroscopic hardness is about 36 (V.D.H.). In the dendritic alloy the lead phase plays a larger part in determining the overall physical properties of the alloy, and the macroscopic hardness is about 25 (V.D.H.). It is clear, however, that in both alloys the lead phase is considerably softer than the average hardness of the alloys.

The frictional properties of these alloys were examined and compared with those of pure copper, pure lead, and thin films of lead deposited on a copper substrate. The friction apparatus described in Chapter IV was used. The lower surface was in the form of a flat plate and the upper surface a hemisphere of radius of curvature of about 0·3 cm.

Frictional behaviour of steel, copper, and lead

The friction of steel on copper or lead depends upon the experimental conditions, but for a curved steel slider on copper or on lead it was of the order of $\mu = 0{\cdot}9$. An examination of the steel surface after sliding showed the characteristic local adhesion and welding of the softer metal on to the hard one. The friction of copper on copper and of lead on lead was somewhat higher and varied from $\mu \sim 1$ to $\mu \sim 2$, and the surfaces showed that considerable seizure and tearing had occurred.

Thin films of lead on copper

The friction of a steel slider rubbing on a copper surface coated with a thin lead film depends on the dimensions of the slider and the thickness of the film. With a hemispherical slider possessing a radius of curvature of 0·3 cm. and a lead film of thickness 3×10^{-3} cm. the sliding was smooth and the friction somewhat less than $\mu = 0{\cdot}3$. Keeping the curvature of the slider fixed, a series of friction measurements was made with lead films of varying thickness. The relation between track width and friction for films of various thickness is shown in Fig. 49.

It is seen that a film of lead 10^{-6} cm. thick (curve II) causes very little reduction in the friction. As the thickness of the film is increased, the friction for any given track width decreases and reaches a minimum when the film is 10^{-3} cm. thick (curve V). Further increase in film thickness produces no further change. The frictional force is now governed by the shear strength of the lead and the width of the track and is not influenced by the substrate (except in so far as this may affect the track width). When similar measurements are made on solid lead, the points lie on curve V (Fig. 49). It is clear that after the film has reached a thickness of about 10^{-3} cm. the friction is due to the interaction between the steel and lead. This minimum film thickness of 10^{-3} cm. for lead films on copper is greater than that observed when the films are deposited on steel. It is also greater than that observed with indium films on steel. As pointed out earlier, the minimum film thickness for complete 'lubrication' is influenced by the hardness of the metal substrate. With softer metal substrates, such as copper, which are more readily deformed, thicker films are necessary. It is also influenced by

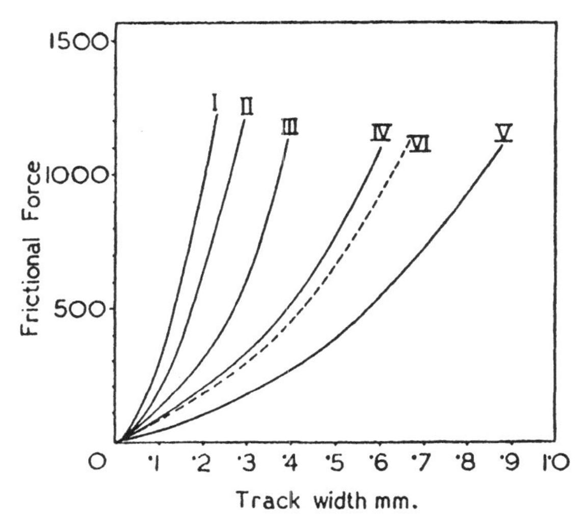

FIG. 49. Frictional behaviour of lead films deposited on copper substrates. I, clean copper; II, lead film 10^{-6} cm. thick; III, lead film 10^{-5} cm. thick; IV, lead film 10^{-4} cm. thick; V, lead film 10^{-3} cm. thick and more, and solid lead; VI, copper–lead alloy. Friction in gm.

the fineness of the original surface finish and by the state of cleanliness of the substrate, since this affects the strength of adhesion of the deposited layer.

Copper-lead alloys

The friction behaviour of both alloys at room temperature was similar. Steel on the alloys gave smooth sliding and a coefficient of friction of about $\mu = 0{\cdot}18$. The tracks were smooth grooves showing little tearing. Some signs of smearing of extruded lead could be detected. The alloys on steel gave intermittent motion, the maximum value of the friction being about $\mu = 0{\cdot}3$. The tracks showed that some metal from the upper contact had been welded on to the steel surface, but the extent of this was considerably less than the welding observed with pure copper or lead.

The main frictional results for pure metals, a thin lead film, and the copper–lead alloys are summarized in the following table. Where the motion is intermittent, the maximum coefficient of friction is given.

TABLE XIV

Coefficient of friction μ_{max}

Lower flat surface	*Upper curved surface*			
	Steel	*Copper*	*Lead*	*Copper–lead alloys*
Steel . . .	1·0	0·7	0·9	0·3
Copper . . .	0·9	2·0	1·0	..
Lead . . .	1·2	1·3	1·6	..
Lead film on copper .	0·18	..	..	..
Copper–lead alloys .	0·18	..	..	..

It is at once apparent that the friction of the copper–lead alloys does not lie between the values of their constituents. The friction of steel on copper, for example, is about $\mu = 0{\cdot}9$, and for steel on lead $\mu = 1{\cdot}2$. With steel on the copper–lead alloys, however, the friction is about $\mu = 0{\cdot}18$. The alloys and the pure copper have approximately similar yield pressures at room temperature, so that the area of contact A is nearly the same for a given load. Nevertheless, the friction of the alloys is very much less than that of their pure constituents.

The relation between the friction F and the track width d for the alloy is plotted on the broken curve (curve VI) in Fig. 49. It will be seen that these results lie very close to those obtained when a lead film 10^{-4} cm. thick is artificially deposited on to copper. This at once suggests that the lead in the alloy is extruded during sliding and forms a thin 'lubricating' film of effective thickness between 10^{-4} cm. and 10^{-3} cm. on the copper. This is borne out by a microscopic examination of the track, which showed traces of lead smeared over the surface.

Effect of wear on the friction

The values of the friction after repeated traversals of the same track are summarized in Table XV for the various combinations of metals.

With pure metals the friction is not appreciably affected by wear. The coefficient of friction after the slider has passed over the same track 20 times is similar to that observed on the first or second run. This is to be expected since the area of contact is essentially the same and the metallic junctions formed in each successive run are still of the same nature as in the first run. With the thin metallic films, as we have seen, the films are gradually worn away. As metallic contact occurs with the underlying substrate which has a much higher shear strength than the film materials, the friction rises. The track also shows the breakdown of the film and the tearing of the underlying metal; the behaviour is similar to that shown for indium films in Chapter V, Plate XIV.

TABLE XV

Effect of repeated sliding over same track

Surfaces	Coefficient of friction μ_{max}		
	1st run	2nd run	20th run
Steel on copper . . .	0·9	0·9	0·8
Copper on steel . . .	0·7	0·8	0·85
Steel on lead . . .	1·2	1·4	1·4
Lead on steel . . .	0·9	1·0	0·8
Steel on alloy . . .	0·18	0·17	0·26
Alloy on steel . . .	0·3	0·28	0·28
Steel on lead film 10^{-4} cm. thick deposited on copper .	0·18	0·18	0·27

The wear results with the copper–lead alloys are similar to those observed with lead films artificially deposited on to copper and again bear out the view that the hard copper matrix is lubricated by a thin lead film. At first the extruded lead offers a reasonably effective lubricating film. The next few runs over the same track may produce a more uniform smearing of the lead, and the friction may even decrease. The value of μ at this stage is about 0·17. With continued traversals over the same track, however, the lead film is worn away, the alloy is gradually exhausted of most of the lead in the vicinity of the track, and the friction rises to a value of about $\mu = 0{\cdot}28$. This is borne out by an examination of the track. An interesting confirmation of this view is to be found in the values obtained for the friction of a curved slider of bearing metal on steel. In this case, all the wear is concentrated on a very small area of the alloy which may quickly be depleted of lead, and the friction rises rapidly to a value of $\mu = 0{\cdot}28$. This is almost identical with the value observed with steel on the bearing metal, after continuous running over the same track; i.e. it corresponds to the friction between steel and bearing metal surfaces when most of the lead between the surfaces has been exhausted.

It is interesting to note a slight difference in the wear properties of the copper–lead alloys. The final value of the friction for the fully worn surfaces was approximately the same for both alloys ($\mu = 0{\cdot}3$). However, with the non-dendritic alloy, the first marked increase in the friction occurred after the 6th run, whilst with the dendritic alloy it did not take place until the 20th run. It is apparent that the extruded lead film on the surface of the dendritic alloy is more readily replenished from the bulk of the alloy than is the case with the non-dendritic alloy.

Effect of temperature on friction

The effect of temperature on the friction of bulk metals and of thin metallic films has been discussed in the previous chapter. With single-phase metals the friction shows little change over a wide range of temperatures. With thin metallic films on hard substrates, however, the frictional force decreases steadily with temperature until the melting-point of the film is reached. For example, with lead films on

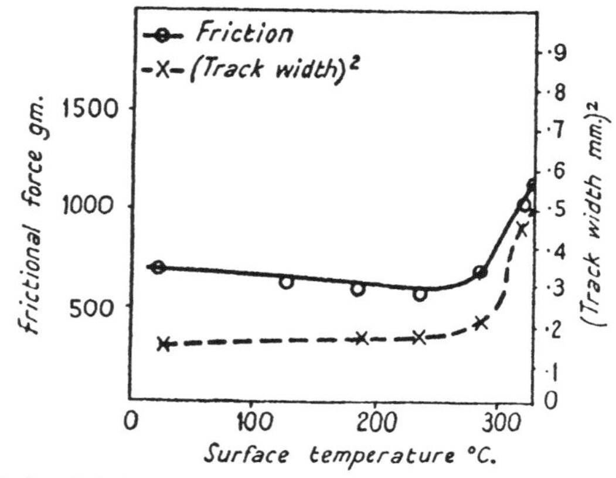

FIG. 50. Frictional behaviour of non-dendritic copper–lead bearing alloy as function of temperature. Full line: frictional force; broken line: square of track width which is a measure of the area of contact. Over the first 250° C. the alloy retains its hardness and there is a slight decrease in friction as the temperature rises. At about 280° C. the alloy softens rapidly (large increase in track width) and there is a marked increase in friction.

a copper substrate, the temperature coefficient of friction between 20° and 300° C. is about $\gamma = -0{\cdot}0016$.

The influence of temperature on the frictional properties of the non-dendritic copper–lead alloy is shown in Fig. 50. The upper curved slider was of steel. The full curve refers to the frictional force and the broken curve to the square of the track width, which is a measure of the area of contact. As the temperature is raised, the friction decreases in an approximately linear manner up to 250° C., with a temperature coefficient of about $\gamma = -0{\cdot}0015$. The track width up to 250° C. increases very slightly, involving an increase in area of contact of about 20 per cent. over this range. Thus the temperature coefficient of friction calculated on the basis of a constant area of contact yields a value of approximately $\gamma = -0{\cdot}0018$. This value of γ is about the same as that obtained with a thin film of lead deposited on copper. This again supports the view that the hard copper matrix of the alloy is lubricated by a thin film of extruded lead. Above 250° C. the alloy begins to soften markedly, the area of contact increases, stick-slips set in, and the friction μ_{max} rises. At 327° C. (the melting-point of lead) the friction rises to almost twice its value at room temperature.

With the dendritic alloy, the behaviour is similar except that there is little or no initial decrease in friction as the temperature is raised (see Fig. 51). Indeed, from 60° C. upwards both the friction and the

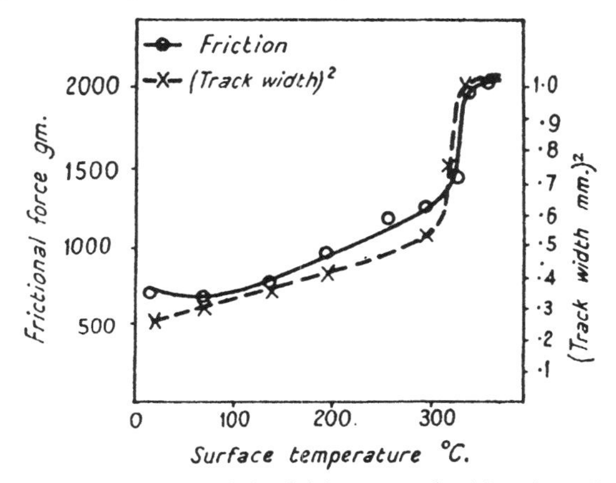

FIG. 51. Frictional behaviour of dendritic copper–lead bearing alloy as a function of temperature. Full line: frictional force; broken line: square of track width which is a measure of the area of contact. There is a continuous softening of the alloy as the temperature is raised (steady increase in track width) and this is accompanied by an increase in friction. Above 300° C. there is a further rapid softening of the alloy and a large increase in friction.

area of contact show a rapid increase with temperature. This behaviour is due to the fact that the lead forms a continuous network between the copper dendrites, so that the softening of the alloy as a whole occurs at a relatively low temperature and is more marked than with the non-dendritic alloy. Nevertheless, as we may see from inspection of Fig. 51, the friction as a function of the area of contact shows little change with temperature up to the melting-point of lead, so that the behaviour is again consistent with the view that the surfaces are lubricated by a thin film of extruded lead. Above 300° C. there is very marked softening of the alloy and a large increase in friction, and at 327° C. the friction is more than three times its value at room temperature.

Role of thin films in bearing alloys

We see, therefore, that the frictional behaviour of a copper–lead alloy resembles very closely that of a copper surface on which a very thin film of lead has been deposited. The actual value of the friction of the alloy is the same as that of a copper surface on which a lead film 10^{-4} cm. thick has been artificially deposited. The temperature coefficient of friction is the same in each case. The increase in friction with wear is also very similar except that, in the case of the alloy, there is evidence that the potential supply of lead is greater since it may continue to be expressed from the alloy during the sliding process. With a thin film of lead deposited on copper the supply is, of course, limited to the amount which is actually present on the surface at the beginning of sliding.

These results show the important part which may be played by thin surface films of metal in reducing the friction and wear of bearing alloys and support the view that in bearing alloys of the copper–lead type the anti-frictional and anti-seizure properties are due, primarily, to the spreading of thin films of the soft low-melting constituent over the surface of the harder constituent.

As we saw in the last chapter, the effectiveness of metallic-film lubrication is due to the fact that it provides a surface for which both the area of contact A and the shear strength s are small. At higher temperatures the metallic film softens (s decreases) without an appreciable increase in A occurring, so that there is a decrease in friction as the temperature is first increased. It is clear, however, that as we approach the melting-point of the metallic film the surface-tension properties of the molten film become important. The film must spread readily on the harder substrate. As we saw, for example, with films of lead and mercury on copper and silver substrates, unless the soft metal readily wets the hard one, it is at high temperatures of sliding relatively ineffective as a lubricant.

In this account of the theory of action of the copper–lead alloys we have only considered the lubricating properties of the extruded lead film itself. We have not considered the effect of adding ordinary lubricants to the surface. As we shall see in later chapters, the whole behaviour is profoundly modified by the presence of lubricants and by the extent to which the surface layers of the various metals are capable of adsorbing the lubricant films. Nevertheless, even in the presence of a lubricant, metallic contact between the moving surfaces may readily occur, so that the basic frictional properties of the alloys themselves are of primary importance.

Comparative behaviour of the dendritic and non-dendritic alloys

The measurements described in this part of the chapter show that there is very little difference between the frictional properties of the dendritic and non-dendritic alloys at room temperature. Both have a coefficient of friction of about $\mu = 0 \cdot 18$, and when the rubbing surfaces are denuded of extruded lead the friction in both cases rises to about $\mu = 0 \cdot 3$. With the dendritic alloy, however, the increase of friction with wear is less rapid than with the non-dendritic alloy. This is probably due to the greater ease with which the extruded film of lead is replaced from the *bulk* of the dendritic alloy. As the micro-structure shows, the pockets of lead in this alloy are all interconnected, so that potentially a greater supply of lead is available to any particular area of the surface which may be in need of it. With the non-dendritic alloy, however, the pockets of lead are isolated, so that a local exhaustion of the lead may occur and a seizure on to the copper surface may take place.

This difference in the distribution of the lead and in its availability at the surface is shown by the comparative wear behaviour described above. It is also directly demonstrated by the 'sweating' out of the lead. When the non-dendritic alloy is heated, the lead appears as a fine mist of droplets over the surface. When the dendritic alloy is heated, the lead collects in one or two large drops, by a surface-tension effect. The dendritic alloy thus has a three-dimensional supply of lead to the surface, while with the non-dendritic alloy it is restricted to the surface layers.

It is clear, therefore, that, quite apart from the surface wetting and spreading effects discussed in the previous section, the *availability* of the lead phase is of great importance in the practical operation of the bearing. For this reason we may expect the dendritic alloy to be more resistant to seizure than the non-dendritic alloy under certain conditions of sliding and wear.

On the other hand, the non-dendritic alloy is harder than the dendritic alloy and has better mechanical properties. This is particularly marked at higher temperatures, where the dendritic alloy softens very much more rapidly than the non-dendritic alloy. As a result, although it has a greater available supply of lead, its friction is higher than that of the non-dendritic alloy.

Summarizing, we may say that, under mild conditions of sliding, both alloys have similar frictional properties. However, the dendritic alloy has a more readily available supply of lead and may therefore be less likely to seize. On the other hand, under severe conditions of sliding, the non-dendritic alloy will possess mechanical properties superior to those of the dendritic alloy, whilst at high running temperatures its frictional properties will also be better.

WHITE-METAL BEARING ALLOYS

The structure of typical 'lead-base' and 'tin-base' bearing alloys is different from that of the copper–lead alloys in that they contain hard crystallites dispersed in a soft matrix. It was usually considered that the hard particles play a fundamental part in the action of these alloys, and to investigate this point special 'matrix' alloys were made up which possessed approximately the same composition and structure as the matrix material of the alloys but which contained no hard particles. A comparison was then made of the properties of each bearing alloy and of its corresponding matrix alloy, along three main lines: first, the hardness and general plastic properties, secondly the frictional properties when clean, and thirdly the frictional properties in the presence of lubricant films (Tabor, 1945).

Lead-base alloys: structure and hardness

The lead-base bearing alloy had the following composition: antimony 15·0 per cent., copper 0·5 per cent., tin 6·0 per cent., and lead 78·5 per cent. The micro-structure is shown in Plate XV. 3, and it is seen that the alloy consists of a lead–tin–antimony eutectic (possessing a fine duplex structure) in which are dispersed the hard cuboids of tin–antimony compound. The dark areas surrounding the cuboids consist of a lead-rich solid solution. The matrix consists of the eutectic and the solid solution.

This alloy was compared with a matrix alloy possessing the following composition: antimony 10 per cent., lead 87 per cent., tin 3 per cent., and copper 0. The micro-structure is shown in Plate XV. 4, and it will be seen that it consists of dendrites of lead-rich solid solution set in a duplex matrix of lead–tin–antimony eutectic. There are no hard particles present, and the structure approximates to that of the matrix material of the lead-base bearing alloy.

Microhardness measurements were made of the various phases of these alloys, and the results, together with some macroscopic hardness measurements, are summarized in Table XVI. As a matter of interest the hardness values of the non-dendritic copper–lead alloy shown in Plate XV. 1 are included.

Two conclusions follow from these hardness measurements. First, that the hard crystallites in the bearing alloy contribute little to its overall hardness. Thus the matrix material has a hardness of 19, whilst the overall hardness of the alloy, in spite of the presence of numerous hard cuboids of hardness 110, is only 21. This view is further confirmed

TABLE XVI

Hardness values of lead-base alloys

Material	*Macroscopic hardness Vickers*	*Microhardness in equivalent Vickers units*
Pure lead . .	4	. .
Lead-bearing alloy .	21	Matrix 19; cuboids 110
Lead-matrix alloy .	18	Lead-rich solid solution 18; duplex Pb—Sn—Sb eutectic 18
Steel . . .	166	. .
Copper–lead alloy .	35	Copper matrix 56; lead phase 8

by hardness measurements at elevated temperatures. As the temperature is raised the hardness decreases, but the hardness–temperature characteristic of the bearing alloy is almost identical both in trend and in absolute values with that of the matrix alloy (Fig. 52). We may

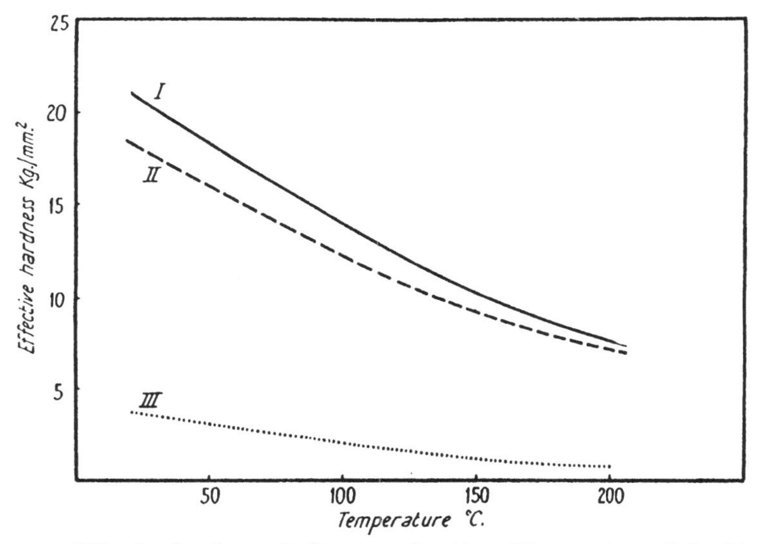

FIG. 52. Effective hardness of alloys as a function of temperature. I, lead-base bearing alloy; II, lead-base matrix alloy; III, lead.

conclude, therefore, that the hardness of the bearing alloy at room temperature and at elevated temperatures is determined essentially by the properties of the matrix material.

The second conclusion concerns the two-phase nature of the matrix alloy. One phase consists of a lead-rich solid solution and the other of a duplex Pb–Sn–Sb eutectic. Both, however, have the same hardness, so that both phases yield equally under an applied load. As we shall see below, this alloy, from the point of view of its plastic and frictional properties, does in fact behave as a homogeneous material. This also applies to the matrix material of the bearing alloy.

Unlubricated surfaces

With a hemispherical steel slider rubbing on a plane surface of the alloys the motion was smooth and the coefficient of friction approximately the same for the bearing alloy ($\mu = 0{\cdot}4$–$0{\cdot}45$) and for the matrix alloy ($\mu = 0{\cdot}35$–$0{\cdot}4$). For a hemispherical slider of the alloy rubbing on a plane surface of steel the motion was intermittent and the friction was again similar for both alloys. There was no indication that the presence of the hard particles produced any marked change in the friction.

This observation is readily understandable if we examine the surface damage of the bearing alloy when the steel slider has passed over the surface once. The nature of the damage is shown clearly if the alloy is polished and lightly etched before the friction experiment. Part of the track is shown in Plate XVI. 1 *a* side by side with the unworn portion of the surface, and it is seen that the hard particles have been pushed under the surface during sliding. This effect is shown in a more striking manner by a taper section recently made by Mr. E. Rabinowicz of the track formed in a similar lead-base bearing alloy when a steel slider passes over the surface once. The size of the crystallites in this alloy was relatively large, and for this reason a much larger load was used in order to give a track appreciably larger than the size of the crystallites. In order to emphasize the behaviour of the hard particles the alloy was vigorously etched before sliding took place so that the crystallites should be well exposed above the surface of the matrix material. The micrograph is shown in Plate XVI. 1 *b*. It is clear that the hard crystallites have been buried under the surface of the alloy and the matrix material smeared over the track as a result of the sliding process. It is evident from these figures that the steel slider is rubbing on the matrix material and not on the hard particles, so that the metallic junctions which are sheared during sliding consist of the matrix material itself.

A similar examination of the track formed in the surface of the matrix alloy shows that there is no preferential smearing of one phase over the other. The alloy as a whole welds on to the steel slider and is torn and smeared during the run, so that from the frictional point of view the alloy behaves as a homogeneous material. Since the matrix material of the bearing alloy has a similar composition and structure, we may expect that it too will behave as a homogeneous material.

The view that the frictional properties of the bearing alloy are determined essentially by the properties of the matrix material itself is supported by a number of other observations. First, the friction is almost independent of temperature from 20° to 200° C. and is very similar for both alloys (see Fig. 53). This behaviour is characteristic of

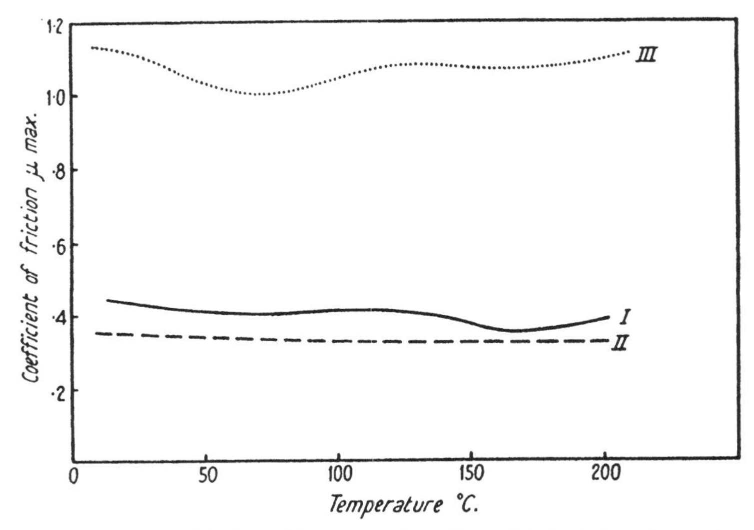

FIG. 53. Variation of friction with temperature for unlubricated surfaces. I, steel on lead-base bearing alloy; II, steel on lead-base matrix alloy; III, steel on lead.

pure metals and of homogeneous substances in general (see, for example, the result for lead in Fig. 53), since the increase in area of contact which occurs as the temperature is raised is compensated by a reduction in the shear strength of the junctions formed between the surfaces. Secondly, repeated sliding over the same track produces a gradual increase in friction which is very similar for both alloys (see Table XVII). This is probably due to the cumulative pick-up of the alloy on the steel, since the upper limit of the friction corresponds roughly to the friction of the alloys sliding on themselves. Thirdly, the wear properties of both alloys are similar. This was measured by repeatedly passing a slider made of the alloy over a steel surface and determining the loss in weight of the slider. The results are given in the last column of Table XVII, and it is seen that after traversing a track 7·5 cm. long 400 times at a load of 4 kg. the wear of the bearing alloy was 11 mgm. whilst that of the matrix alloy was 14 mgm. This difference is not significant.

TABLE XVII

Friction of lead-base alloys. Unlubricated

Surfaces		Coefficient of friction				Wear after traversing 3,000 cm. of track mgm.
		1st run	20th run	100th run	400th run	
Steel on	Bearing alloy	0·45	0·7	..	..	...
	Matrix alloy	0·4	0·8	..	..	..
On steel	Bearing alloy	0·55	0·8	1·0	1·0	11
	Matrix alloy	0·65	1·1	1·2	1·2	14

Lubricated surfaces

Similar measurements were made of the friction between the alloys and steel surfaces in the presence of a refined mineral oil. The frictional behaviour of both alloys was again similar. It is interesting to examine the track formed in the bearing alloy when the steel slider traverses the same track a large number of times. The track formed for the first run corresponding to a coefficient of friction of $\mu = 0{\cdot}12$ is shown in Plate XVII. 1, and it is seen that the arrangement of the hard cuboids has been little disturbed by the passage of the slider. If this is compared with Plate XVI. 1*a* for the unlubricated surface it is seen that the presence of the oil film has greatly reduced the amount of smearing and surface damage. The central position of the track formed during the 100th run is shown in Plate XVII. 2, and it is seen that the hard crystallites have been almost completely obliterated by the repeated passage of the slider. The coefficient of friction is $\mu = 0{\cdot}11$. The fact that the friction for the first run, when the crystallites are still clearly visible in the track, is nearly the same as for the 100th run, when the crystallites have been almost completely wiped over by the smeared matrix material, is strong evidence for the view that the friction is determined essentially by the properties of the matrix material itself.

Experiments in which stearic acid was used as the lubricant again show a close parallelism between the properties of the bearing alloy and the matrix alloy. These results are summarized in Table XVIII.

The role of the matrix and the hard particles

These results show repeatedly the similarity between the properties of the lead-base bearing alloy and those of the lead-base matrix alloy. The friction when clean and lubricated, the behaviour with respect to temperature and wear, the types of track formed, and the nature of wear are all markedly alike.

TABLE XVIII

Friction of lead-base alloys. Lubricated

Surfaces		Coefficient of friction			
		Mineral oil			Stearic acid
		1st run	20th run	100th run	1st run
Steel on	Bearing alloy	0·12	0·12	0·11	0·08
	Matrix alloy	0·14	0·15	0·16	0·07
On steel	Bearing alloy	0·32	..	..	..
	Matrix alloy	0·35	..	..	..

A survey of the main results for the lead-base alloys (summarized in Tables XVII and XVIII) indicates that, in general, the friction of the lead-base bearing alloy is less by a few per cent. than that of the corresponding matrix alloy. This difference may be due to the presence of the hard crystallites in the bearing alloy. On the other hand, it is extremely difficult to reproduce in a separate matrix alloy the complex phase relationships existing in the matrix material of the bearing alloy. This difference in friction may therefore be due to differences between the constitution and structure of the matrix alloy and the matrix material of the bearing alloy. This difference in frictional behaviour, however, is small.

When sliding takes place, the pressure and motion of the steel surface impress the hard crystallites deeper into the matrix and cause an extrusion and smearing of the matrix material over the surface of the alloy. Welding takes place between the steel surfaces and the matrix material, and the shearing takes place within metallic junctions composed of the matrix material itself. When clean surfaces are used, this smearing and tearing of the matrix material over the surface of the track completely obliterates the arrangement of the hard crystallites. When a lubricant is used, the seizure and smearing is far less, and this process is much more gradual. For both clean and lubricated surfaces the hard crystallites play no appreciable part in the frictional behaviour of the bearing alloy.

Tin-base alloys: structure and hardness

A similar series of experiments was carried out on a tin-base bearing alloy with the following composition: antimony 6·5 per cent., copper 4·2 per cent., tin 89·2 per cent., and nickel 0·1 per cent. The microstructure (see Plate XVII. 3) shows that it consists of a matrix of a tin-rich solid solution, in which are embedded numerous hard needles of copper–tin compound. This alloy was compared with pure tin and with a tin-base matrix alloy containing 7 per cent. antimony, no nickel, and less than 0·4 per cent. copper. This alloy contained no hard particles and consisted essentially of a tin-rich solid solution. Its hardness, compared with that of the tin-base bearing alloy, is given in Table XIX.

TABLE XIX

Hardness of tin-base alloys

Material	*Macroscopic hardness (Vickers)*	*Microhardness in equivalent Vickers units*
Pure tin . .	7	7
Tin-bearing alloy .	24	Matrix 20 Needles 100
Tin-matrix alloy .	18	18

It is apparent that the matrix alloy is similar in hardness to the matrix material of the bearing alloy.

Unlubricated surfaces

Frictional measurements of steel sliding on the alloys and of the alloys sliding on steel showed that the motion was always intermittent. The values of the coefficient of friction during the 'stick' are summarized in Table XX.

TABLE XX

Friction of tin-base alloys. Unlubricated

Surfaces		*Coefficient of friction μ max.*		*Wear after traversing 3,000 cm. of track mgm.*
		1st run	*20th run*	
Steel on	Bearing alloy	0·7–0·8	0·65–0·8	. .
	Matrix alloy	0·7–0·8	0·85	. .
On steel	Bearing alloy	0·7	0·8	0·2
	Matrix alloy	0·8	0·85	0·3

It is seen that the friction of the two tin alloys is similar. Further, there is no appreciable increase in friction for the first 20 runs, and even after 400 runs the increase in friction was not marked. This may be due to the fact that the pick-up of the tin alloys on the steel surfaces is relatively small. For example, a photo-micrograph of the steel surface after the tin-matrix alloy has passed over the same track 400 times (total track length 3,000 cm.) shows relatively slight smearing of the alloy over the steel, so that the lapping marks in the steel surface are still visible (Plate XVII. 4). This behaviour is in marked contrast to that of the lead-base alloy, where the pick-up of the alloy on the steel almost completely obliterates the lapping marks in the steel surface (Plate XVII. 5). Corresponding to this large amount of metallic transfer the wear of the lead-base alloys is very much heavier than that of the tin-base alloys (see Chap. XIV).

The effect of temperature on the frictional properties of the tin-base alloys is similar to that of the lead-base alloys. Over a temperature range of 20° to 200° C. the coefficient of friction for both alloys remains approximately constant.

Lubricated surfaces

The main friction results are shown in Table XXI for surfaces lubricated with a mineral oil and with pure stearic acid. In all cases the motion was smooth. It is seen that the behaviour of the bearing alloy is similar to that of the matrix alloy.

TABLE XXI

Friction of tin-base alloys. Lubricated

	Coefficient of friction			
	Mineral oil			Stearic acid
Surfaces	1st run	20th run	100th run	1st run
Steel on Bearing alloy	0·13	0·13	0·11	0·07
Steel on Matrix alloy	0·13	0·13	0·12	0·09

It is evident from these results that for both lubricated and unlubricated surfaces the frictional properties of the tin-base bearing alloy are similar to those of the tin-base matrix alloy. The actual values of the friction, the frictional behaviour for repeated traversals of the same track, the types of track formed, and the nature of wear are all markedly alike. This again shows that the frictional properties of the bearing alloy are determined mainly by the matrix itself. The agreement is not as close as with the corresponding lead-base alloys and this is discussed more fully in the original papers.

Comparison of lead-base and tin-base bearing alloys

A point of some interest is the comparison between the lead-base and tin-base alloys. The frictional properties of the lead-base bearing alloy when clean and when lubricated, at room temperature and at elevated temperatures, are generally slightly better than those of the tin-base bearing alloy. However, its wear on steel surfaces is very much heavier.

Its hardness (measured in Vickers units) is also about 15 per cent. less. Apart, therefore, from the question of mechanical properties which may play a vital role in the practical behaviour of a bearing, the lead-base bearing alloy should, in practice, be at least as satisfactory as the tin-base alloy from the point of view of its frictional behaviour. Under severe conditions, however, the rate of wear of the lead-base bearing alloy may be very much higher.

Silver–Lead Bearings

This bearing is not an alloy but consists of silver on which a very thin layer of lead has been deposited. A thick layer of silver is cast or electro-deposited on to a steel shell which serves as a backing and gives added strength to the bearing. The silver has good mechanical properties in that it has an appropriate hardness and ductility and is resistant to fatigue. It also has a very good thermal conductivity and so is effective in carrying away heat from the surface. As we shall see later, however, silver gives a high friction and is very difficult to lubricate. For this reason a very thin film of lead (a few thousandths of an inch thick) is deposited on it. This serves a double role. It enables the fatty acids in the oil to lubricate the surface (see Chap. X), and it also gives metallic film lubrication in the manner that has been described.

Under high temperatures and oxidizing conditions the chemical attack of the lead by the lubricating oil may be too rapid, and in order to prevent this a small amount of indium or of tin is alloyed with the lead film. The presence of these metals in the lead film render it resistant to excessive corrosion by the fatty acids in the oil. These bearings are effective under very severe conditions and provide a good practical example of metallic film lubrication.

Effect of Temperature Changes on Bearing Alloys

Under practical conditions, the frictional heat liberated during running may raise the temperature of the surface layer of a bearing to high values, and this may have a marked effect on the mechanical and frictional properties of the bearing. In addition, however, to this localized surface heating, the bearing as a whole may be gradually warmed up and cooled down as the engine is started or stopped, or as the temperature of the oil changes, and we may inquire what effect these thermal cycles may have on the bearing alloy. If, as is frequently the case, the alloy is bonded on to another metal, we may expect that the difference in the thermal expansion of the two metals may set up thermal stresses near the boundary. In the case of a typical lead-base alloy the coefficient of thermal expansion is $24 \times 10^{-6}/°C.$, while for steel it is c. $12 \times 10^{-6}/°C$. If the bearing is heated and cooled through 100° C. or so, the difference in expansion may cause deformation and cracking of the softer metal. A characteristic example of this is shown in Plate XVIII. 1. This 'bimetallic expansion effect' is of considerable practical importance since it may lead to premature failure of the bearing.

Another effect of considerable interest has recently been described by Boas and Honeycombe (1947). They have found that if certain metals such as tin, cadmium, or zinc, unattached to any other metals, are heated and cooled, deformation and cracking occur. The deformation becomes more marked the larger the number of heating-and-cooling cycles to which the metals are subjected. With lead, however, no such deformation is observed. This effect has been explained in terms of the thermal expansion of the metals themselves. Pure lead is a cubic metal and its coefficient of thermal expansion is the same in all directions in the crystal. Tin, cadmium, and zinc are non-cubic metals and their coefficients of thermal expansion differ along the various crystal axes. In a polycrystalline specimen of an anisotropic metal the orientation of any two neighbouring grains will in general be dissimilar, so that when the metal is heated the expansion on two sides of the grain boundary is different. The stresses set up are usually sufficient to produce plastic deformation and slip in the crystals, and the amount of deformation will increase with the number of cycles. This is shown very clearly in Plate XVIII. 3, 4, 5, and 6 for tin and cadmium specimens when heated between 30° and 150° C. through 50 and 200 cycles. This behaviour is therefore a fundamental property of non-cubic polycrystalline metals themselves. It follows that it is very difficult to prepare any anisotropic crystalline solid in a strain-free state since the preparation usually involves some heating or cooling. Even if the solid is initially free from strain, these strains are readily set up by small changes in temperature.

These observations have been extended to a study of the behaviour of lead-base and tin-base bearing alloys. With lead-base alloys the matrix is essentially isotropic and there is no appreciable deformation produced in the mass of the alloy by cyclic heating (see Plate XVIII. 1). With tin-base alloys, however, the matrix is markedly anisotropic and marked deformation readily occurs (see Plate XVIII. 2). Indeed, as Boas and Honeycombe (1947) have shown, the deformation by anisotropic expansion is generally far more marked for tin-base alloys than that produced by the 'bimetallic expansion effect' described above. It is found, however, that increasing the quantity of hard particles reduces the 'anisotropic effect' in tin-base alloys, since the hard particles have a thermal expansion which lies between the two principal coefficients of expansion of the matrix. Further, the hard particles appear to have a 'stiffening' effect on the alloy as a whole. Unfortunately there is a limit to this, since the alloy becomes increasingly hard and brittle, and therefore less suitable for most bearing applications.

It is clear from this discussion that there is a marked difference in the behaviour of lead-base and tin-base alloys on cyclic thermal treatment. The lead-base alloys suffer only from the 'bimetallic effect' at the interfaces where they are bonded to the steel backing. The tin-base alloys, however, show marked deformation within the alloy due to the anisotropic effect and, if this is sufficiently acute, it may lead to failure by 'thermal fatigue'.

The Role of the Soft Constituent in Bearing Alloys

The results described in this chapter show one feature that is common to both the copper–lead and the white-metal bearing alloys. Both types of alloys contain a soft low-melting constituent in which the actual shearing, during sliding, takes place. With the copper–lead alloys the shearing takes place in the extruded lead film; with the white-metal alloys, in the lead-base or tin-base matrix material. It is clear that this characteristic will prevent excessive seizure, since the high local temperatures developed under severe conditions of running will readily cause a local softening or melting of the low-melting constituent at the regions of momentary contact. The molten or plastic material will then be carried round to a cooler portion of the bearing.

The major difference between these two types of bearing alloys lies in the manner in which the load is supported. In the non-dendritic copper–lead alloy the hard copper matrix takes the load, so that the area of contact A is relatively small, whilst the shear strength s of the junctions formed in the lead film is also small. As a result, the frictional force, given by $F = As$, is also small, and the coefficient of friction for the unlubricated surfaces has a value of about $\mu = 0{\cdot}2$. With the white-metal alloys, however, the hard constituent plays practically no part in supporting the load, and from the frictional point of view the alloys behave as homogeneous materials of the same composition and properties as their matrices. The friction is thus characteristic of most homogeneous materials, and the coefficient of friction for the unlubricated surfaces has a value lying between about $\mu = 0{\cdot}6\text{–}0{\cdot}9$.

The dendritic copper–lead alloy is of considerable interest in this connexion, since it occupies an intermediate position between the non-dendritic copper–lead alloy and the white-metal bearing alloys. In the dendritic alloy the lead phase forms a continuous network between the copper dendrites and, strictly speaking, constitutes the matrix of the alloy. However, as an examination of the micro-structure shows, about 75 per cent. of the area of any surface consists of the hard copper phase.† As a result, the major portion of the load is still borne by the copper phase, and at room temperatures the friction is of the same order as that obtained with the non-dendritic alloy.

At higher temperatures the basic difference between the non-dendritic copper–lead alloy and the white-metal alloys is again apparent. With the non-dendritic copper–lead alloy the lead phase softens as the temperature is raised, whilst the copper matrix retains very largely its original hardness. As a result, s decreases whilst A scarcely changes, so that there is a gradual decrease in friction as the temperature increases. This proceeds until a temperature is reached at which there is a very rapid softening of the alloy as a whole (above 300° C.) and a correspondingly rapid increase in the friction. With the white-metal alloys there is a continuous softening of the matrix of the alloy as the temperature is raised, so that s decreases and A increases proportionately. As a result, the friction remains almost independent of temperature almost up to the melting-point of the alloy. Here again the dendritic copper–lead alloy occupies an intermediate position: the softening of the lead phase continues as before with a corresponding decrease in s. However, the softening of the lead network between the copper dendrites leads to a relatively rapid decrease in the overall hardness of the alloy, with a corresponding increase in A. As a result, the friction rises rapidly, and the coefficient of friction approaches that of the lead-base white-metal alloy.

It is clear that the hard copper matrix in the non-dendritic copper–lead alloy plays a well-defined part in the frictional mechanism of the alloy. It imparts greater strength and hardness to the alloy and by supporting the thin extruded lead surface-film produces a relatively low friction even for unlubricated surfaces. The dendritic alloy lacks the superior physical and mechanical properties of the non-dendritic alloy,

† In contrast to this, the hard particles of the white-metal alloys occupy only about 15 per cent. of the area of any surface.

but its supply of lead is very much more readily available at any surface. With the white-metal bearing alloys, however, the hard particles contribute little to the overall hardness or frictional properties of the alloys. It is possible that, as the white-metal type of bearing alloy was developed empirically, the hard particles are simply a relic of an historical accident. Nevertheless, it should be borne in mind that the frictional and wear properties are not the only factors which determine the suitability of an alloy for use as a bearing metal. For example, the hollows and crevices which may develop in the softer phase, particularly at the boundaries of the hard particles, may serve as minute reservoirs for the lubricating oil. Again, it is possible that many of the mechanical properties, such as the hardness and its variation with temperature, the compressive strength and the resistance to mechanical fatigue, may all play a vital part in the practical performance. For example, as we saw above in the work of Boas and Honeycombe, the presence of hard particles in a softer matrix may 'stiffen' the whole alloy and increase its resistance to 'thermal fatigue'. It is possible, therefore, that some of the significant differences between the mechanical properties of the white-metal bearing alloys and their corresponding matrix alloys may be due, in some measure, to the presence of the hard particles. It is clear, however, that in the white-metal alloys described in this chapter the *basic frictional properties* are determined essentially by the matrix material itself and that the hard particles play little part.

REFERENCES

H. Bassett (1937), *Bearing Metals and Alloys*. London, Edward Arnold & Co.

W. Boas and R. W. K. Honeycombe (1947), *Proc. Roy. Soc.* A **188**, 427 ; *J. Inst. Metals*, **73**, 433.

F. P. Bowden and D. Tabor (1943), *J. Appl. Phys.* **14**, 141.

D. Tabor (1945), *J. Appl. Phys.* **16**, 325.

VII

FRICTION OF CLEAN SURFACES: EFFECT OF CONTAMINANT FILMS

All things and everything whatsoever however thin it be which is interposed in the middle between objects that rub together lighten the difficulty of this friction. LEONARDO DA VINCI

Influence of Surface Films

In the previous chapters we have dealt with friction experiments carried out on metal surfaces which have been cleaned in the atmosphere. It is clear, however, that when solids are cleaned in air their surfaces are still covered by a thin film of oxide, water vapour, and other adsorbed impurities. The contaminant film is usually at least several molecular layers in thickness, and any complete theory of friction must take this film into consideration. As we shall see, even the thinnest surface films may have a profound effect on the friction.

The grosser surface contaminations may be removed by lapping or polishing and even more effectively by scraping with a degreased diamond tool (see Chap. X). None of these methods can, however, remove the last remaining adsorbed film from the surface. The only really effective method is to heat the metal surfaces *in vacuo*, and some earlier investigations along these lines have been described by Jacob (1912), Shaw and Leavey (1930), and Holm (1931). In what follows we shall describe some work carried out by Bowden and Hughes in 1939 and recently by Mr. J. E. Young on the friction of metals outgassed in high vacuum. The influence of surface films will also be discussed in relation to the effect of interfacial potential on friction. It is also of interest in this connexion to discuss some work carried out by Savage on the friction of graphite *in vacuo*.

The friction apparatus used by Dr. Hughes is shown in Fig. 54. The friction surfaces consist of the wire or cylinder XY and the larger cylinder C which hangs on it. The method consists essentially of projecting the suspended cylinder C along XY and photographically estimating its deceleration. From this the frictional force between the surfaces can be calculated. The cylinder is projected by the spring S which is released by the electromagnet M. The lower surface is degassed by passing a suitable heating current through it, while the cylinder is lifted off the lower surface on to the molybdenum rails R and heated by bombarding with 2,000- or 5,000-volt electrons from the filament F. Both metals are kept at a temperature just below that at which excessive evaporation sets in during the final stages of degassing. In these experiments the pressure in the envelope was maintained below 10^{-6} mm.

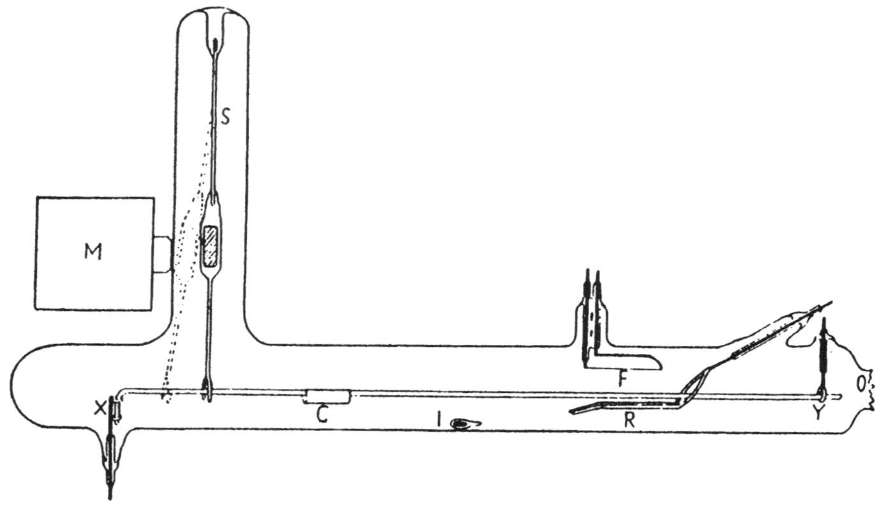

FIG. 54. Apparatus used for measuring the friction between outgassed metal surfaces. XY, stationary surface; C, moving surface; R, rails, and F, filament for supporting and heating surface C; S, spring for propelling C; M, magnet.

during the friction measurements, and, except when otherwise stated, the friction was measured immediately after the degassed surfaces had been allowed to cool to room temperature.

Effect of adsorbed gases on metallic friction [*A*]

The main results for two different combinations of metals are shown in Figs. 55 (*a*) and (*c*). It is seen that the coefficient of friction of the surfaces at the beginning of the experiments lies around $\mu = 0{\cdot}5$. After prolonged heating *in vacuo* at bright red heat and subsequent cooling the friction increases to values lying between $\mu = 4{\cdot}5$ and $\mu = 6$.

These values show that the removal of the surface films leads to a very great increase in the friction. This is due largely to the ease with which the metal surfaces can weld together in the absence of contaminant films. It would also seem that the shearing action of the sliding process itself produces an increase in the area over which metallic junctions are formed. These results, therefore, support the more detailed theory of metallic friction developed in Chapter V.

If the clean surfaces are allowed to stand at room temperature in a vacuum of 10^{-5} to 10^{-6} mm. a steady decrease in friction occurs. This decrease takes place in a few minutes and is presumably due to the gradual contamination of the surfaces by residual gases in the apparatus. Since rapid contamination occurs under these conditions, it is clearly impossible under ordinary atmospheric experimental conditions in the

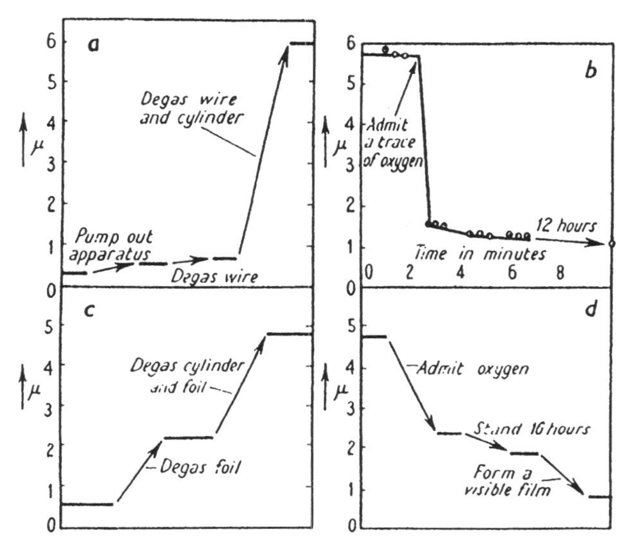

FIG. 55. (*a*) and (*c*) Effect of removing the adsorbed film of oxygen and other contaminants from metal surfaces: (*a*) nickel on tungsten; (*c*) copper on copper. The friction rises by a factor of 10 or more. (*b*) and (*d*) Effect of deliberately adding a trace of oxygen to clean outgassed metals: (*b*) nickel on tungsten; (*d*) copper on copper. There is a rapid reduction in friction.

laboratory to produce surfaces which, from the frictional point of view, are really clean.

The effect of deliberately admitting a trace of oxygen to the surfaces is shown in Figs. 55 (*b*) and (*d*). It is seen that there is a sudden large reduction in friction followed by a slower reduction which continues with time. On the other hand, the admission of pure hydrogen or pure nitrogen has little effect on the friction of the clean surfaces.

These results have been generally supported in recent work carried out by Mr. J. E. Young (1949). His apparatus is shown in Fig. 56. The envelope and the whole of the moving parts are made of fused silica to facilitate degassing of the surfaces. Its main point of difference from Dr. Hughes's apparatus is that the upper surface is dragged slowly over the lower surface instead of being projected along it. Further, the load is appreciably higher, being of the order of 15 gm., whereas in Dr.

Hughes's apparatus it is less than 1 gm. In addition the region of contact between the surfaces is more clearly defined as it occurs between a small curved protrusion and a flat surface. The degassing is carried out by high-frequency induction heating as shown in the figure.

With nickel surfaces as prepared 'clean' in the atmosphere the coefficient of friction is about $\mu = 1{\cdot}4$. When the surfaces are heated in a vacuum at 1,000° C. they are gradually freed of contaminant films. If

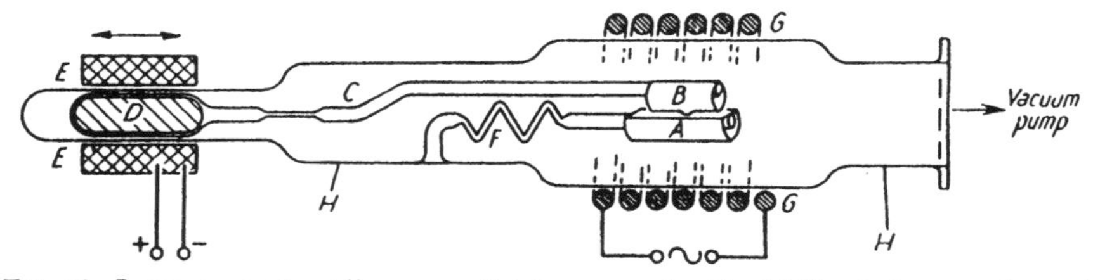

FIG. 56. Later apparatus (diagrammatic) for measuring the friction between outgassed metal surfaces. The specimens A and B are in the form of hollow cylinders to facilitate heating by high-frequency induction. A, fixed lower surface with flattened face; B, moving upper surface with 'pip' to give localized contact; C, connecting links of silica; D, sealed bulb of soft iron; E, electromagnet to move D and hence the surface B; F, silica spring device for measuring frictional force on A; G, induction heating coil; H, silica envelope connected to vacuum system.

friction measurements are carried out *in vacuo* after the surfaces have been allowed to cool to room temperature, it is found that the friction increases steadily as the outgassing becomes more thorough and readily reaches a value of about $\mu = 9$. On admitting air or a trace of oxygen the friction falls gradually to a low value and there is a further slow decrease with time. On the other hand, hydrogen even after a prolonged period of time produces no appreciable reduction in friction. If friction measurements are carried out with surfaces which are even more thoroughly outgassed, such large-scale seizure occurs on contact that the surfaces can only be separated by prizing them apart and the friction is too high to be measured ($\mu \sim 100$).

The fact that adsorbed films often take a finite time to produce their maximum reduction in friction has also been observed by Hughes when caproic acid vapour is admitted to clean gold surfaces. The friction of the degassed gold is about $\mu = 4$. On admitting caproic acid vapour at room temperature the friction falls immediately by about 10 per cent., in the course of the next 16 hours it falls to a value of about $\mu = 2$.

These experiments again show the marked effect of small quantities of adsorbed films on the friction of clean metals. With oxygen there is a large reduction in friction, and on tungsten the main effect is extremely rapid (Fig. 55 (*b*)) presumably due to the chemisorption of oxygen on the metallic surface. As Roberts (1935) has shown, the chemisorption here is almost instantaneous. The further increase in thickness of the adsorbed layer with time has relatively little effect on the friction. On other metals, however, the reduction in friction by oxygen is considerably slower and may proceed over many hours or days. Here it would seem that the effect is due essentially to the gradual growth of the thick oxide film, which provides increased protection of the surfaces and produces a corresponding reduction in friction. A similar time effect is observed with caproic acid on gold.

These results with specific contaminant films again support the view that metallic friction is due to the shearing of welded metallic junctions formed between the contacting surfaces. In the absence of contaminant films, as we have seen, these junctions are formed very easily and the friction is very high. The experiments also show that extremely small quantities of surface contamination may produce a marked reduction in the intimacy of contact, and hence in the frictional force.

Simple considerations of the kinetic theory show that in a vacuum of the order of 10^{-7} mm. of Hg the number of molecules striking the surface would, in a minute or so, be sufficient to form a monolayer. Even under the conditions described above, we cannot expect that the surfaces, after they have cooled, will be entirely free from adsorbed gas. Nevertheless the experiments show that when the surfaces are placed in contact under these conditions the metal atoms on each surface are able to fit together in a lattice and a comparatively large-scale 'crystal growth' can occur across the interface.

Influence of oxide films on friction [*A*]

The oxidation of most metals is a fast process which is only retarded by the protection provided by the oxide layer itself (Evans, 1946). Under ordinary room-temperature conditions exposed metal surfaces will soon be oxidized. Most freshly cleaned metals (lapped, ground, etc.) will acquire a layer of oxide between 10 and 100 A thick in about 5 minutes or less (cf. Mott). This applies to copper, iron, aluminium, nickel, zinc, chromium, and numbers of other similar metals. This oxidation will go on at even very low oxygen pressures, and copper will oxidize as fast at 10^{-3} mm. Hg pressure of oxygen as at atmospheric pressure (cf. Mrs. Garforth). During the process of oxidation the oxide layer formed may either take on the parameter of the parent metal (copper to about 100 A thick of oxide) or a completely independent oxide structure may be formed (aluminium) which is bonded to the metal at the coincident points of the two lattices. According to Frank, if the atomic lattice of the metal and of the oxide match to within 10 to 15 per cent. the oxide will take up the metal lattice parameter. If the match is poorer than 15 to 20 per cent. the oxide structure will form immediately. In the intermediate region of match from 10 to 20 per cent. patches of both structural types may be present.

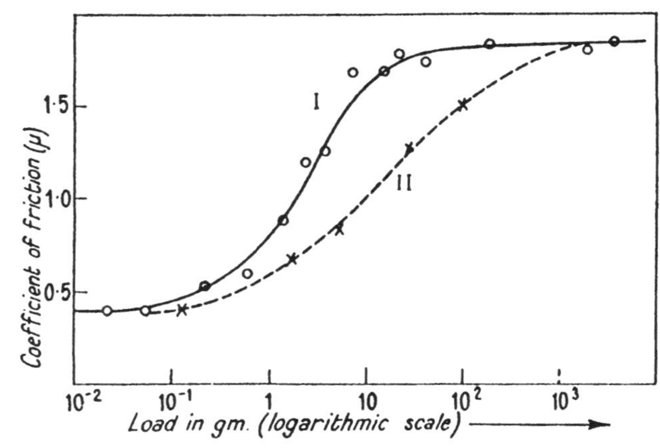

Fig. 57. Friction of copper on copper as a function of load. Curve 1: electrolytically polished copper. At loads above 100 gm. the coefficient of friction μ is sensibly independent of load and $\mu \approx 1{\cdot}6$. As the load falls the oxide films present on the copper surface are incompletely ruptured and the friction decreases. The value of the friction at very small loads ($\mu = 0{\cdot}4$) is essentially the friction of copper oxide. If the thickness of the oxide film is deliberately increased by oxidizing the copper surface, the effect is similar but the low friction persists up to a higher load (Curve II).

As the oxide layer on copper increases in thickness the internal compressive force will increase. At some thickness of about 200 A the oxide layer probably breaks up and re-forms and oxidation may then continue. This makes for a weak layer as opposed to the much stronger Al_2O_3 layers where the original arrangement is permanent.

A series of experiments which shows the importance of oxide films in ordinary friction measurements has recently been carried out by Dr. Whitehead (1950). With copper sliding on electrolytically polished copper he finds that the coefficient of friction is independent of load and has a value of about $\mu = 1{\cdot}6$ for loads above about 100 gm. With decreasing load the friction falls off and reaches a steady value of $\mu = 0{\cdot}4$ at loads below about 10 gm. (see Fig. 57, curve I). There is also a marked change in the appearance of the surface damage as is shown in Plate XIX. 1, 2, 3, and 4. For the lower loads where the friction is small the track has the appearance of a relatively smooth groove, which is comparable with that observed on lubricated surfaces. As the load is increased the track shows more tearing and plucking, and at loads giving a friction of $\mu = 1{\cdot}6$ the surface damage is very heavy and is characteristic of that observed with unlubricated metals. It is clear that this effect is due to the gradual break-through of the surface oxide film. At the lower loads the friction is essentially that of copper oxide sliding on copper oxide and this is sheared more easily than the metal. At the higher loads the oxide film is ruptured and considerably more metallic adhesion occurs. This is confirmed by the fact that if the copper surface is deliberately oxidized after polishing to produce a thicker oxide film the low friction persists up to a higher load (Fig. 57, curve II).

The behaviour of aluminium is in marked contrast to this (see p. 99). The friction remains constant at a high value of $\mu = 1{\cdot}2$ from 10^4 to 10^{-2} gm. Further, the oxide film is broken up and there is characteristic metallic welding even at the smallest loads (Plate XI). This difference in behaviour might be attributed to the 'fit-misfit' nature of the oxide layers; or to their state of tension or compression. The evidence, however, suggests that the main factor is the relative mechanical properties of the oxide and the underlying metal. If the oxide forms a very hard film on the surface of a soft metal as with aluminium (Mohs's hardness of Al ≈ 2, of $Al_2O_3 \approx 9$) its behaviour may be analogous to that of a thin film of ice on mud. The slightest load will deform the substrate so that break-through of the surface will occur. If, however, the mechanical properties of both metal and oxide are similar as with copper (Mohs's hardness of Cu ≈ 3, of oxides $\approx 3{\cdot}5$) the surface film will deform with the underlying metal; break-through will not occur and the friction and surface damage will be small. (Whitehead 1950.)

Influence of Temperature on Friction of Clean Metals

Bowden and Hughes (1939) have carried out some experiments on the effect of temperature on the friction of clean metals. The apparatus used was that described in Fig. 54, and the results show at once that unless elaborate precautions are taken to degas the metals very thoroughly the results are erratic and irreproducible. This is presumably due to the influence of incompletely removed surface films superimposed on any purely thermal effect.

If the surfaces are first thoroughly outgassed and a good vacuum subsequently maintained the results are reproducible and reversible. Typical results for several metals are shown in Fig. 58. In general the friction decreases as the temperature is raised, but the effect is slight, and even at 1,000° C. the friction has fallen to only about half its value.

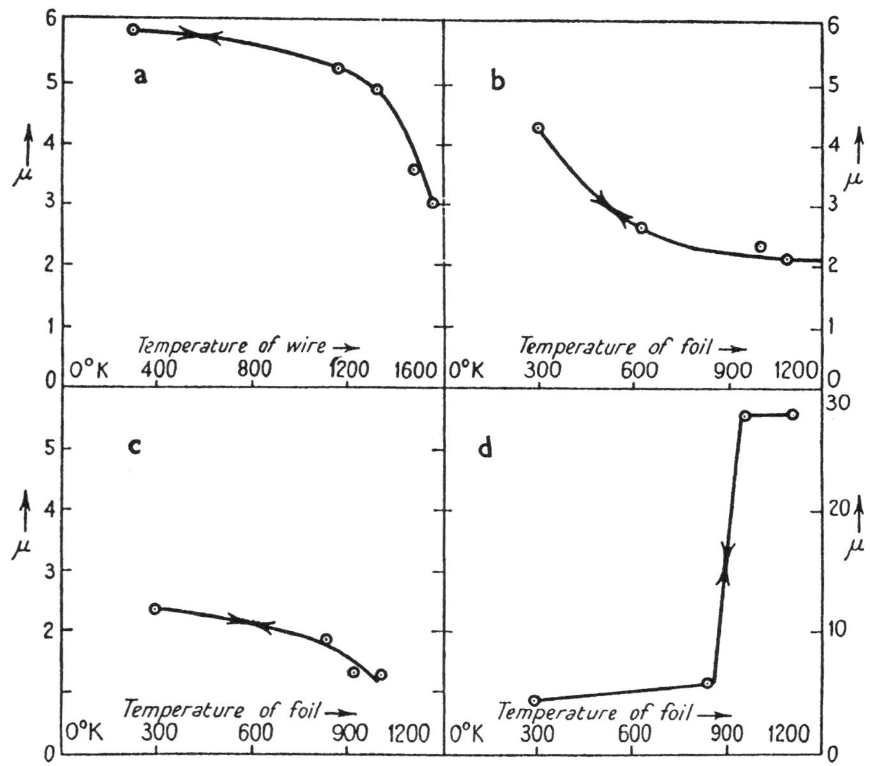

FIG. 58. The effect of temperature on the friction between clean outgassed metal surfaces: (*a*) nickel sliding on tungsten; (*b*) nickel on nickel; (*c*) copper on copper; (*d*) gold on gold. In general the friction decreases somewhat as the temperature is raised. With gold there is a sudden large rise in friction when the surfaces begin to soften (*c*. 600° C.) and large-scale welding occurs.

The behaviour with gold is exceptional as the friction shows little change until a temperature of about 600° C. is reached, when the friction rises suddenly to a value of about $\mu = 20$. This is probably due to the pronounced softening of gold which occurs at this temperature, so that the metal flows readily and the sliding surfaces would weld together over a large area. Similar results are obtained by Mr. J. E. Young with outgassed nickel surfaces when the friction is measured at 1,000° C.

The fact that the measurements recorded in Fig. 58 are reversible, i.e. that the friction returns to its original value when the surfaces are cooled down, is important since it shows that the frictional decrease is due to the temperature alone and not to any irreversible change in the state of the surface. The fact that the friction actually decreases as the temperature is raised also suggests that the surfaces are reasonably clean, for in the presence of contaminating films a high temperature would diminish their concentration and so cause an increase in the friction.

The results show that the decrease in the coefficient of friction for most metals is only a few per cent. over a temperature rise of 100° C. This small effect is in harmony with the theory of metallic friction described in Chapter V, and may be accounted for as follows. The area of contact over which welded metallic junctions are formed depends on the load and the yield pressure of the metal. At high temperatures the yield pressure diminishes so that, for a given load, the total cross-section of the metallic junctions is increased. However, the shear strength of the metal will similarly fall off, so that the force to shear the junctions, which is the frictional force, should remain approximately the same. For this reason the friction of clean metals is not markedly dependent on the temperature unless the temperature is so high that marked thermal softening occurs. This behaviour may be contrasted with that of a solid which retains its bulk hardness up to the melting-point, so that a soft or molten surface film can be formed on top of a relatively hard substrate. In this case, as we saw in earlier chapters, the friction may fall to a very low value.

Influence of Interfacial Potential on Friction

Effect of electrodeposited hydrogen and oxygen

It is clear that the adsorption of a layer of gas on the metal surface may have a profound effect on the friction, and in this connexion it is interesting to consider the friction of metals which are immersed in an electrolyte. If an inert metal such as platinum, for example, is immersed in an acid electrolyte it is possible, by suitably varying the interfacial potential, to bring about the electrodeposition of either hydrogen or of oxygen. When the interfacial potential is in the region of $+1{\cdot}0$ volt on the hydrogen scale (the metal being positive) the electrode surface becomes covered with a monolayer of oxygen, and if the potential is made more positive still, oxygen is evolved, and we are in the region of oxygen overpotential. If the interfacial potential is reduced below about $+1{\cdot}0$ volt, the monolayer of oxygen is removed and, in the region of 0 volt on the hydrogen scale, a monolayer of hydrogen is deposited (Bowden, 1929). If the potential is made more negative still, hydrogen is evolved and we are in the region of hydrogen overpotential. In the intermediate region of potential, that is, round about $0{\cdot}3$ to $0{\cdot}6$ volt on the hydrogen scale, the electrode surface should be reasonably free from either oxygen or hydrogen.

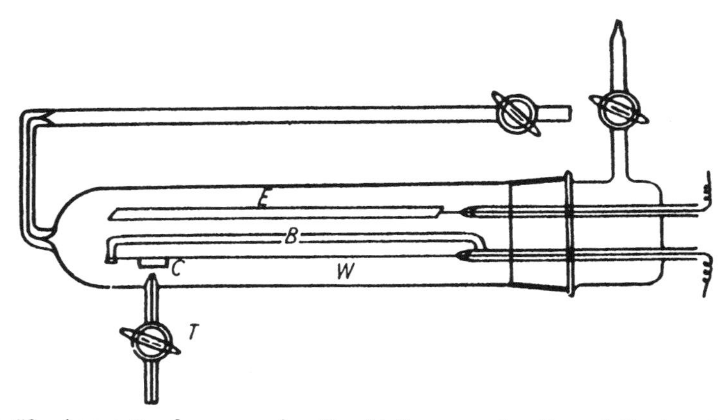

FIG. 59. Apparatus for measuring the friction as a function of the interfacial potential. *B*, glass bow supporting platinum wire *W* (stationary surface); *C*, platinum cylinder (sliding surface) suspended on *W*; *E*, electrode; *T*, tap leading to reference electrode. The coefficient of friction is determined from the angle of inclination of *W* necessary to cause *C* to slide.

It would clearly be of interest to investigate the friction of metal surfaces in an electrolyte at different interfacial potentials. Several workers have reported changes in the friction between a metallic surface and some other material such as glass or chalk as the metal surface was polarized. Such effects were first demonstrated by Edison (1877–9) and subsequently studied by Koch (1879), Krouchkoll (1882), and Waitz (1883). It has also been shown (Bastow, 1936; Clark, 1940) that the friction of immersed solids is dependent upon the hydrogen-ion concentration of the solution. The results, however, cannot be easily interpreted since the interfacial potentials of the surfaces are not clearly defined.

Measurements of the friction of platinum, as the interfacial potential is varied under controlled conditions, have been made recently by Dr. G. C. Barker (1947) and the investigations are being continued by Dr. L. Young (1949). One method used is illustrated in Fig. 59. The friction is measured between a stretched platinum wire *W* and a platinum cylinder C (weighing about 2 gm.) which can slide on it. The platinum wire is stretched taut by the glass bow B. These are contained in a glass cell which can be freed from gas and filled with an electrolyte. A large platinized platinum electrode E serves as the polarizing electrode through which a current can be passed to the wire and cylinder, and the interfacial potential of the wire and cylinder can be measured against a reversible hydrogen electrode immersed in a similar electrolyte. Electrical connexion to the hydrogen electrode is made through the tap T. The whole cell can be rotated so that the wire W can be tilted through 180°. The angle at which sliding occurs gives a measure of the friction.

It was found that great care must be taken to free the system from traces of impurity, but when this was done results similar to those shown in Fig. 60 were obtained. Curve 1 shows the results obtained by Dr. L. Young for the static friction of platinum on platinum in pure dilute sulphuric acid (0·1 N). It will be seen that the changes in friction as the potential is varied are very great indeed. In the region of +1·0 volt, which corresponds to oxygen deposition, $\mu = 0{\cdot}7$. As the potential is decreased and the oxygen is removed, the friction rises, and in the region of 0·3 volt has reached a value of $\mu = 3{\cdot}4$. As the potential is reduced below this, and we come into the region of hydrogen deposition, the friction falls to a value of $\mu = 2{\cdot}3$. At still more negative potential, that is, in the region of hydrogen overpotential, there is a further decrease, but the friction is still high ($\mu = 2$) compared with the values obtained in the region of oxygen deposition ($\mu = 0{\cdot}7$). (It will be seen that in the region of oxygen overpotential, i.e. as the surface becomes more positive than +1·0 volt, there is a small increase in the friction. The reason for this is more complex and will not be discussed here.)

The change in friction with potential is accompanied by a corresponding change in the damage to the surface. Plate XIX. 5 and 6 show the tracks formed on a platinum surface when a small platinum slider passes over it once under a load of 20 gm. Plate XIX. 5 shows the damage when the interfacial potential is *c.* 1·0 volt and the friction is a minimum. The surface damage is slight. Plate XIX. 6 shows the corresponding damage when the potential is *c.* 0·3 volt and the friction is a maximum. In this region there is profound damage, welding, and seizure of the metals. It is frequently found that the surfaces stick together at this potential so that the coefficient of friction cannot be measured.

If a trace of impurity is present, these high frictions are not observed.

Curve II shows the results obtained by Barker when the solution is contaminated with a trace of H_2S. The friction stays at a low value over the whole potential range. This suggests that the nature of the film adsorbed on the surface can be more important in determining the friction than the value of the interfacial potential itself.

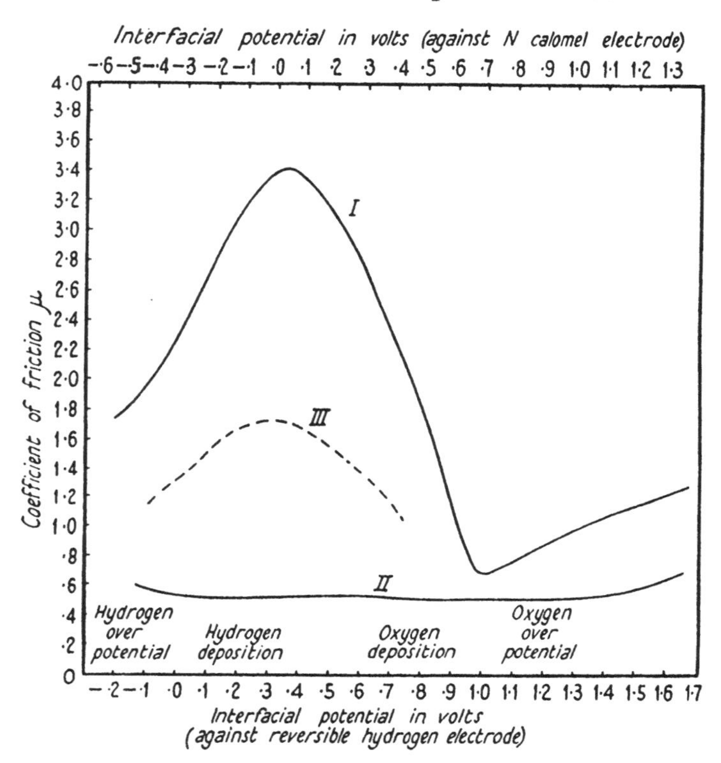

FIG. 60. Effect of interfacial potential on friction of platinum surfaces in dilute sulphuric acid. Curve I shows that the friction is a maximum in the range where neither hydrogen nor oxygen is deposited and a minimum where oxygen is present. The maximum friction corresponds to the potential at which the surface tension of platinum is a maximum. This is shown in curve III, taken from results by Gorodetskaja and Kabanov, where the surface tension is plotted in arbitrary units. Curve II is the friction of platinum surfaces in sulphuric acid when the solution is contaminated with a trace of H_2S. The friction is scarcely affected by the interfacial potential.

These results are consistent with the observations on the friction of outgassed metals. The accumulation of a monolayer of oxygen on the surface reduces the friction to a low value. A film of hydrogen is much less effective as a lubricant than the corresponding film of oxygen. In the intermediate region, where neither is present, the friction is very high indeed and seizure may readily occur when the surfaces are placed in contact. Apparently water itself is a very poor lubricant.

Friction and surface tension

It is well known from electrocapillary curves that the surface tension at a mercury surface in contact with an electrolyte is dependent upon the interfacial potential. If, for example, a mercury surface is placed in contact with a dilute solution of sulphuric acid containing a few mercury ions, the surface of the mercury will be positively charged. These charges on the surface will mutually repel one another and will reduce the surface tension. If the interfacial potential is made less positive by applying a polarizing current (or in any other way), the surface tension will rise as the positive charge on the surface is reduced and will reach a maximum when the surface charge is zero. Further reduction in the interfacial potential will result in the accumulation of electrons on the mercury surface and the mutual repulsion of these will again decrease the surface tension. Corresponding to the charge on the surface, positive or negative ions will be attracted from the solution to form the electrical double layer and these also will influence the surface tension. With a mercury surface in 0·1 N H_2SO_4 the maximum of the electrocapillary curve would be approximately $+0{\cdot}56$ on the N calomel electrode, and at this point there should be effectively no charge on the surface.

In the case of platinum we cannot measure the surface tension direct, but Gorodetskaja and Kabanov (1934) have measured the contact angle between a gas bubble and a platinum surface immersed in 0·1 N Na_2SO_4 solution and have shown that it is a maximum (corresponding to maximum surface tension between the platinum and the electrolyte) when the interfacial potential is *c.* 0 on the calomel scale, i.e. about $+0{\cdot}28$ on the hydrogen scale. Their results are plotted on curve III, Fig. 60. It will be seen that the position of maximum surface tension corresponds roughly to the position of maximum friction.

Since the surface tension is a measure of the attractive force between the metal atoms and the friction is due to the attraction between the metal atoms on the two adjacent surfaces (with consequent localized welding), we might expect to find a correlation between them and it is interesting to see that this is so.

In general, we have two effects operating. The first is the presence on the surface of gas or other films which may reduce the friction. If these films are removed when the interfacial potential is changed there will be a corresponding change in the friction (cf. oxygen). If they are not removed and the films are effective lubricants, the friction may remain at a low value even when the interfacial potential is changed (cf. H_2S, Fig. 60, curve II). In addition to this, there will be the effect of the electrical charge on the surface. If there is a concentration of positive charge (or of electrons) on the surface, the electrical repulsion will reduce the surface tension at the metal electrolyte interface. At the same time the repulsion between the charges will reduce the attraction and adhesion between the two metal surfaces and hence reduce the friction. At the point where this surface charge is reduced to zero the surface tension will be a maximum and the adhesion and friction between the two surfaces will also be a maximum. This relation between the interfacial potential and the friction and adhesion of the surfaces is of considerable interest in connexion with a number of other physical phenomena, such as the variation of strength properties of solids in the presence of surface active materials or when immersed in electrolytes (e.g. Rehbinder and Wenström, 1937; Andrade, 1949).

Friction of Graphite [*A*]

It has long been known that the friction of graphite is low even in the absence of lubricants, and it has generally been considered that this is due to the lamellar structure of graphite. According to this view the structure imposes marked anisotropy of physical properties: in a direction normal to the lamellae the graphite is strong and resists compression, whilst in the plane of the lamellae it is weak and hence shearing occurs readily. As we have seen in Chapter V, these properties provide the necessary conditions for a low coefficient of friction.

Recent work by Savage (1948) in America has shown, however, that this explanation of the action of graphite is not satisfactory. Most of his experiments have been carried out with graphite sliding on copper in order to simulate the conditions operating between commutator brushes and slip rings. A few experiments with graphite sliding on graphite, described below, yield, however, very similar results. Savage finds that for graphite surfaces freed from contaminant films by heating in a vacuum the friction is higher, and the wear very heavy. In addition the abraded graphite dust possesses very high adsorptive power; for example it adsorbs more hydrogen at room temperature than charcoal at $-195°$ C. The same behaviour was observed with all types of graphite examined under these conditions. This suggests that the wear is due to fragmentation, with tearing apart of the crystals across the lamellae, rather than to flaking along the lamellae. Such a mechanism would account for the high friction and wear, whilst the exposed valence arms would explain the high adsorptive power of the graphite dust.

The high friction is still observed with decontaminated graphite after hydrogen or nitrogen is admitted to the surface. If, however, certain organic vapours or water vapour, are admitted at pressures as small as 6–7 mm. the friction is reduced to the usual low value and the wear is entirely eliminated. Oxygen has a similar lubricating effect at pressures above about 600 mm.

It would appear from Savage's work that the low friction and wear normally associated with graphite surfaces are due to the presence of adsorbed layers. Under normal conditions this adsorbed layer is provided by atmospheric moisture, and is hence self-forming and self-repairing. Savage has offered an explanation for the fact that, although contaminated graphite has a finite though very small coefficient of friction, its rate of wear is infinitesimally small. He suggests that this sliding is essentially a surface process so that, by applying the earlier theory of Tomlinson (1929), he is able to express the friction in terms of the surface tension of the adsorbed moisture film. Further work, however, must be done before this can be finally accepted. In particular it is necessary to know whether adsorbed vapours have any effect on the physical properties of graphite and whether, by penetrating the lattice, they are able to accentuate the anisotropic properties of it.

Nevertheless, the results provide striking evidence for the important part played by small quantities of contaminant films in reducing the friction between sliding solids.

REFERENCES

E. N. da Costa Andrade (1949), *Nature*, **164**, 536.
G. C. Barker (1947), Ph.D. dissertation, Cambridge.
S. H. Bastow (1936), Ph.D. dissertation, Cambridge.
F. P. Bowden (1929), *Proc. Roy. Soc.* **A 125**, 446.
—— and T. P. Hughes (1939), ibid. **A 172**, 263.
—— and J. E. Young (1949), *Nature*, **164**, 1089.
N. Cabrera and N. F. Mott (1948–49), *Reports on Progress in Physics*, **12**, 163.
R. E. D. Clark (1940), *J. Soc. Chem. Ind.* **59**, 216.
T. A. Edison (1877), *Telegr. Journal*, **5**, 189; (1879), ibid. **7**, 332.
U. R. Evans (1946), *Corrosion, Passivity, and Protection.* Arnold & Co.
F. C. Frank and J. H. van der Merwe (1949), *Proc. Roy. Soc.* **A 198**, 216.
F. Garforth (1949), to be published.
A. Gorodetskaja and B. Kabanov (1934), *Phys. Zeit. Sowjet.* **5**, 418.
R. Holm *et al.* (1931), *Wiss. Veröff. Siemens Konzern.* **10** (4), 20.
C. Jacob (1912), *Ann. Phys. Lpz.* **38**, 126.
K. R. Koch (1879), *Wied. Ann.* **7**, 92.
M. Krouchkoll (1882), *Compt. rend.* **95**, 177.

P. Rehbinder and E. Wenström (1937), *Bull. Acad. Sci. U.R.S.S. Ser. Phys.* **4**, 531.
J. K. Roberts (1935), *Proc. Roy. Soc.* **A 152**, 445.
R. H. Savage (1948), *J. Appl. Phys.* **19**, 1.
P. E. Shaw and E. W. L. Leavey (1930), *Phil. Mag.* **10**, 809.
G. A. Tomlinson (1929), ibid. **7**, 907.
K. Waitz (1883), *Wied. Ann.* **20**, 285.
J. R. Whitehead (1950), Ph.D. Dissertation, Cambridge; *Proc. Roy. Soc.* **A 201**, 109.
L. Young (1949), Ph.D. Dissertation, Cambridge.

VIII
FRICTION OF NON-METALS

IT is interesting to consider the friction of non-metals and to compare their behaviour with that of metals. As we saw earlier, the characteristic frictional properties of metals are due largely to their ability to flow plastically and to weld together under load. The welded junctions so formed are sheared during the sliding process. With many non-metals, such as glass and certain plastics, flow and welding may occur as with metals. With other materials, particularly those which are markedly crystalline in structure, it is difficult to envisage plastic flow and subsequent welding. Nevertheless many non-metals, even of a crystalline nature, have frictional properties similar to those of metals and it is possible that other mechanisms analogous to the welding process are responsible for the frictional behaviour observed. In other cases, as we shall see, the friction is very different from that of metals.

The coefficient of friction of most metals exposed to the air lies within a relatively small range of values ($\mu = 0{\cdot}5$ to about $1{\cdot}5$), and as we saw in Chapter V this is due to the fact that with metals we usually deal with polycrystalline specimens which are comparatively isotropic in their 'strength' properties. With some non-metals we have to work with single crystals and these are often appreciably anisotropic. Consequently we may expect these materials to possess exceptional frictional properties. Materials such as diamond and sapphire have, in fact, exceptionally low coefficients of friction, and the same is true of graphite. It is clear that the low coefficient of friction is not always due to anisotropic 'strength' properties. For example with graphite, which has a marked lamellar structure, recent work of Savage (1948) suggests that the low coefficient of friction is due to surface films rather than to the intrinsic structural properties of the graphite. This raises one interesting point in the investigation of non-metallic friction. With metals, although it is difficult to prepare the surfaces in a reproducible state, we do in general know the nature of the surface films or contaminant layers that are present. With non-metals we know considerably less about the nature of these surface layers and the frictional measurements are often variable and irreproducible.

Crystalline solids [*A*]

Some frictional measurements carried out by Mr. R. Hutchison (unpublished) some years ago, on crystals of $NaNO_3$, KNO_3, and NH_4Cl, showed that for the materials sliding on themselves the coefficient of friction was of the order of $\mu = 0{\cdot}5$. He did not observe any clear distinction between electrovalent and covalent solids; for example the friction of sulphur on sulphur was not very different from that of rock-salt on rock-salt. He concluded that the friction depended more on the bulk mechanical properties of the solid than on its chemical nature.

More recent experiments by Hutchinson and Rideal (1947) have confirmed these results and have shown that Amontons's law holds over an appreciable range of loads. Sliding is accompanied by considerable fragmentation and surface damage of the solids. In the presence of surface films of long-chain polar compounds the friction is reduced to about $\mu = 0{\cdot}12$, a value which is close to that observed with metals (see Chap. IX). Although these materials are relatively brittle it is apparent that there is marked adhesion between the surfaces during sliding and considerable surface damage. The coefficient of friction for both clean and lubricated sliding is of the same order as that observed with metals.

Sapphire and diamond [*A*]

Sapphire and diamond are very hard crystalline materials with exceptional frictional properties. The frictional behaviour of sapphire depends somewhat on the crystal plane exposed to the rubbing process, but the effect is small and the friction is always low. For sapphire sliding on itself the friction is about $\mu = 0{\cdot}2$, and the damage produced on the surface is not generally visible. However, more refined methods such as electron microscopy (see Chap. IV, Pl. X. 2) show that some surface damage does occur. For a steel slider on a sapphire surface the coefficient of friction is about $\mu = 0{\cdot}12$ and Dr. Tingle finds that mineral or vegetable oils have little effect on the friction. Silicones, however, produce a marked *increase* in friction, in some cases the coefficient reaching a value of $\mu = 0{\cdot}25$. It is suggested that for unlubricated sliding the friction occurs between the sapphire and a protective oxide film on the steel slider: this film is continuously re-formed during sliding so that the friction remains low. In the presence of silicone, however, the oxidation is strongly inhibited, so that once the oxide is worn off the slider, the sliding takes place on a relatively clean steel surface. The friction now reaches a value comparable to that of sapphire on sapphire. This view is supported by the observation that when a *sapphire* slider rubs on a steel surface the coefficient of friction is scarcely affected by the presence of silicones and remains at the low value of about $\mu = 0{\cdot}15$.

In specific practical applications the oxide film on the steel surface may play a very important part. Shotter (1937), for example, finds that in a jewelled bearing the ferric oxide formed at the tip of the steel pivot acts as an abrasive and increases the wear of the sapphire cup. For this reason unlubricated sliding is unsatisfactory. Similarly Stott (1937) finds that in the absence of lubricants the accumulation of oxide detritus in the cup may lead to an actual increase in the friction. This effect is presumably due to a 'clogging' action of the detritus.

The friction of diamond is exceptionally low. For unlubricated surfaces of diamond on diamond the coefficient may be as low as $\mu = 0{\cdot}05$ and the surface damage is extremely small. When metals slide on diamond the friction is of the same order, and the amount of metallic transfer is again very small. For these reasons diamond cutting-tools are very effective in the machining of hard metals. In addition, since diamond has an extremely high melting-point and retains its hardness to very high temperatures, it is very effective as a polishing powder.

The low friction of sapphire and diamond is probably not due to appreciable anisotropy of physical properties. The strength properties do vary somewhat according to the crystallographic direction, but the effect is not large. The same applies to the friction. As mentioned above, there is only a small variation of friction with the crystal face exposed. It is evident that when surfaces slide on sapphire or diamond the adhesion is small, but it was not clear whether this is an intrinsic property of the material or whether it is due to adsorbed surface films. Recent experiments by Mr. J. E. Young (unpublished) show that when the surface films are removed from diamond by heating in a vacuum, the friction of diamond sliding on diamond may rise to a high value.

Carbon, graphite, and molybdenum disulphide [*A*]

It has long been known that the friction of carbon, especially when it is in the form of graphite, is low. For hard non-graphitic carbon surfaces, Mr. Burns found that the friction of steel on carbon and of carbon on carbon was approximately $\mu = 0{\cdot}16$. This was reduced by about 20 per cent. in the presence of a lubricant. With graphite the friction is considerably less ($\mu \approx 0{\cdot}1$) and it remains low up to very high temperatures. The amount of wear is also very small. For these reasons lubricants containing graphite have proved very effective under conditions where more conventional lubricants are unsuitable or ineffective.

The more recent work by Savage (1948) on outgassed graphite has already been discussed in Chapter VII.

Another material of considerable interest is molybdenum disulphide (MoS_2). This possesses a laminar structure and has the added advantage of being stable at high temperatures. Experiments by Mr. K. Shooter show that the friction of MoS_2 on itself is low. Further, a thin film on metal surfaces gives a very low coefficient of friction ($\mu = 0{\cdot}05$) which is maintained up to elevated temperatures. A study of the frictional behaviour of MoS_2 and of its performance at high speeds of sliding has been made by Johnson, Godfrey, and Bisson (1948).

Mica [*A*]

The friction of mica is of considerable interest since it also possesses a well-defined lamellar structure. It is well known that freshly cleaved faces of mica adhere very strongly to one another and the adhesion decreases with exposure of the mica to the atmosphere. Derjaguin and Lazarev (1934) found a corresponding change in the friction. For freshly cleaved mica the friction was high ($\mu = 1$), but with prolonged exposure it fell to about $\mu \doteq 0{\cdot}4$. If liquids were added to the surface the friction was even lower, of the order $\mu = 0{\cdot}2$. It is clear from these results that with mica there are strong adhesive forces between freshly cleaved crystal faces. These forces are comparable with the interlattice forces, since when sliding occurs portions of mica are plucked out on either side of the original plane of cleavage. Recently Dr. Courtney-Pratt (1950) has made the interesting observation that occasionally the mica molecule is cut in half during cleavage (see also Tolansky, 1948).

Plastics [*A*]

The frictional properties of plastics are markedly dependent on the type of plastic considered. Recently Mr. Shooter and Mr. Thomas (1949) have carried out some frictional measurements on four types of plastic summarized in Table XXII. The friction was measured on the apparatus described in Chapter IV, Plate V, between a hemispherical slider and a flat surface. Over the range of loads used (1 to 4 kg.) Amontons's law was obeyed. With polystyrene the friction was determined for three samples of molecular weight 130,000, 90,000, and 66,000 and the results were essentially the same. With perspex the friction was determined for an unplasticized specimen containing titanium oxide filler and for a plasticized specimen containing no filler. The friction was again essentially the same for both specimens. The main results are given in

TABLE XXII

Polymer	Chemical formula	Characteristics	Brinell hardness
Teflon	$[-CF_2-CF_2-]_n$		1–2
Polythene	$[-CH_2-CH_2-]_n$	Alkathene grade 20 (I.C.I.)	1–2
Polystyrene	$\left[-CH_2-CH(C_6H_5)-\right]_n$	Molecular weight (a) 130,000 (b) 90,000 (c) 66,000	20–25
Perspex	$\left[-CH_2-C(CH_3)(COOMe)-\right]_n$	(a) plasticized (b) unplasticized containing titanium oxide pigment.	25–30

Table XXIII. It is seen that the coefficients of friction for polystyrene and Perspex are of the same order as for metals. The values for Teflon and polythene are very much lower.

TABLE XXIII

Polymer	Polymer sliding on polymer	Polymer sliding on steel	Steel sliding on polymer
Teflon	0·04	0·04	0·10
Polythene	0·1	0·15	0·2
Polystyrene	0·5†	0·3	0·35
Perspex (a)	0·8†	0·5†	0·45†

† Intermittent motion.

The variation of the friction with temperature for the four plastics is summarized in Table XXIV, and it is seen that, as with metals, the friction is not markedly dependent on the temperature. With Perspex and polystyrene the plastics became too soft for reliable friction measurements above 80° C. With Teflon, on the other hand, there is little sign of thermal softening even at 200° C. and the friction remains low throughout.

These results show that the frictional properties of Teflon are exceptionally good. With unlubricated metals the coefficient of friction is of the order of $\mu = 1$, whilst with Teflon it is about $\mu = 0{\cdot}04$. Even in the presence of the best boundary lubricants, metal surfaces rarely give a friction as low as $\mu = 0{\cdot}04$. Indeed, the friction of Teflon on Teflon is comparable with that of ice on ice. The friction is unaffected by lubricants, and the friction and mechanical properties remain unchanged up to a temperature of almost 300° C.

TABLE XXIV

Sliding surfaces	*20° C.*	*50° C.*	*80° C.*	*100° C.*	*150° C.*	*200° C.*
Polystyrene on polystyrene	0·5†	0·65†	0·65–0·7†	..	..	..
Perspex on Perspex	0·8†	..	0·85†	..	..	..
Teflon on Teflon	0·04	0·04	0·04	0·04	0·04–0·05	0·04–0·05
Teflon on steel	0·04	0·04	0·04	0·04	0·04	0·04
Steel on Teflon	0·09	0·09	0·10	0·10	0·11	0·14

† Intermittent motion.

Hanford and Joyce (1946) have suggested that in Teflon the comparatively large fluorine atoms screen the positive charge on the carbon atoms. On this view the interaction of the negative charges on the fluorine atoms of neighbouring molecules results in a low molecular cohesion. This effect will also occur in polythene, but to a much less extent, since the hydrogen atoms are much smaller than the fluorine atoms and consequently have less screening power. The presence of aromatic groups as in polystyrene and polar groups as in Perspex should still further increase the molecular cohesion. The friction of polymer sliding on polymer depends upon the adhesion of the two surfaces and this in turn depends upon the molecular cohesion. One would therefore expect the friction to increase in the order (i) Teflon, (ii) polythene, and (iii) polystyrene and Perspex. This is observed. The results so far obtained are insufficient to warrant a more detailed discussion. Here again it would be desirable to determine the frictional properties when adsorbed surface films have been removed.

Frictional welding of plastics

It is of interest to consider the rise in temperature of the sliding surfaces produced by frictional heating. The frictional measurements on the plastics were carried out at a load of about 2 kg. and a sliding speed of 0·01 cm./sec. For polythene and Teflon, where the sliding is smooth, the temperature rise of the slider can be calculated using the equations given in Chapter II and the value is of the order of 10°. Polystyrene and Perspex, on the other hand, give intermittent motion and the relative velocity of the two surfaces during slip is of the order of 1 or 2 cm./sec. This higher sliding velocity combined with the low thermal conductivity of these materials (10^{-4} compared with 1–10^{-1} c.g.s. units for metals) might be expected to give rise to considerable increases in the temperature at the area of contact. Indeed, using Jaeger's equations for high-speed sliding, calculations indicate that the rise of temperature will be of the order 100–500° C. The equation, however, assumes that thermal equilibrium has been attained: more detailed considerations show that in the very small time during which slip occurs the actual rise in temperature is only a small fraction ($\sim$ 10 per cent.) of the temperature resulting after a long time. It is probable, therefore, that even during intermittent motion, under the conditions of these experiments the frictional heating will not be sufficient to produce appreciable thermal softening of the plastic. At steady speeds of sliding of the order of a few cm./sec., however, marked thermal softening may be expected.

The occurrence of thermal softening as a result of frictional heating may be demonstrated in the following manner. A rod of polystyrene is held in a lathe chuck and rotated at a uniform speed. A flat plate of the same material is pressed against the end of the rod with a known force. With suitable loads and speeds thermal softening occurs at the interface and the rubbing surfaces become tacky. At this stage the rotation is stopped and the surfaces allowed to cool. The bond formed at the interface has the strength of the bulk material, i.e. on breaking the joint the break does not necessarily occur along the joint. This technique forms the basis of the industrial process of friction welding (Freres, 1945).

The load and speed required to produce thermal softening can be calculated by using the equation of Jaeger for the temperature rise T at the interface between a cylinder rubbing on a flat plate of the same material (Chap. II, equation (11)). Experiments show that there is good agreement between the calculated loads and speeds, and those necessary to produce thermal welding.

For rods of diameter about 1 cm., polystyrene, plasticized Perspex, and polythene can be welded quite easily at a load of 5 kg. and a speed of 500 r.p.m. Under these conditions unplasticized Perspex containing titanium oxide pigment cannot be welded, much of the frictional energy being dissipated in chip formation or wear, unless the surfaces are pre-heated by immersion in boiling water. An alternative technique for friction welding this material is to increase the speed to 3,000 r.p.m. with a corresponding reduction of the load.

With Teflon it was not possible to form a thermal weld even under the most severe conditions of load and speed. Although much wear and chipping of the rubbing surfaces occurred, the dissipation of the frictional energy involved did not prevent the temperature of the interface rising above 300°. The occurrence of temperatures of this order was shown by the onset of thermal decomposition. This resistance to seizure and the low coefficient of friction suggest that Teflon may find many important applications as an 'anti-friction' and 'anti-welding' material in bearings and other sliding mechanisms.

Tungsten carbide [*A*]

Tungsten carbide is a very hard material which has been widely used as a material for cutting-tools. The carbide as prepared commercially consists of very fine crystals bonded together (sometimes in the presence of titanium or tantalum carbide) by a metal such as cobalt. The resulting material consists of a sintered mass of tungsten carbide, titanium carbide, and cobalt, of which the cobalt is the softest constituent. Frictional measurements of some typical 'tungsten carbide' specimens have recently been made by Mr. Shooter. He finds that for the carbide sliding on itself (unlubricated) the friction is low, $\mu = 0{\cdot}2$. For a curved slider of tungsten carbide sliding on steel $\mu = 0{\cdot}6$, while for a curved steel slider on the carbide $\mu = 0{\cdot}45$.

The friction of commercial tungsten carbide sliding on itself is reduced to about $\mu = 0{\cdot}12$ in the presence of a wide range of lubricants, as Barwell and Milne (1948) have recently shown, and the friction is not appreciably different for steel sliding on the carbide. These experiments have all been carried out at very slow sliding speeds where the frictional heating is small. At higher speeds of sliding, such as those occurring between the cutting face of a tool and the chip in a machining operation, much higher surface temperatures will be developed. Under these conditions the frictional and wear behaviour may be substantially different from that described above. The wear or 'cratering' of cutting tools in actual machining operations is described in Chap. XI, pp. 240–4.

Glass [*A*]

The early experiments of Hardy showed that the coefficient of friction of clean glass on glass is about $\mu = 0{\cdot}9$ and that fragments of glass are flaked off the surface during sliding. These fragments are doubly refracting showing that they have been heavily deformed. Here again the results suggest that strong glass–glass junctions are formed at the points of contact and that these junctions are sheared during the sliding process. The view that the frictional mechanism is similar to that of metals is supported by the fact that Amontons's law is obeyed. Further, some experiments of Shaw and Leavey (1930) show that the friction remains essentially constant over temperatures ranging from 20° to 300° C.

The formation of welded junctions is also observed when metals slide on glass. Wooster and Macdonald (1947), for example, have shown that titanium, when slid over glass, adheres to the surface. Adhesion of titanium is also observed on quartz, tourmaline, topaz and other jewels. A photomicrograph by Mr. J. E. Young of a glass surface, after a titanium slider has passed over it three times, is shown in Plate XX. 1. There is marked transfer of titanium on to the glass and similar metallic pick-up is observed with steel and other metals. The nature of the damage produced in the surface of glass when sliding takes place is shown in another photomicrograph taken by Mr. J. E. Young (Plate XX. 2). In this experiment tungsten carbide was used as the slider and the adhesion and friction were both small. It is seen, however, that the stresses produce intermittent cracking of the glass which extends far beyond the boundaries of the track. This may be contrasted with the slip-lines beyond the track which are observed with metals (Chap. IV, Pl. XI; see also Preston, 1922.)

Rubber [*A*]

The friction of rubber has been investigated by Roth, Driscoll, and Holt (1942). They find that Amontons's law is obeyed over a fairly wide range of loads. On glass and steel surfaces the friction of rubber is higher on smooth surfaces than on rough. As with most metals the friction at first falls with increasing speed, but at higher velocities the friction increases with speed. At a sliding speed of about 10 cm./sec. the friction of rubber on steel is as high as $\mu = 4$. However, other workers find that for a wide range of solids sliding on rubber, the friction is of the order of $\mu = 1$. It is, of course, clear that standardization of surface cleanliness will be very difficult with such a material. The compounded ingredients appear to have little effect since Roth *et al.* find that the friction depends mainly on the nature of the rubber matrix rather than on the fillers incorporated in the rubber.

Fibres [*A*]

The friction of fibres is of considerable practical importance. In the spinning of cotton, for example, the friction between the fibres raises a number of technical problems. In certain operations the high electrostatic charge which is generated between the fibres may produce a large electrostatic force between them. Since the mechanical loads applied to the fibres are usually very small, this electrostatic attraction may play a predominant part in the friction between the fibres.

The problem of frictional electricity is not discussed in this book.

It is rather odd that so old a physical phenomenon as the charging of a piece of amber when it is rubbed should still be so little understood. It is not even certain how much the charge is due to 'contact potential' and how much to the frictional rubbing. It is probable that both effects come in. Most of the experimental observations which have been made are complicated by the presence of surface films on the solids and these naturally have a profound influence on the size and magnitude of the charge (see, for example, Richards, 1920, 1923). If the rubbing speeds are appreciable, the local high temperatures developed at the points of rubbing contact will play an important part, particularly with non-conductors such as amber or 'catskin and vulcanite'. These temperatures would be sufficiently high to cause a local chemical decomposition of the solids.

Apart from the possible electrostatic effects there will, of course, be the ordinary frictional interaction between the fibres. With wool fibres the frictional properties play a fundamental part in shrinking and 'felting'. The wool fibre, like all natural fibres (hair, nails, horn, hides), has a very fine scale-like structure at the surface. Measurements have shown that in general the friction from the tip to the root μ_2 (that is against the scales) is greater than the friction μ_1 from the root to the tip. Consequently when wool fibres are rubbed together there is a tendency for them to travel preferentially in the direction of smaller friction and it has been suggested that this process leads to a knotting or 'felting' of wool. This is supported by the general observation that most treatments which reduce felting in practice do in fact reduce the difference between the two coefficients of friction.

Although it is clearly very difficult to prepare wool fibres of the same degree of cleanliness, there is a surprisingly broad basis of agreement in the frictional measurements of various workers. Most measurements are made with one wool fibre (or a bundle of fibres) sliding over another, or with a horn-like material (which also possesses scales) sliding on wool. Over the small range of loads used (about 0·1 to 1 gm.) it is found that Amontons's law is approximately true. For cleaned wool rubbing on horn, for example, $\mu_2 = 0{\cdot}8\text{–}1{\cdot}0$ and $\mu_1 = 0{\cdot}4\text{–}0{\cdot}7$, whilst for greasy fibres $\mu_2 = 0{\cdot}6\text{–}0{\cdot}8$ and $\mu_1 = 0{\cdot}3\text{–}0{\cdot}4$. If the 'directional coefficient' δ is defined as

$$\delta = \frac{\mu_2 - \mu_1}{\mu_1 + \mu_2},$$

it is found that both the friction and the directional coefficient are profoundly affected by the presence of liquids and especially by the pH.

For example, the directional coefficient is considerably increased by wetting the fibre, the increase being appreciably greater in acid and alkaline solutions than it is in natural media. As mentioned above, there is a corresponding change in the felting; the rate of felting is greater in acid and alkali than in neutral solutions. The effect of pH on the friction of fibres has been used by Clark (1940) as a means of measuring the end-point of certain titrations.

Other liquids have specific effects on the frictional properties. For example, Mercer (1945) has shown that alcoholic KOH and $SOCl_2$ may increase the friction but the two coefficients become much more nearly equal so that the directional coefficient is reduced. On the other hand, chlorine and bromine reduce both the friction and the directional coefficient. All these substances are effective in reducing felting. However, mercuric acetate, which is a good antifelting agent, does not produce an appreciable change in the directional coefficient (Lipson and Mercer, 1946). It would seem that, although the directional coefficient is often associated with the felting, it is not the only factor.

A number of workers have attempted to explain the directional friction properties of wool fibres. Thomson and Speakman (1946) coated wool fibres with very thin metal films and found that the directional friction effect still remained. Photomicrographs showed that the scale-like structure was still visible. Thicker metallic films, however, eliminated the effect. This suggests that the directional properties are due to a ratchet-like action of the scales. This view has been strongly supported by Mercer, who has compared the frictional properties of a wool fibre with its structure as revealed by the electron microscope (Mercer and Rees, 1946). It is found that the scale normally protrudes from the surface as a well-defined ratchet, but that after treatment with $SOCl_2$, which reduces the directional effect, the projecting tip of the ratchet has been eaten away. On the other hand, Martin (1944) has suggested that the directional effect is due to an asymmetrical molecular field at the surface of the fibre. Recently Makinson (1948) has developed a more quantitative treatment of the ratchet theory which is of wider interest since it can be extended to the general effect of surface roughness on friction.

Suppose one surface consists of a series of scales or faces making a small angle θ_1 with the general surface (Fig. 61 (*a*)), and the other body rests at a number of well-defined points on these faces. If the true coefficient of friction at each face is μ, then the angle of friction λ, that is the angle of tilt which would just cause sliding, is given by $\mu = \tan\lambda$.

On account of the inclination of the faces the apparent coefficient of friction when sliding occurs from left to right will be

$$\mu_1 = \tan(\lambda+\theta_1). \tag{1}$$

The friction from right to left would similarly be $\mu_2 = \tan(\lambda-\theta_1)$, but this ignores the interaction between the surfaces on the steep faces of the scales.

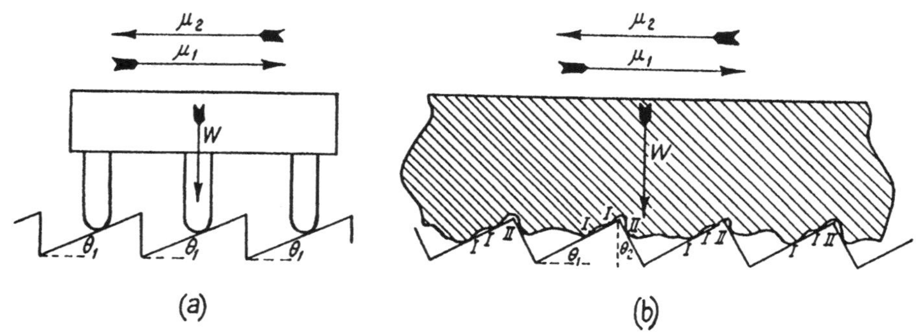

FIG. 61. Effect of asperities on friction of fibres.

Consider the more typical type of contact between a wool fibre and a piece of horn under a given load W (Fig. 61 (*b*)). The load may be assumed to be distributed over contacts of type I on the flat face of a scale and of type II between the steep face of a scale and an asperity or a fissure on the horn. From geometric considerations we may expect that the number of type I contacts will be much larger than type II. For motion with the scales the simple relation given earlier is still approximately valid, so that again $\mu_1 = \tan(\lambda+\theta_1)$. For motion against the scale (since θ_2 is generally smaller than λ) true sliding cannot occur at contacts of type II. Rupture or deformation of the scale or of the asperity on the horn must occur: suppose this requires a total force F for all the type II contacts. The value of F will probably depend more on the geometry of the surfaces than on the load. If m is the fraction of the load W borne by all the contacts of type II, the fraction borne by all the contacts of type I will be $(1-m)$. We may now express the coefficient of friction μ_2 as the sum of two terms. The first is the sliding component on the face of the scales and is equal to $(1-m)\tan(\lambda-\theta_1)$; the second is the ploughing or tearing term representing the force to tear off the tips of the scales and is equal to F/W. Hence the friction in sliding against the scales will be given by

$$\mu_2 = (1-m)\tan(\lambda-\theta_1)+F/W. \tag{2}$$

In equation (2) the unknowns are m, λ, θ_1, and F. From microscopic examination it is known that θ_1 is approximately 5°. Hence if we measure μ_1 we can, from equation (1), calculate λ. From geometric considerations (see above) we may assume that m is small and lies between, say, 0·01 and 0·1. If we measure μ_2 we may now calculate F. It turns out that the value of F obtained does not depend very critically on m although F may be relatively large. This indicates that contacts of type II do not contribute much to the support of the load but they do contribute a large part to the friction against the scales. A check on the method is obtained by lubricating the fibre. This alters λ but should not affect m or F. In fact Mrs. Makinson's measurements show that the values of F obtained for 'clean' fibres agree very well with those obtained for lubricated fibres. In some cases the lubricant layer was 5×10^{-5} cm. thick, so that asymmetric molecular fields could scarcely be effective. Experiments also show that the directional coefficient is greatly reduced by squeezing the scales flat or by chemical treatment which etches away the tips of the scales or weakens them in shear. These observations directly support the ratchet theory.

Surface Irregularities and the Friction of Metals

We may now consider how far this treatment is applicable to the sliding of metals. For simplicity let us consider a two-dimensional model in which a soft metal rests on a harder metal with asperities tilted in one direction at an angle θ_1 to the surface (Fig. 62 (*a*)). The coefficient of friction on the sloping faces is again taken to be equal to μ and the applied load is W. If now we try to slide the upper surface from left to right there will be, in addition to the friction on the sloping faces, a component of the weight W, since we must raise the upper surface in order to allow sliding. Here again the exact treatment shows that the coefficient of friction μ_1 is given by equation (1), which may be expanded to give

$$\mu_1 = \frac{\mu+\tan\theta_1}{1-\mu\tan\theta_1}. \tag{1 a}$$

It is interesting to note that if θ_1 is small and μ is not too large this reduces to

$$\mu_1 = \mu+\tan\theta_1 \tag{1 b}$$

which is the relation suggested by Ernst and Merchant (1940). This value of μ_1 may be considered as the sum of the ordinary friction component μ and the component $\tan\theta_1$ of the weight. The summation is, of course, only valid for small values of θ_1.

These equations may be expected to apply if the asperities are large compared with the area over which metallic junctions are formed. If, for example, the metal flows so that the asperities are completely covered by flowed metal (Fig. 62 (*b*)) it is clear that the frictional force given by equation (1 a) or (1 b) will be greater than the force required to shear the metal in the plane BB'. Consequently shearing will occur

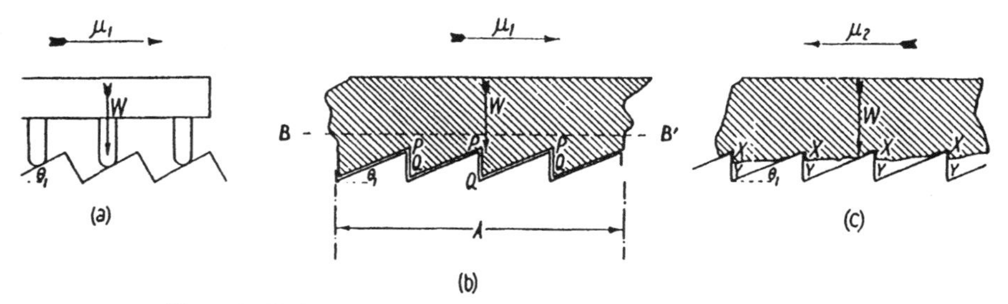

FIG. 62. Effect of surface asperities on the friction of metals.

in this plane and the upper surface will not be raised along the asperities. The friction will now be independent of the surface roughnesses and will in fact revert to the original value μ.

The same arguments apply for sliding from right to left. If the regions of contact are small compared with the surface roughnesses, the frictional component will be approximately $\mu - \tan\theta_1$ if, as above, θ_1 is small. To this we must add the force to shear off the ratchets XY (Fig. 62 (*c*)). It is clear that this part of the frictional force corresponds to the ploughing term in the derivation given in Chapter V. We may calculate the ploughing term in a very simple way if the surface asperities are completely filled as in Fig. 62 (*b*). If the overall area of contact is A, the area to be ploughed out is the sum of all the faces PQ and equals $A\tan\theta_1$. But, as we saw in Chapter I, $A = W/p$, where p is the yield pressure of the softer metal. Hence the area to be ploughed out is $(W/p)\tan\theta_1$. Since the ploughing pressure is approximately equal to p, (see Chap. V) the ploughing force becomes $W\tan\theta_1$, so that it adds to the frictional coefficient a term $\tan\theta_1$. Hence the total friction becomes

$$\mu_2 = (\mu - \tan\theta_1) + \tan\theta_1 = \mu. \tag{2 b}$$

The reason for this result is simple. The model we have just considered is equivalent to shearing across the plane $B'B$, and this, of course, gives the value $\mu_2 = \mu$.

It is clear from these considerations that the separation of the friction into a sliding and ploughing term is possible only when the two processes are more or less independent. This occurs, for example, in the sliding of a hard hemisphere on a softer metal (as in Chap. V). It also probably occurs in the sliding of fibres where the regions of contact are small compared with the size of the scales. In the model considered in Fig. 62 (*b*), however, this separation is artificial and the sliding process becomes almost independent of the surface asperities. It is probable that these conditions apply to most sliding operations between finely ground or polished metal surfaces. For this reason, over a wide range of surface finish, the friction of metals is nearly independent of the degree of surface roughness.

REFERENCES

F. T. Barwell and A. A. Milne (1948), *VIIth Int. Cong. App. Mechanics* (London).
R. E. D. Clark (1940), *J. Soc. Chem. Ind.* **59**, 216.
J. S. Courtney-Pratt (1950), *Research*, in press.
B. Derjaguin and W. Lazarev (1934), *Kolloid Zeit.* **69**, 11.
H. Ernst and M. E. Merchant (1940), *Conf. Friction and Surface Finish*, M.I.T. 76.
R. N. Freres (1945), *Ind. Plastics* (Dec.).
W. E. Hanford and R. M. Joyce (1946), *J.A.C.S.* **68**, 2082.
E. Hutchinson and E. K. Rideal (1947), *Trans. Faraday Soc.* **43**, 435.
R. L. Johnson, D. Godfrey, and E. E. Bisson (1948), N.A.C.A. Tech. Note No. 1578.
M. Lipson and E. H. Mercer (1946), *Nature*, **157**, 134.
K. R. Makinson (1948), *Trans. Faraday Soc.* **44**, 279.
A. J. P. Martin (1944), *J. Soc. Dyers and Col.* **60**, 325.
E. H. Mercer (1945), *Nature*, **155**, 573.
—— and A. L. G. Rees (1946), *Austral. J. Exp. Biol. Med. Science*, **24**, 175.
F. W. Preston (1922), *Trans. Opt. Soc.* **23** (3), 141.
H. F. Richards (1920), *Phys. Rev.* **16**, 290.
—— (1923), ibid. **22**, 122.
F. L. Roth, R. L. Driscoll, and W. L. Holt (1942), *Bureau Stand. J. Res.* **28**, 439.
R. H. Savage (1948), *J. Appl. Phys.* **19**, 1.
P. E. Shaw and E. W. L. Leavey (1930), *Phil. Mag.* **10**, 809.
K. V. Shooter and P. H. Thomas (1949), *Research*, **2**, 533.
G. F. Shotter (1937), *Inst. Mech. Eng. Discussion on Lubrication*, **2**, 140.
V. Stott (1937), ibid. p. 145.
H. M. S. Thomson and J. B. Speakman (1946), *Nature*, **157**, 804.
S. Tolansky (1948), *Multiple-beam interferometry of surfaces and films*, O.U.P.
W. A. Wooster and G. L. MacDonald (1947), *Nature*, **160**, 260.

IX

BOUNDARY FRICTION OF LUBRICATED METALS

. . . it is necessary to know the nature of the contact which this weight has with the smooth surface where it produces friction by its movement, because different bodies have different kinds of friction; because if there shall be two bodies with different surfaces, that is that one is soft and polished and well greased or soaped, and it is moved upon a smooth surface of a similar kind, it will move much more easily than that which has been made rough by the use of lime or a rasping-file.

LEONARDO DA VINCI

IN hydrodynamic or fluid lubrication the surfaces in relative motion are separated by a lubricant layer of appreciable thickness and under 'ideal' conditions there is no wear of the solid surface. The resistance to motion is due entirely to the viscosity of the interposed layer. In practice, however, it is often impossible to obtain fluid lubrication, particularly if the sliding speeds are low or the loads are high. In such cases the thick lubricant layer breaks down and the surfaces are separated by lubricant films of only molecular dimensions. Under these conditions, which Hardy (1936) referred to as boundary conditions, the friction is influenced by the nature of the underlying surface as well as by the chemical constitution of the lubricant. The bulk viscosity plays little or no part in the frictional behaviour.

Boundary lubrication is of great importance in engineering practice. It governs the behaviour of most sliding mechanisms, and in the case where fluid lubrication breaks down, it determines whether serious wear or seizure will take place. The coefficient of friction for unlubricated metal surfaces is about 1·0, whilst the value for surfaces lubricated with boundary films is of the order of 0·05–0·15. This is considerably less than for clean surfaces but is much higher than for fluid lubrication.

FLUID LUBRICATION

In this chapter we shall describe the main characteristics of boundary lubrication and the chief properties of boundary films. Before doing so, however, we may mention briefly some of the more recent trends in fluid lubrication. As Reynolds (1886) showed in his classical hydrodynamic treatment of the problem, fluid lubrication occurs when the pressures developed in a converging film of liquid of given viscosity are sufficient to keep the solid surfaces apart (see, for example, Hersey 1938). In recent years, however, with the development of mechanisms operating under more extreme conditions it has become evident that Reynolds's simple theoretical picture is rarely applicable in its original form. In particular, when bearings are run at heavy loads the thickness of the lubricant film may become comparable with the size of the surface asperities. The simple hydrodynamic theory is no longer valid and the detailed theoretical treatment presents difficulties which have not yet been satisfactorily solved. In addition, the viscosity of the lubricant may be considerably affected by the running conditions themselves. For example, Everett (1937) has shown that the viscosity of most lubricants is modified and usually increased by pressure. As the hydrodynamic pressures developed in modern bearings may be very high, the effective viscosity of the lubricant wedge may be higher than the bulk viscosity. On the other hand, Blok (1946) has shown that the temperatures developed in the lubricant film as a result of the dissipation of viscous energy may appreciably *reduce* the effective viscosity of the lubricant wedge. Finally, with some lubricants, particularly of the synthetic polymer type, the high rates of shear may produce thixotropy in the film, or may lead to a breakdown of the lubricant molecule, and thus to a reduction in the viscosity of the lubricant. All these factors may seriously modify the hydrodynamic behaviour of the system.

In general, the viscosity will be affected simultaneously by the pressure, temperature, and rate of shear. In practice, however, the most marked effect is usually due to temperature. This may be seen from the fact that, even with a high-grade conventional type lubricant, a temperature rise of 10° C. may reduce the viscosity by one-half. For this reason, a great deal of applied research has been devoted to the development of additives which improve the viscosity index of a lubricating oil. Most of these additives are extremely long-chain organic polymers such as polybutenes, polyethylenes, vinyl polymers, polystyrenes, methacrylates, etc. Another line of great promise has been the development of synthetic lubricants, containing silicon, which in themselves show a particularly small variation of viscosity with temperature. These compounds consist of relatively short-chain and ring polymers and their molecules show no tendency to break down physically under high rates of shear. They are, therefore, well suited theoretically for use as hydrodynamic lubricants. In practice, however, silicones have two main disadvantages. Firstly, although they are far more stable than the corresponding hydrocarbons, extremely severe conditions may lead to chemical decomposition, one of the products of which is likely to be silica (SiO_2). This may have very serious abrasive action on the rubbing surfaces. A second and far more serious disadvantage is that they are very poor boundary lubricants. Thus if for any reason thick-film lubrication fails, they are unable to prevent seizure and wear of the rubbing surfaces. For this reason, silicones have been mainly used as hydraulic fluids and as damping fluids. However, by suitable chemical modifications, silicone products have recently been prepared which have slightly better boundary lubricating properties than straight mineral oils. As further progress is made in this direction, it is probable that silicones and other synthetic polymers, such as polyethylene oxides, will find increasing use in a number of practical applications.

In parallel with the development of lubricants possessing improved lubricating properties, there has been an interesting development in the *design* of hydrodynamic bearings. In the past it has generally been considered that the essential condition for fluid lubrication is the existence of an oil *wedge* in which a hydrodynamic pressure can be developed sufficient to keep the two sliding surfaces completely separated. This formed the basis of the lubrication of journal bearings and of the operation of the Michell thrust bearing, and of the more recent Michell pad bearing. Recently Fogg (1945) has shown that a thrust bearing in which the oil film is of uniform thickness may also function very effectively under conditions of hydrodynamic lubrication. It has been suggested that this type of bearing functions by the formation of a 'thermal wedge', but the detailed theory of its action is still the subject of discussion. Nevertheless the Fogg bearing has already been used with considerable success.

Boundary Lubrication by Long-chain Compounds

Influence of chain length

The transition from hydrodynamic to boundary lubrication is relatively gradual. As the sliding speed is decreased (or the load increased) the wedge of lubricant separating the surfaces becomes thinner and thinner and the number of surface asperities penetrating the film becomes correspondingly greater. At the tips of these asperities the lubrication is of a boundary nature, so that the amount of boundary lubrication increases gradually whilst the amount of fluid lubrication decreases. This type of 'mixed' lubrication extends over a very wide range of experimental conditions, and experiments by Beeck, Givens, and Smith (1940), Forrester (1946), Kenyon (1946), and others show that even at sliding speeds of a few cm./sec. an appreciable fraction of the load is supported by a hydrodynamic film. For this reason investigations of boundary lubrication necessitate the use of experimental methods where the pressures are extremely high and the sliding speeds are also very low. These conditions are satisfied in the friction apparatus described in Chapter V, where the pressure between the surfaces is of the same order as the yield pressure of the metals and the sliding speed is of the order of 0·01 cm./sec.

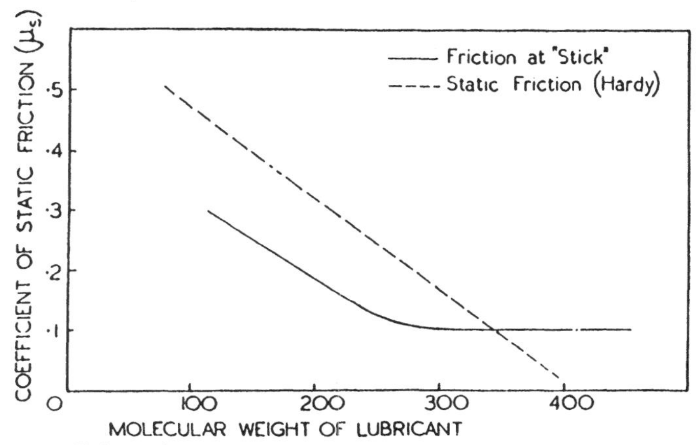

FIG. 63. Coefficient of friction as a function of molecular weight. Straight-chain paraffins on steel surfaces.

Experiments were carried out with homologous series of paraffins, alcohols, and fatty acids. The lubricants when liquid were applied as a thin smear; when solid they were applied in a volatile solvent. The friction results are somewhat complicated by the fact that for paraffins and alcohols the motion was intermittent for compounds which were liquid at room temperature and smooth for solid lubricant films. With fatty acids (except for the short-chain acids which gave marked corrosion) the motion was smooth, even when the acid was liquid. As with intermittent motion on unlubricated surfaces, the friction at the stick corresponds to the static friction and refers to conditions far more reproducible and precise than those occurring during the rapid 'slip'. In what follows, therefore, whenever the motion is intermittent we shall consider the maximum friction at the stick. In Figs. 63, 64, and 65 the coefficient of friction μ_s has been plotted against the chain length for paraffins, alcohols, and fatty acids on steel respectively. It is seen that in all cases there is a steady decrease in friction with increasing chain length. Similar results for the static boundary friction were obtained by Hardy and are shown in the dotted lines. It is seen that, according to Hardy, at a sufficiently long chain length the coefficient of friction approaches zero; the results given here, however, show that the friction reaches a lower limit of about $\mu = 0{\cdot}07$ however long the chain may be. In agreement with Hardy's observations it was found that essentially

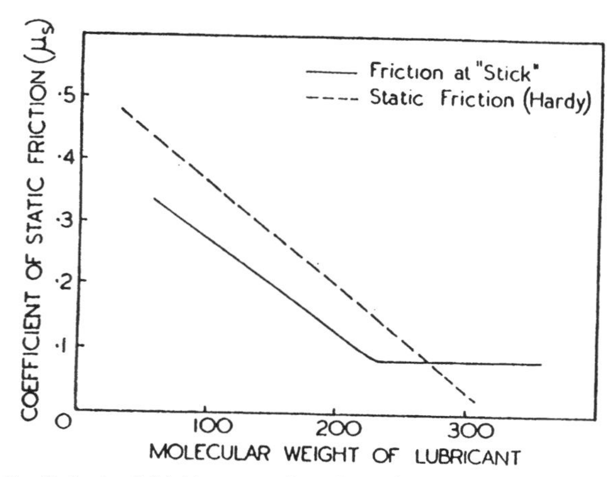

FIG. 64. Coefficient of friction as a function of molecular weight. Saturated straight-chain alcohols on steel surfaces.

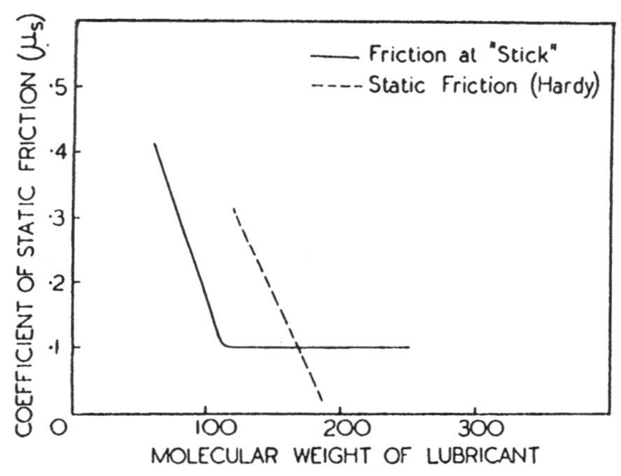

FIG. 65. Coefficient of friction as a function of molecular weight. Saturated fatty acids on steel surfaces.

the same results were obtained whatever the surface finish of the metal surfaces.

With these lubricants there is a fairly direct correlation between the friction and the amount of surface damage. Using paraffins, alcohols, and fatty acids with chain lengths sufficiently long to give the minimum value of the friction, the track formed is relatively light and the surface shows very small surface damage. With shorter chain molecules where the friction increases, the track formed is of approximately the same width but it is more deeply grooved and there is heavier surface damage. It is important to note, however, that even under the most favourable conditions some surface damage is always observed. This is shown very clearly in Chapter IV, Plate IX, where the electrographic surface analysis and the radio-active tracer technique show the existence of metallic pick-up and surface damage through the lubricant film.

Effect of temperature [*A*]

The friction between metal surfaces sliding under conditions of boundary lubrication does not, in general, vary continuously with the temperature. There is usually a sharp rise or transition in the friction at a certain temperature, with a corresponding increase in the wear and damage of the surfaces. Provided the heating has not been too high to cause appreciable oxidation of the lubricant, these changes are reversible on cooling.

For pure paraffins and alcohols the transition is well defined and occurs at the bulk melting-point of the lubricant. For example, docosane (m.p. 43° C.) gives a low friction and smooth sliding until a temperature of 43° C. is reached, when there is a sudden rise in friction. Similarly, cetyl alcohol (m.p. 49·5° C.) gives a transition temperature of 50° C. These transition temperatures are independent of the nature of the underlying metal, and it would seem that the breakdown occurs at the melting of the boundary film.

The behaviour of these lubricant films is thus similar to that observed in the lubrication of surfaces by thin metallic films. In both cases, effective lubrication is provided until the melting-point of the film is reached. It is clear that in these cases the main function of the lubricant film is to diminish the amount of intimate metallic contact between the sliding surfaces. With hydrocarbon films the *lateral* adhesion between the molecules in the boundary film is of primary importance in protecting the surfaces. When the temperature is raised and melting occurs, the lateral adhesion is diminished and breakdown of the film takes place. The increased metallic contact which now occurs through the lubricant film leads to an increase in the friction and the amount of surface damage.

The behaviour of fatty acids is significantly different. These also show a transition effect, but in general the transition temperature (although dependent somewhat on the experimental conditions) is very much higher than the bulk melting-point of the fatty acid (Tabor, 1940, 1941). For example, lauric acid (m.p. 43° C.) will lubricate zinc surfaces effectively up to a temperature of about 110° C. This is shown in Fig. 66 where the friction is plotted against the temperature and shows a rapid increase above 100° C. This transition is accompanied by increased surface damage and wear, as may be seen in Plate XX. 3 *a* and 3 *b* for zinc surfaces below and above the transition temperature

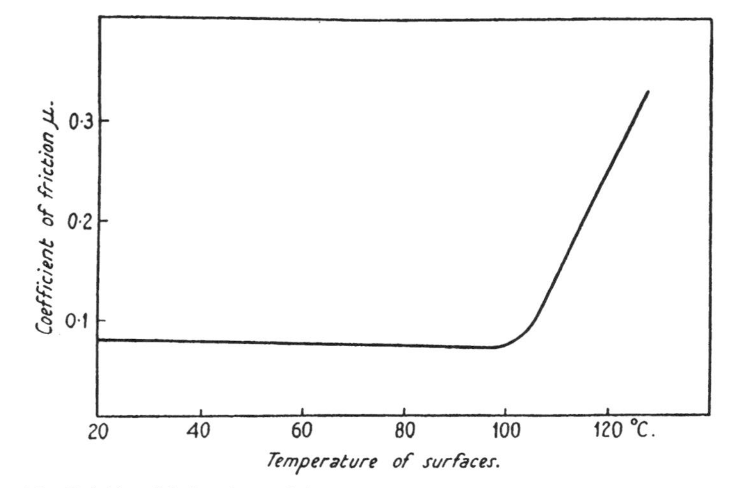

FIG. 66. Frictional behaviour of zinc surfaces lubricated with 1 per cent. solution of lauric acid in paraffin oil, as a function of temperature. The friction increases rapidly above 110° C.

(Gregory, 1943). The effect is reversible, and on cooling the surfaces the friction falls to its original low value and the surface damage becomes very slight.

The effect of temperature on the breakdown of the lubricant film is also shown very graphically by some experiments recently carried out by Mr. E. Rabinowicz using the radio-active technique described in Chapter IV. The lower surface was of copper and the upper surface was a radio-active copper slider. The lubricant was palmitic acid and the friction was recorded as the temperature of the surfaces was raised to 150° C. After the run the lower surface was placed against a photographic plate and produced a record of the amount of radio-active copper transferred from the slider. The results are shown in Plate XX. 4, where the friction trace has been reproduced on such a scale that there is point-to-point correspondence between the friction and the track. It is seen that the friction at room temperature is low, the motion smooth,

Mg stearate 200c
stearic acid 69°
Zn stearate 120-130
Cu stearate 112
Cd stearate 134

and the amount of pick-up very small.† At 11(
rapid increase in friction and a large increase ir
Each large 'stick' in the friction above this te
with a large amount of metallic transfer. I
reversible, and on cooling the surface the frict
small. At higher temperatures or under more
versible effects may occur largely as a result
reversible and irreversible changes which occ
are heated are of considerable practical importa
in greater detail in a later chapter.

Fatty acids in solution

Several workers have observed that the addition of a small trace of a fatty acid to a non-polar mineral oil or to a pure hydrocarbon can bring about a considerable reduction in the friction and wear. This is demonstrated very strikingly in Fig. 67 for cadmium surfaces lubricated with a highly refined paraffin oil. The friction is very high and at the 'stick' reaches a value of about $\mu = 0{\cdot}6$. At the point A a small amount of lauric acid was added to the surfaces and it is seen that there is a sudden drop in friction to a value of about $\mu = 0{\cdot}07$. The tracks show a corresponding decrease in the amount of surface damage.

Gregory (1943) has investigated the minimum concentration of fatty acid necessary to provide effective lubrication. Using cadmium surfaces he found that a 1 per cent. solution of lauric acid in paraffin oil gave a coefficient of friction of about $\mu = 0{\cdot}05$. A concentration of 0·01 per cent. gave a coefficient of friction of about $\mu = 0{\cdot}1$, whilst a 0·001 per cent. solution gave a large friction ($\mu = 0{\cdot}45$) comparable with that observed with pure paraffin oil. With this dilute solution, however, the friction decreased slowly with time and after 12 hours it had fallen to a value of about $\mu = 0{\cdot}26$. (See also Hardy, 1936.)

This investigation has been extended by Gregory to estimate the thickness of the fatty acid film responsible for effective boundary lubrication. A small quantity of 0·1 per cent lauric acid in paraffin oil was placed on a clean cadmium surface and gave a low coefficient of friction. The area of the cadmium surface covered by the oil was increased in stages using a micro-pipette until the newly covered regions gave increased friction and surface damage. The total area covered by

† The radio-activity of the slider in this experiment was not sufficient to show appreciable metallic pick-up at room temperature. With a more strongly irradiated slider, however, some metallic interaction is observed through the lubricant film (see Chap. IV, Pl. IX. 3 *b*).

the lubricant at this stage was measured. From these observations it was possible to calculate the amount of lauric acid which was just sufficient to give good boundary lubrication. The thickness was found to be of the order of 1 to 2 molecular layers. Thus the effective lubrication observed with the fatty acid is due to a boundary film, 1 or 2 molecules thick.

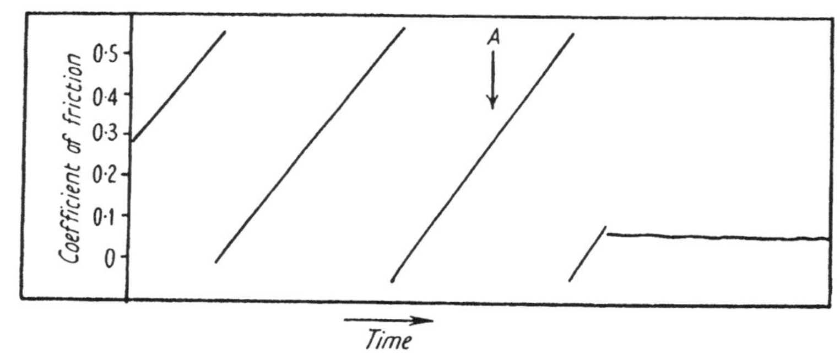

FIG. 67. Frictional behaviour of cadmium surfaces lubricated with a highly refined paraffinic oil. A drop of lauric acid is placed on the surfaces at the point *A* and there is a marked change in the friction and in the character of the motion.

These results emphasize the extremely marked effect of very small quantities of fatty acids on the boundary properties of lubricants. There is little doubt that the effectiveness of many mineral oils as boundary lubricants is due to the presence of minute quantities of fatty acids or similar materials. (See Southcombe and Wells, 1920.)

LUBRICATING PROPERTIES OF MONOLAYERS AND MULTILAYERS

It is interesting to consider the frictional behaviour of metal surfaces covered with thin films of molecular dimensions and of known thickness. In this investigation (Bowden and Leben, 1939) the films were deposited by the Langmuir–Blodgett technique. Although there is some uncertainty as to the exact structure of these films (see, for example, Bikerman, 1939), this technique is a very convenient one for the deposition of films of known and controllable thickness.

In these experiments the surfaces were of stainless steel and two substances were used as lubricants. One was a long-chain fatty acid (stearic acid) and the other a large flat polar molecule (cholesterol). The fatty acid film was first spread as a monolayer on tap water and then picked up in the standard way on the metal surface by successive dippings. Later work has shown that the film deposited in this way consists not of pure acid but of a mixture of stearic acid and calcium

The film, however, is close packed and regularly oriented e polar group in the water surface. The cholesterol, which is a lat molecule containing a complex ring system and a single OH .p, was spread in the same way on water. It forms a stable con- nsed film with the OH group in the water and the molecule oriented .ormally to the surface. As we shall see, although the cholesterol film is very stable and well oriented it is far less effective as a boundary lubricant than the stearic acid film.

The films were deposited either on the upper or lower surface. In the first set of experiments the small upper curved slider was covered with a known number of films while the lower surface was unlubricated. The lower surface was then started moving at a speed of 1·0 cm./sec. and the friction recorded from the beginning of the sliding.

In the second set of experiments a known number of films was deposited on the lower plate while the upper surface was kept unlubri- cated. The surfaces were allowed to slide together and the friction recorded at intervals. The upper slider was then run repeatedly over the same track on the lower surface and the friction during each run observed. The process was continued for 100 runs, or until the friction had risen to a very high value and the surfaces were badly torn.

Stearic acid films

1. *Films on upper surface only.* Fig. 68 shows the friction results obtained with various numbers of molecular films on the upper surface and these results are summarized in Fig. 69. It is seen that in every case immediately after motion began the friction was low at about $\mu = 0{\cdot}1$. The friction, however, soon began to rise, the rate of increase in friction being more marked the fewer the number of films on the upper surface. Photomicrographs of the tracks formed in the lower surface showed a corresponding increase in the amount of tearing and surface damage as sliding proceeded. In these experiments the wear on the upper slider was concentrated on one small area of the surface and it is clear that this gradual increase in friction and surface damage is associated with the removal of the protective film from this region. Experiments carried out at much slower speeds of sliding (0·001 cm./sec.) showed that the rate of wear of the film per cm. of track was approximately the same as when the surface speed was 1 cm./sec.

2. *Films on lower surface.* When two surfaces slide together under these conditions fresh portions of the film are continually coming under the upper slider, and consequently during any particular run over the

surface no change in the friction could be observed. When only one film was present on the surface (Plate XXI. *a*) the coefficient of friction

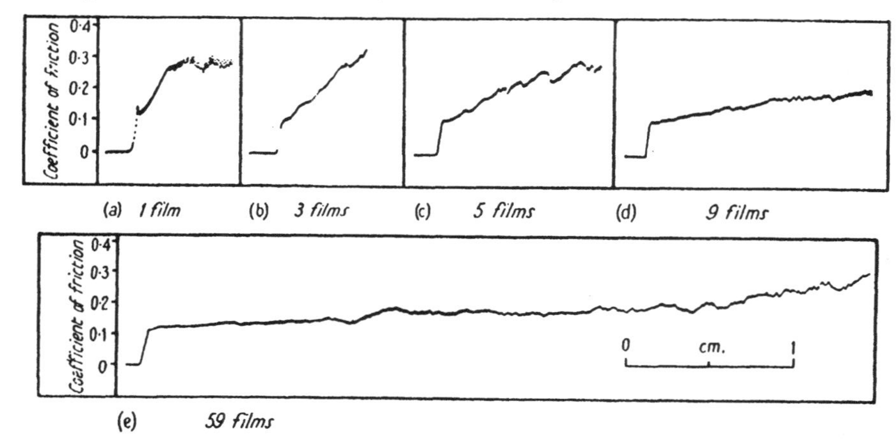

Fig. 68. Friction traces of stainless steel sliding on stainless steel when the upper surface or slider is covered with stearic acid films of various thicknesses. The horizontal ordinate corresponds to the distance traversed by the slider. The monolayer is worn away very rapidly compared with the polymolecular layer.

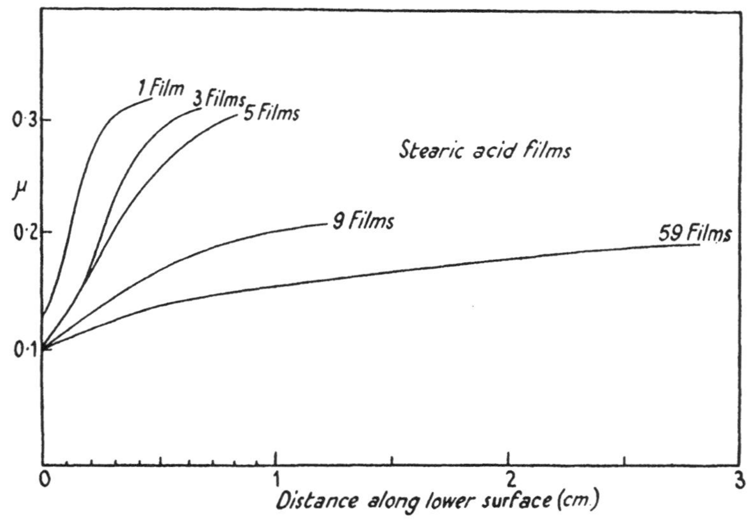

Fig. 69. Wear of stearic acid films deposited on upper surface or slider (stainless steel surfaces). Results of Fig. 68 plotted on a single figure. The thicker films wear much more slowly than the thin films.

during the first run was low ($\mu = 0{\cdot}1$) and was similar to that observed when a large number of films (e.g. 53) was present (Plate XXI. *b*). The track formed during the first run is shown in Plate XXI. *c*, and it is seen that there is very little surface damage. However, on repeatedly running over the same track, the friction soon began to rise and eventually attained the high value characteristic of unlubricated surfaces (see Fig. 70). At the same time the wear between the surfaces increased greatly, and the appearance of the track after 20 runs is shown in Plate XXI. *d*. It is clear that as a result of the repeated sliding the

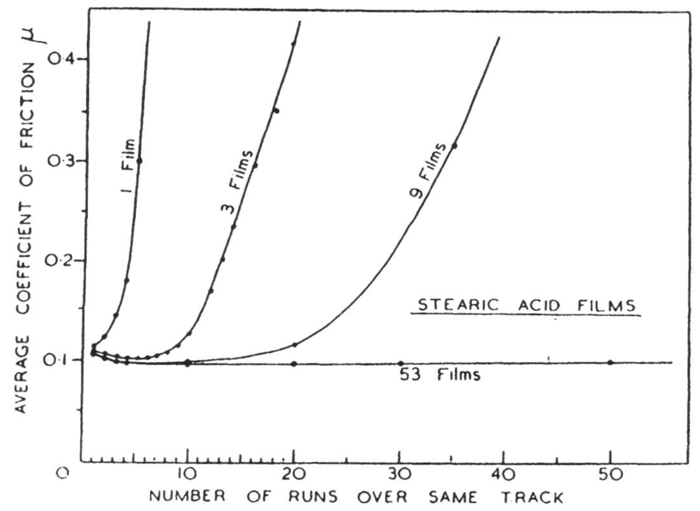

FIG. 70. Wear of stearic acid films deposited on the lower surface (stainless steel surfaces). A single molecular layer produces the same reduction in friction as a thick film but it is worn away far more rapidly.

protective film is rapidly worn off the lower surface, with a corresponding increase in the friction and in the amount of surface damage.

Similar results were obtained in the presence of other thicknesses of film, but the curves given in Fig. 70 show that the thicker films wear away less rapidly. Eventually, when a sufficiently large number of films was present, no rise in friction was observed even after 100 runs over the lower surface. The coefficient of friction during the first run on a film 53 molecules thick was $\mu \approx 0{\cdot}1$ (Plate XXI. *b*) and the corresponding surface damage, which was slight, is shown in Plate XXI. *e*. These results show that both the friction and surface damage are not very different from those obtained during the first run on only one film. The track formed after 100 runs on the 53 films is illustrated in Plate XXI. *f*, and it will be seen that there was very little tearing of the surfaces. This result stands in marked contrast to the extensive wear observed after only 20 runs on the single film, Plate XXI. *d*.

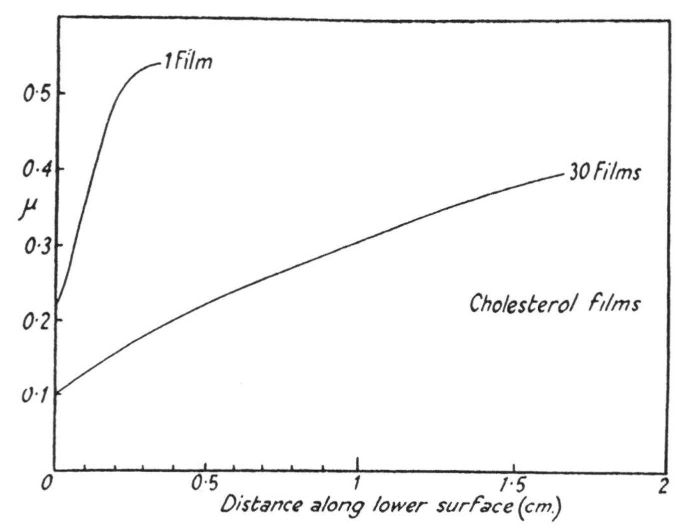

FIG. 71. Wear of cholesterol films deposited on upper surface or slider (stainless steel surfaces). The films are worn away more rapidly than stearic acid (see Fig. 69).

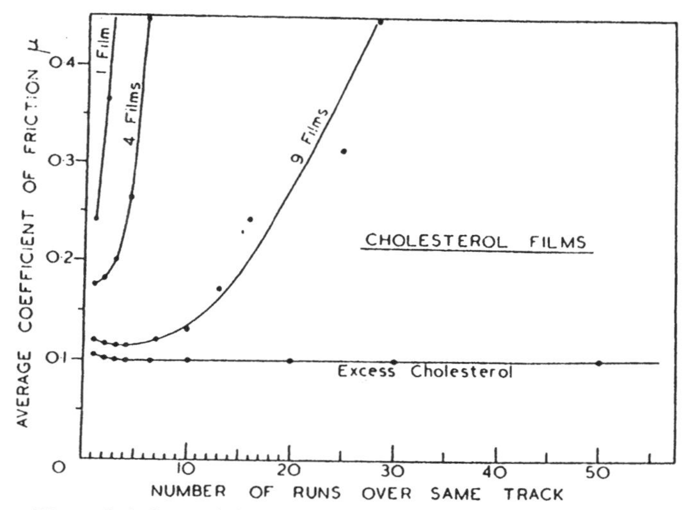

FIG. 72. Wear of cholesterol films deposited on the lower surface. (Stainless steel surfaces.) A single molecular layer does not produce as great a reduction in friction as a thick film and the wear-resisting properties are inferior to those of stearic acid (see Fig. 70).

Cholesterol films

The experiments were repeated with cholesterol films, and the main friction results obtained are plotted in Figs. 71 and 72. It will be seen that in many ways the behaviour was similar to that observed with 'stearic acid' films. With films on the upper surface only, a rapid rise in the friction and wear was observed during the sliding, and the rate of this rise was dependent on the number of films present. When the films were placed on the lower surface the friction and wear remained unchanged during each run, but, as before, showed an increase during subsequent runs. The rate at which the lubricating layer wore off again depended on the number of molecular films of which it was composed.

One important difference in the behaviour of these two lubricants will be seen, however. With 'stearic acid' a unimolecular film was able to reduce the friction almost to the same value as that observed when many films were present. With cholesterol, however, the friction for a unimolecular film was comparatively high and the minimum value was only attained when several films had been deposited on the surface.

Wear properties of lubricant layers

Three interesting conclusions follow from these experiments. The first is that the rate of wear of the film is roughly independent of the speed of sliding, for speeds ranging from 0·001 to 1 cm./sec. It is clear that the rise in friction depends essentially on the relative distance moved by the surfaces and not on the time during which they have been sliding together.

The second conclusion is that as the thickness of the lubricant film increases, there is a gradual decrease in the rate at which this film wears off the surface. This effect has been generally observed by other workers, especially Langmuir (1934) and Claypoole (1939).

The third conclusion is that substances which give the same coefficient of friction are not necessarily equally efficient as lubricants. Thus a comparison between Fig. 72 for cholesterol films and Fig. 70 for stearic acid films shows that with cholesterol the lubricant film wears off more rapidly than with stearic acid. It is apparent, therefore, that stearic acid is a better boundary lubricant than cholesterol under these conditions. Similar differences have been observed between various fatty acids. Dacus, Coleman, and Roess (1944) have recently developed a special apparatus for measuring the rate at which a lubricant film is worn away during sliding. Their results show that with monolayers of fatty acids there is a marked increase in the 'durability' of the lubricant film with increasing chain length.

Minimum film thickness for effective lubrication [*A*]

The direct experiments of Gregory quoted earlier in the chapter show that in the lubrication of cadmium surfaces by lauric acid the lubricating film is of the order of 1–2 molecules thick. It is, of course, difficult to carry out systematic experiments on the minimum film thickness for effective lubrication by any method other than that involving the use of the Langmuir trough, since, as has already been pointed out, this provides the most direct means of depositing films of known thickness. Although there is some evidence that the deposited films are less robust than adsorbed monolayers (see Chap. X), experiments carried out with deposited films generally confirm conclusions derived from other experiments. For example, the results described above show that with stearic acid films deposited on steel surfaces from the Langmuir trough a single monolayer produces a coefficient of friction almost as low as stearic acid applied in bulk. This result directly confirms the early observations of Langmuir (1920), who was the first to show that a monolayer of fatty acid deposited on glass surfaces was sufficient to reduce the friction from about $\mu = 1{\cdot}0$ for clean glass to about $\mu = 0{\cdot}1$. Similar results have been obtained by Isemura (1940), Hughes and Whittingham (1942), Frewing (1942), and other workers with films deposited from the Langmuir trough.

In some cases, however, a monolayer of fatty acid is insufficient to provide adequate lubrication even for a single run. Some recent experiments, for example by Gregory and Spink (1947), and especially by Greenhill (1949), on molecular layers of stearic acid deposited from the Langmuir trough, are summarized in Table XXV.

TABLE XXV

Lubrication of metal surfaces by layers of stearic acid and metal stearates deposited from the Langmuir trough (pH 9·5)

Metal	*No. of layers for effective lubrication*	
	Stearic acid	*Soap* (Cu *or* Ag *stearate*)
Platinum	> 10	7–9
Stainless steel	3	1
Silver	7	3
Nickel	3	3
Cobalt	..	1
Copper	3	3

It is seen that with some metals 1 or 3 monolayers give good lubrication. With others, such as platinum, at least 7 molecular layers must be present for effective lubrication. These results fall in line with the earlier investigations of Bowden and Hughes (see Chap. VII) on the friction of outgassed gold. In the absence of all surface films the friction was high. When a trace of caproic acid vapour was admitted to the surface, a slight decrease in friction occurred which decreased with time as a thicker film of lubricant formed on the surface.

With materials other than fatty acids or metallic soaps, the film thickness necessary for effective lubrication is usually considerably more than a single molecular layer. For example, as we saw above, with cholesterol on stainless steel surfaces, 9 molecular layers are necessary. Similar observations have been made by Isemura for long-chain alcohols and esters deposited on glass surfaces.

It is clear from these results that the lubricating properties of boundary films depend on the nature of the underlying surfaces as well as on the lubricant itself. In many cases the first monolayer of the lubricant is responsible for the lubrication observed. This single layer is usually unable to provide adequate protection for continuous traversals of the same track; for effective lubrication in such cases a multilayer of lubricant should be present and the thicker the film the slower the rate at which it will be worn away. The best protection is afforded by an excess of lubricant on the surface. In some cases a single molecular layer is unable to give effective boundary lubrication even for the first run; a comparatively thick film is necessary to produce smooth sliding and low friction. The rate of wear of such films is usually relatively high.

Lubricating Properties of Silicones and of Fluorinated Hydrocarbons (Fluorolubes)

It is interesting to consider briefly the boundary lubricating properties of non-hydrocarbon lubricants such as the silicones and the fluorolubes. The silicones consist of long-chain polymers containing chains, rings, or three-dimensional networks of silicon and oxygen atoms.

```
       R1      R1                  R1     O     R1
       |       |                     \  /   \  /
  —O—Si—O—Si—O—                     Si       Si
       |       |                     /  |       |  \
       R2      R2                  R2   O       O   R2
                                         \     /
                                           Si
                                          /  \
                                        R1    R2
```

These substances may be obtained in various ranges of viscosity according to the nature of the polymer and its length, and the viscosity is very much less dependent on temperature than is the case with petroleum oils. In addition the silicones are more stable to heat, oxidation, and chemical attack, and they are in fact capable of providing anti-corrosion protection to most metal surfaces. For these reasons silicones are of considerable interest as hydrodynamic lubricants or hydraulic fluids. Unfortunately their boundary properties are very poor. This is shown by some recent measurements by Dr. Tingle (1948), who determined the friction between metal surfaces lubricated with silicones ranging in viscosity from 20 centistokes (at 20° C.) to 1,000 centistokes (at 20° C.). With copper surfaces the friction for the lubricated surfaces was $\mu = 1{\cdot}4$ and there was very heavy tearing of the surfaces. The friction was not markedly dependent on the viscosity of the silicone lubricant. With steel surfaces the friction was somewhat lower, of the order of $\mu = 0{\cdot}4$ to $\mu = 0{\cdot}8$, but the lubrication was always worse than that of a highly refined mineral oil such as medicinal paraffin. Only one silicone, which was stated to be a copolymer of dimethyl and diphenyl siloxane, gave reasonably satisfactory lubrication. The coefficient of friction at room temperature on steel surfaces was about $\mu = 0{\cdot}2$ and the surface damage was less marked. At 100° C., however, the lubrication had rapidly deteriorated and the friction was high, $\mu = 0{\cdot}5$. Under certain conditions, for example with steel sliding on sapphire, as we saw in Chapter VIII, silicones may actually produce an increase in the friction if they inhibit the oxidation of the surfaces during sliding.

It is clear from these results that silicones are poor boundary lubricants. This is to be expected from their structure. As we shall see later, effective lubrication is provided only if the lubricant film can react with the surface to form a surface layer with strong lateral adhesion between the molecules. On account of their marked chemical stability the silicones are unable to react with the metal surface. In addition, the lateral cohesive forces between the silicone molecules are relatively weak.

It is possible by suitably 'tailoring' the silicone molecule to provide it with a means of attaching itself more firmly to the metal surface. In this way more effective lubrication may be obtained. For example, Gregory and Newing (1948) have shown that polymers formed by the hydrolysis of alkyl chlorosilicones possess very good lubricating properties on copper, silver, and other metals. Similarly, Hunter *et al.* (1947) have shown that dichlorosilanes effectively lubricate glass surfaces. There are, however, complicating factors which militate against the use of these materials in practice. It is evident that there is still considerable scope for further work in the development of silicones which have good boundary properties while still retaining the desirable bulk properties of silicones. An alternative approach which has already met with some success is the development of good boundary lubricants which can be dissolved in small quantities in the silicone.

Another type of non-hydrocarbon lubricant of considerable interest is the class known as 'fluorolubes'. These consist of hydrocarbon molecules in which most, or all, of the hydrogen is replaced by fluorine. These compounds often possess high chemical and thermal stability and are non-inflammable. We are indebted to Professor M. Stacey, who is largely responsible for the development of these compounds, for supplying us with some of them. Dr. Tingle has recently examined their lubricating properties and finds that in general they are effective on steel surfaces over a temperature range of 20°–150° C. For example, a typical fluorolube possessing an average molecular weight corresponding to the formula $C_{21}F_{44}$ gave a value of $\mu = 0{\cdot}1$ at 20° C. and a value of $\mu = 0{\cdot}15$ at 100° C. A typical chlorolube, made by the action of chlorine on ethyl-benzene under intense ultra-violet light, gave even better results. For example, on steel surfaces $\mu = 0{\cdot}12$ at 20° C. and $\mu = 0{\cdot}05$ at 200° C. It is possible that at higher temperatures the chlorolubes react with the surface to form the metallic chloride which helps to provide good lubrication (see Chap. XI). It is doubtful whether this occurs with the fluorolubes since the molecules are far too stable. It would seem that the fluorolubes in themselves are intrinsically fairly good boundary lubricants. This may be connected with the observation described in Chapter VIII that polymerized fluorinated hydrocarbons which are solid at room temperature (e.g. Teflon) have exceptionally good frictional properties with values of μ of the order of $\mu = 0{\cdot}05$. However, until more is known of the structure and composition of the fluorolubes it would be difficult to provide a more complete explanation. The results, however, suggest that under conditions where extreme chemical inertness is required, the fluorolubes may find many practical applications as boundary lubricants.

Influence of Load and Speed on Friction of Lubricated Surfaces

In the experiments described in the earlier part of this chapter the frictional measurements on paraffins, alcohols, and fatty acids were carried out at a constant speed (about 0·01 cm./sec.), and a constant load (about 4 kg.). It is interesting to consider the way in which the frictional behaviour depends on the load and speed.

Effect of load

Frictional measurements were carried out at loads ranging from 500 to 6,000 gm. Exactly as for unlubricated surfaces, it is found that the main effect of increasing the load is to increase the magnitude of the frictional force. If the motion is smooth, as for fatty acids, the frictional force is directly proportional to the load, so that Amontons's law is accurately obeyed. A typical result for pelargonic acid on steel is given

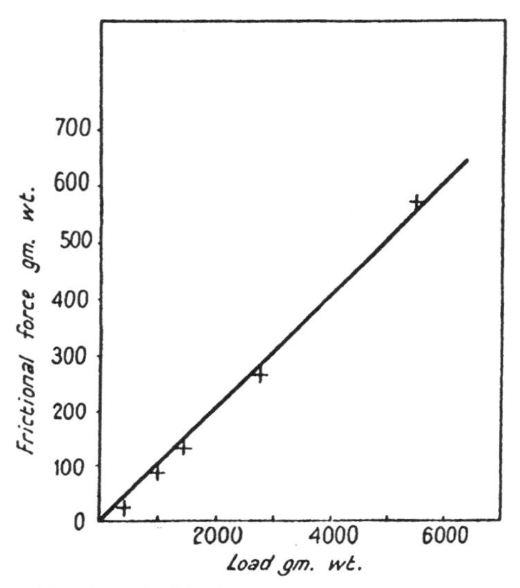

Fig. 73 (a). Effect of load on the friction of steel surfaces lubricated with pelargonic acid. The frictional force is directly proportional to the load.

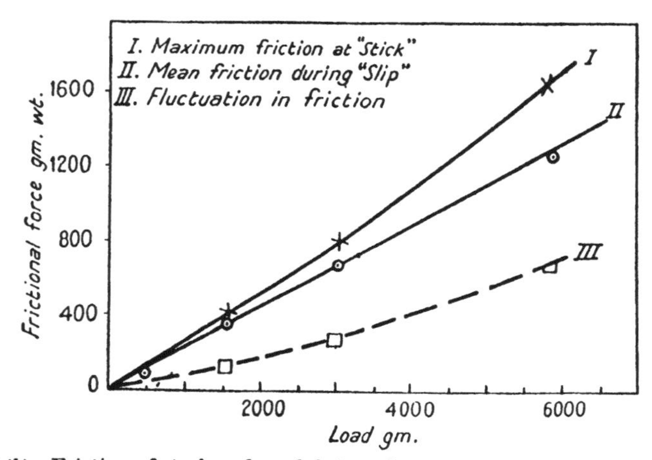

Fig. 73 (b). Friction of steel surfaces lubricated with decane as a function of load. The friction and the fluctuations increase proportionately with the load.

in Fig. 73 (*a*). If the motion is intermittent, the maximum friction, the mean friction, and the size of the fluctuations increase proportionately with the load. A typical result for decane on steel is shown in Fig. 73 (*b*).

These results show that, as with unlubricated surfaces, increasing the load causes the whole friction process to take place on a proportionately increasing scale.

Friction of lubricated surfaces at very light loads

Under most conditions of sliding, penetration of the lubricant film occurs. For this reason it is of considerable interest to examine the friction at very light loads, where the penetration of the film may be

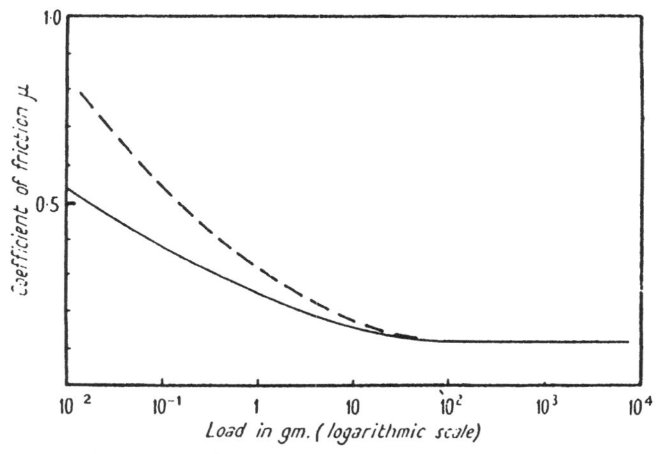

FIG. 74. Friction of copper surfaces lubricated with fatty acids as a function of load. Continuous curve: lauric acid; broken curve: octacosanoic acid. Amontons's law is obeyed for loads above about 10 gm., but at small loads the coefficient of friction steadily increases as the load is reduced. It is seen that at a load of 10^{-2} gm., $\mu > 0{\cdot}5$. For unlubricated surfaces at this load, $\mu = 0{\cdot}4$ (see Fig. 57), so that the friction for the lubricated surfaces is actually higher than for the unlubricated surfaces.

avoided and a change in the frictional behaviour may be expected. These experiments have been carried out by Dr. Whitehead (1950) using the apparatus described in Chapter IV, Fig. 29. He finds that with copper surfaces lubricated with lauric acid the coefficient of friction at a load of 10 gm. is $\mu = 0{\cdot}1$, at 100 gm. $\mu = 0{\cdot}1$, and on a heavier apparatus at a load of 4,000 gm. the friction has again a value of about $\mu = 0{\cdot}1$. Thus Amontons's law holds over a load range of 400 or more. At loads below 10 gm., however, there is a change in the frictional behaviour and a marked deviation from Amontons's law is observed. The coefficient of friction steadily increases as the load is reduced and at a load of 0·01 gm., $\mu = 0{\cdot}5$ for lauric acid, while for octacosanoic acid $\mu = 0{\cdot}8$ (see Fig. 74). With unlubricated surfaces, as we saw in Fig. 57, the coefficient of friction at these loads is of the order of $\mu = 0{\cdot}4$, so that with octacosanoic acid the friction of the lubricated surfaces is actually higher than for the unlubricated surfaces. This exceptional behaviour is probably due to the fact that at these very small loads the deformation of the surface is insufficient to produce penetration of the lubricant film (see Chap. X) and we are measuring the interaction of the hydrocarbon film itself. Nevertheless, for loads above about 10 gm. the frictional behaviour is independent of the load and Amontons's law is closely obeyed.

Effect of speed

When the motion is smooth the friction shows little change when the speed varies from 0·001 cm./sec. to 2 cm./sec. A typical result for pelargonic acid on steel is shown in Fig. 75.

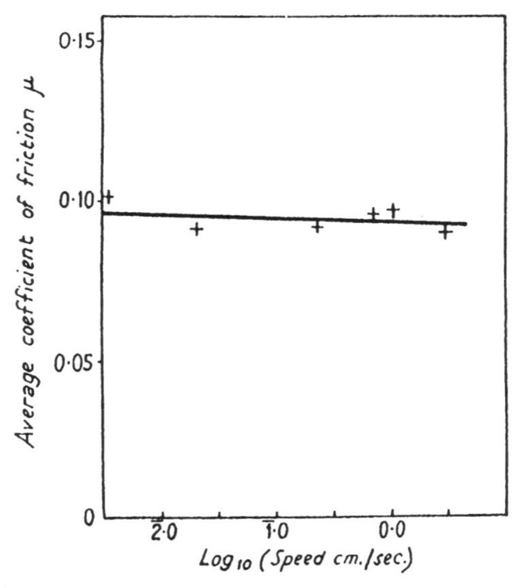

FIG. 75. Friction of steel surfaces lubricated with pelargonic acid. The coefficient of friction is almost independent of the speed.

When the motion is intermittent, as with decane on steel, the behaviour is different (Fig. 76). The friction at the 'stick' decreases relatively rapidly with increasing speed. The mean friction during the 'slip' decreases more gradually. Consequently as the speed increases the friction at the stick approaches the mean friction during slip, with a corresponding decrease in the size of the fluctuations. This result is consistent with the discussion in Chapter V on intermittent motion; as the surface speed increases the processes during stick approach more nearly the processes occurring during slip, so that the fluctuations steadily diminish in size. The converse of this is equally true. If the friction–speed characteristic is flat (as in Fig. 75) the sliding will always be smooth. If it shows a marked falling characteristic, the motion may be intermittent if the elastic constants of the moving parts are suitable (Bristow, 1942). The fact that intermittent motion can occur with lubricated surfaces is of great importance in many practical operations where it may be essential to avoid vibration. The experiments show

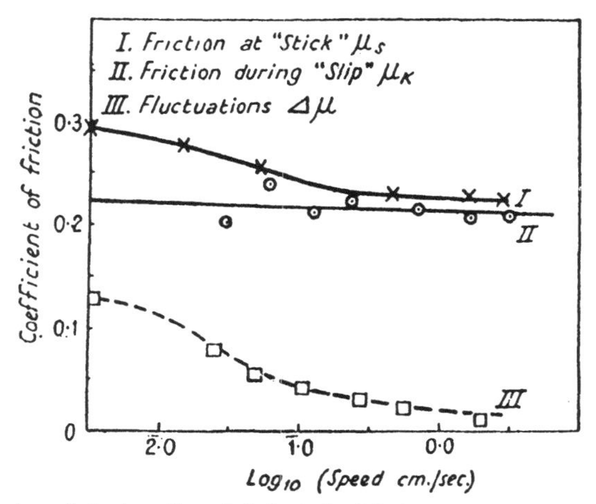

FIG. 76. Friction of steel surfaces lubricated with decane as a function of speed of sliding. The friction during the slip (kinetic friction) is almost independent of the speed, but the friction at the stick (static friction) decreases as the speed increases and gradually approaches the value of the kinetic friction.

that lubricants containing fatty acids are much more likely to provide smooth sliding than lubricants containing only paraffins or alcohols.

There is, however, an additional factor with lubricated surfaces that makes intermittent motion more general than may at first sight appear. If 'slip' does occur the motion is usually relatively rapid. Under these conditions the oil film may be drawn between the sliding surfaces and provide, at least partially, hydrodynamic lubrication. If this occurs the frictional force may be appreciably smaller than the friction at lower speeds where the conditions are predominantly those of boundary lubrication. Thus a relatively rapid change in friction with speed may occur and this will facilitate the onset of intermittent motion. In contrast to 'true' boundary lubrication, the occurrence of quasi-hydrodynamic lubrication of this type depends on the viscosity of the lubricant film, on the area of the sliding surfaces, and on the surface finish.†

† Beeck, Givens, and Smith (1940) find that quasi-hydrodynamic lubrication is also favoured by the presence of polar bodies in the lubricant. This will be discussed in the next chapter.

As we shall see in the next chapter, the viscosity of the lubricating film *between* the sliding surfaces may bear little relation to the bulk viscosity of the lubricant. It may also be profoundly affected by the high rates of shear and the transient temperature flashes occurring during sliding. Further, the area of the surfaces over which the hydrodynamic film may act and the surface finish will depend on the detailed wear processes occurring during sliding itself. It follows that the conditions favouring the incidence of fluid lubrication will be complicated and extremely difficult to estimate quantitatively.† We may therefore expect considerable difficulty in generalizing about the type of motion to be expected between lubricated metal surfaces.

It is apparent from these observations that under boundary conditions the coefficient of friction is sensibly independent of the load over a very wide range, i.e. Amontons's law is obeyed. It is only when the load is very light that deviations from Amontons's law are observed. The effect of speed, however, is more involved. For fatty acids the coefficient of friction does not vary appreciably with speed unless the speed of sliding is sufficient to initiate fluid lubrication. For paraffins and alcohols the friction decreases as the sliding speed is increased but at higher sliding speeds reaches a fairly steady value. The falling friction–velocity characteristic for these substances implies that the sliding may be intermittent. However, the general conditions for intermittent motion in the presence of boundary films are complex and we may expect considerable difficulty in generalizing about the type of motion to be expected.

† An interesting discussion of these and other factors has been recently given by Forrester (1946).

REFERENCES

O. Beeck, J. W. Givens, and A. E. Smith (1940), *Proc. Roy. Soc.* **A 177**, 90.
J. J. Bikerman (1939), ibid. **A 170**, 130.
H. Blok (1946), *VIth Int. Cong. Appl. Mechanics*, Paris.
F. P. Bowden and L. Leben (1939), *Phil. Trans.* **A 239**, 1.
J. R. Bristow (1942), *Nature*, **149**, 169.
W. Claypoole (1939), *Trans. Amer. Soc. Mech. Eng.* **61**, 323.
E. N. Dacus, E. F. Coleman, and L. C. Roess (1944), *J. Appl. Phys.* **15**, 813.
H. A. Everett (1937), *S.A.E. Journ.* **41**, 531.
A. Fogg (1945), *Engineering*, **159**, 138.
P. G. Forrester (1946), *Proc. Roy. Soc.* **A 187**, 439.
J. J. Frewing (1942), ibid. **A 181**, 23.
E. B. Greenhill (1949), *Trans. Faraday Soc.* **45**, 631.
J. N. Gregory (1943), C.S.I.R. (Australia) Tribophysics Division Report A 74.
—— and M. Newing (1948), *Aust. J. Sci. Research* **A 1**, 85. See also M. Newing (1949), Ph.D. Dissertation, Cambridge.
—— and J. A. Spink (1947), *Nature*, **159**, 403.

SIR W. B. HARDY (1936), *Collected Works*, Camb. Univ. Press.

M. D. HERSEY (1938), *Theory of Lubrication*. John Wiley & Sons Inc., New York.

T. P. HUGHES and G. WHITTINGHAM (1942), *Trans. Faraday Soc.* **38**, 9.

M. J. HUNTER, M. S. GORDON, A. J. BARRY, J. F. HYDE, and R. D. HEIDENREICH (1947), *Ind. Eng. Chem.* **39**, 1389.

T. ISEMURA (1940), *Bull. Chem. Soc. Japan*, **15**, 467.

H. F. KENYON (1946), private communication.

I. LANGMUIR (1920), *Trans. Faraday Soc.* **15**, 62.

—— (1934), *J. Franklin Inst.* **218**, 143.

O. REYNOLDS (1886), *Phil. Trans. Roy. Soc.* **177**, 157.

J. E. SOUTHCOMBE and H. M. WELLS (1920), *J. Soc. Chem. Ind.* **39**, 51 T. See also Discussion on Lubrication (1920) *Proc. Phys. Soc. London*, **32**, 1 S, and especially contribution by R. M. Deeley.

D. TABOR (1940), *Nature*, **145**, 308.

—— (1941), ibid. **147**, 609.

E. D. TINGLE (1948), Ph.D. Dissertation, Cambridge.

J. R. WHITEHEAD (1950), *Proc. Roy. Soc.* A **201**, 109.

X

MECHANISM OF BOUNDARY LUBRICATION

Importance of Chemical Attack

It has already been indicated that the boundary properties of lubricant films depend on the nature of the metal surfaces as well as on the composition of the lubricant itself. Nevertheless, little work of a consistent nature has been carried out on the effect of the underlying metal on boundary lubrication. Hardy (1936) made some comparative measurements of the coefficient of static friction μ_s on steel, glass, and bismuth surfaces and found that for any given hydrocarbon, fatty acid, or alcohol, μ_s for glass $>$ μ_s for steel $>$ μ_s for bismuth.

Sameshima (1940) described some similar experiments on steel, glass, and silver surfaces. He found that glass was poorly lubricated whilst steel was well lubricated by fatty acids and alcohols. Silver showed an intermediate behaviour.

A more systematic investigation has been carried out by Dr. Hughes and Dr. Whittingham (1942), who examined the lubricating properties and transition temperatures of stearic acid and a commercial oil on various metal surfaces. In these experiments, however, they used an upper slider of a fixed metal throughout (steel). This introduced a serious uncertainty into the interpretation of the results.

It is clearly more satisfactory to carry out experiments using similar metals for the upper and lower surfaces. An investigation of this type was recently carried out (Gregory, 1943; Bowden, Gregory, and Tabor, 1945) on the large friction apparatus described in Chapter V. The load used was about 4,000 gm. and the sliding speed about 0·01 cm./sec. A pure fatty acid (lauric acid) was used as the lubricant and it was applied as a thin smear. In some experiments it was applied pure in the molten state: in other cases as a dilute solution in paraffin oil.

The results show that the lubricating properties of the fatty acid depend markedly on the nature of the metal. One of the most striking results is that for unreactive surfaces such as nickel, chromium, platinum, silver, and glass, the fatty acid is scarcely more effective as a boundary lubricant than paraffin oil itself. The results are shown in Table XXVI *a*. It is seen that the lauric acid does not lubricate these unreactive surfaces effectively.

However, pure lauric acid below its melting-point, when present as

TABLE XXVI *a*

Surfaces	*Clean*	*Coefficient of friction*		
		Paraffin oil, room temp.	*1% lauric acid in paraffin, room temp.*	*Pure lauric acid just above the m.p.*
Nickel . .	0·7	0·3 intermittent	0·28 intermittent	0·26
Chromium .	0·4	0·3 ,,	0·3 ,,	0·25
Platinum . .	1·2	0·28 ,,	0·25 ,,	0·28
Silver . .	1·4	0·8 ,,	0·7 ,,	0·7
Glass . .	0·9	..	0·4 ,,	0·6

a relatively thick film, gave effective lubrication with a low coefficient of friction of the order of $\mu = 0{\cdot}1$. As soon as it was melted, however, the friction rose to the relatively high values given in Table XXVI *a*, column 5. Here again, therefore, lauric acid is scarcely better as a boundary lubricant than a solid long-chain paraffin.

In contrast, the results for the reactive metals show that very effective lubrication may be obtained with a 1 per cent. solution of lauric acid in paraffin oil (Table XXVI *b*).

TABLE XXVI *b*

Surfaces	*Clean*	*Coefficient of friction*	
		Paraffin oil, room temp.	*1% lauric acid in paraffin, room temp.*
Copper . .	1·4	0·3 intermittent	0·08 smooth
Cadmium .	0·5	0·45 ,,	0·05 ,,
Zinc . .	0·6	0·2 ,,	0·04 ,,
Magnesium .	0·6	0·5 ,,	0·08 ,,
Iron . .	1·0	0·3 ,,	0·2 irregular smooth
Aluminium .	1·4	0·7 ,,	0·3 stick slips

These results led to a study by Gregory of the chemical reactivity of the various surfaces with fatty acids under standard conditions. Lauric acid was used and the metal surface heated up to 150° C. with the acid in contact with it. The results at once showed that the metals examined fall into two distinct classes: (*a*) in which chemical attack is absent or very slight (magnesium, iron, silver, aluminium, nickel, chromium, glass), and (*b*) those for which chemical attack is marked (copper, cadmium, zinc).

A comparison of the chemical reactivity and lubricating properties shows a striking correlation, as may be seen from Table XXVII.

These results show that the boundary lubricating properties of fatty acids are profoundly affected by the nature of the metal surfaces. Those

TABLE XXVII

Efficiency of lubrication with 1 per cent. lauric acid in paraffin oil compared with reactivity of the metal to lauric acid

Metal	Coefficient of friction (20° C.)	Transition temperature °C.	% acid† reacting	Type of sliding at 20° C.
Zinc	0·04	94	10·0	Smooth
Cadmium	0·05	103	9·3	,,
Copper	0·08	97	4·6	,,
Magnesium	0·08	80	Trace	,,
Platinum	0·25	20	0·0	Intermittent
Nickel	0·28	20	0·0	,,
Aluminium	0·30	20	0·0	,,
Chromium	0·34	20	Trace	,,
Glass	0·3–0·4	20	0·0	Intermittent (irregular)
Silver	0·55	20	0·0	Intermittent (marked)

† Estimated amount of acid involved in the reaction assuming formation of a normal salt.

metals which are most readily attacked by the fatty acid are those which are most effectively lubricated. On the other hand, the unreactive metals (and glass) are poorly lubricated and give intermittent motion with a relatively high coefficient of friction. It is not suggested that there is a quantitative relation between the amount of chemical reaction and the coefficient of friction. It is, however, clear that unreactive metals are not well lubricated by fatty acid solutions whilst, in general, reactive metals are. In addition the less reactive metals such as aluminium (and iron) which are not lubricated by a 1 per cent. solution of fatty acid are well lubricated by a more concentrated solution. All these results strongly suggest that, under these conditions of sliding, *lubrication is effected not by the fatty acid itself but by the metallic soap formed as a result of chemical reaction between the metal and the fatty acid.*

This view is supported by an examination of the effect of temperature on the breakdown of lubricant films. As we saw in the last chapter, hydrocarbon films cease to lubricate effectively at their melting-points. With fatty acids on reactive metals, however, the breakdown does not occur at the temperature at which the fatty acids melt but at considerably higher temperatures. This is shown in Fig. 77 for a series of fatty acids on steel surfaces, and it is seen that the breakdown occurs at 50° to 70° C. above the melting-point (Tabor, 1941). The actual value of the breakdown temperature depends on the nature of the metals and on the load and speed of sliding, but as we shall now show, it corresponds approximately to the stage at which the metallic soap film, formed by chemical reaction, softens or melts. This may be shown by comparing the lubricating properties of a fatty acid ($C_nH_{2n+1}COOH$) with those of the corresponding metallic soap ($C_nH_{2n+1}COOM$).

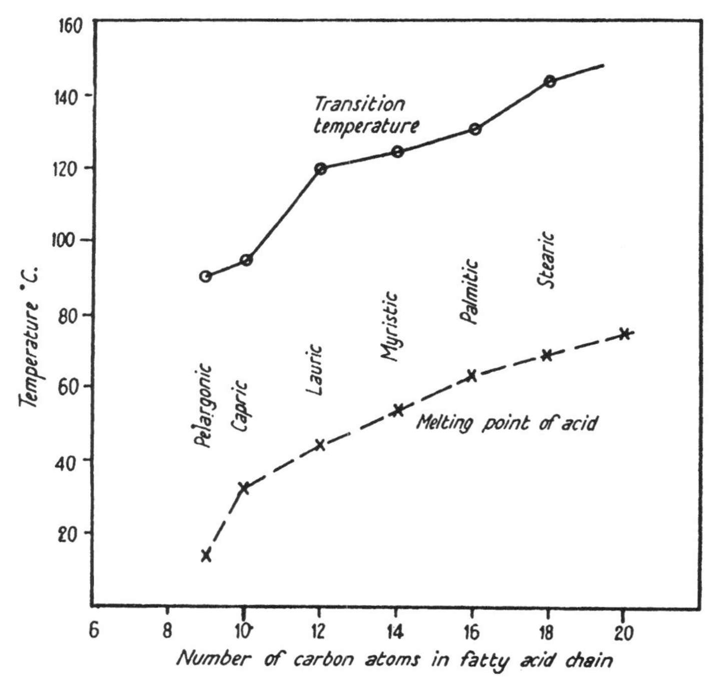

FIG. 77. Breakdown or transition temperature of fatty acids on steel surfaces as a function of chain length. Lubrication ceases to be effective at a temperature considerably above the bulk melting-point of the fatty acid (Tabor 1941).

Lubricating properties of metallic soaps

(*a*) *Lauric acid and metallic laurates.* Results for lauric acid ($C_{11}H_{21}COOH$, m.p. 44° C.) and copper laurate ($[C_{11}H_{21}COO]_2Cu$, softening-point about 100° C.) are summarized in Table XXVIII. It is seen that lauric acid on copper gives the same behaviour as copper laurate whether the soap is deposited on copper or on platinum. Further, the breakdown temperature is very close to the softening-point of the soap. Results for zinc laurate ($[C_{11}H_{21}COO]_2Zn$, softening-point about 120° C.) and for lauric acid on zinc are also included and show a similar behaviour. Results for magnesium are also given, and these again behave in a similar way.

TABLE XXVIII

Surfaces	Lubricant	Coefficient of friction at 20° C.	Breakdown temperature °C.
Copper . .	1% lauric acid in paraffin	0·08, smooth	100
,, . .	Smear of copper laurate	0·08, ,,	100
Platinum . .	,, ,, ,,	0·10, ,,	100
Zinc . . .	1% lauric acid	0·05, ,,	130
,, . . .	Smear of zinc laurate	0·05, ,,	120
Platinum . .	,, ,, ,,	0·10, ,,	130
Magnesium . .	1% lauric acid	0·12, ,,	150
,, . .	Smear of magnesium laurate	0·12, ,,	150
Platinum . .	,, ,, ,,	0·12, ,,	160

(*b*) *Stearic acid and metallic stearates.* Table XXIX summarizes results obtained with stearic acid ($C_{17}H_{35}COOH$, m.p. 69° C.) on copper and cadmium surfaces. The results are compared with the corresponding metallic soaps. In addition results are given for sodium stearate ($C_{17}H_{35}COONa$, softening-point about 260° C.) on steel. As we should expect, lubrication is effective up to about 280° C.

TABLE XXIX

Surfaces	Lubricant	Coefficient of friction at 20° C.	Breakdown temperature °C.
Copper . .	1% stearic acid	0·08, smooth	90
,, . .	Smear copper stearate	0·08, ,,	94
Platinum . .	,, ,, ,,	0·1, ,,	110
Cadmium . .	1% stearic acid	0·05, ,,	130
,, . .	Cadmium stearate	0·04, ,,	140
Platinum . .	,, ,,	0·08, ,,	140
Steel . . .	Smear sodium stearate	0·1, ,,	280

(*c*) *Cetyl mercaptan and corresponding copper and cadmium compounds.* The lubricating properties of cetyl mercaptan ($CH_3(CH_2)_{14}CH_2SH$, m.p. 15° C.) on copper and cadmium were compared with those of cupric cetyl mercaptide ($[C_{15}H_{31}CH_2S]_2Cu$, softening-point about 115° C.) and cadmium cetyl mercaptide ($[C_{15}H_{31}CH_2S]_2Cd$, softening-point about 120° C.) when applied to an unreactive surface such as platinum. The results are summarized in Table XXX. Cetyl mercaptan is not a fatty acid, but it reacts with metals to form compounds which are analogous to the metallic soaps. This substance is commonly considered to be an extreme pressure lubricant (see next chapter), but the results are included here to illustrate the importance of the softening-point of the film formed by chemical reaction.

TABLE XXX

Surfaces	*Lubricant*	*Coefficient of friction at 20° C.*	*Breakdown temperature, °C.*
Copper	1% cetyl mercaptan in paraffin oil	0·1 smooth	130
Platinum	Smear of cupric cetyl mercaptide	0·1 ,,	100
Cadmium	1% cetyl mercaptan in paraffin oil	0·1 ,,	140
Platinum	Smear of cadmium cetyl mercaptide	0·1 ,,	100

It is seen that the lubrication breaks down with the copper compound at approximately the same temperature as with cetyl mercaptan on a copper surface. The same applies to cadmium. The breakdown temperatures correspond roughly to the bulk softening-points of the metallic compounds concerned.

(*d*) *α-mercapto-palmitic acid and cadmium compound.* Similar experiments were also carried out with α-mercapto-palmitic acid

$$(CH_3—(CH_2)_{13}—H.C.SH—COOH, \text{ m.p. } 71° \text{ C.})$$

which reacts with cadmium to form cadmium mercapto-palmitate ($[C_{14}H_{29}.HCSH.COO]_2Cd$, softening-point about 140° C.). The acid is directly analogous to a fatty acid and the cadmium compound to a metallic soap. The results are given in Table XXXI.

TABLE XXXI

Surfaces	*Lubricant*	*Friction at 20° C.*	*Breakdown °C.*
Cadmium	1% mercapto-palmitic acid in paraffin	0·08 smooth	140
,,	1% cadmium mercapto-palmitate in paraffin	0·08 ,,	145
Platinum	Smear of cadmium mercapto-palmitate	0·1 ,,	140

It is seen that the metallic compound gives the same frictional behaviour on both cadmium and platinum. The behaviour is also similar to that of the acid when applied to cadmium surfaces. The temperature of breakdown corresponds approximately to the softening-point of the cadmium compound.

(*e*) *Metallic soaps in presence of paraffin oil.* The results already described indicate that metallic soaps and analogous metallic compounds lubricate metal surfaces up to the stage at which they soften. If, however, there is a relatively weak attachment between the soap and the metal surface, then in the presence of excess paraffin oil the soap may leave the surface at a relatively low temperature due to its increased solubility in the superincumbent oil. As a result, breakdown of the lubricant film occurs at a lower temperature than the softening-point. This is shown in Table XXXII.

TABLE XXXII

Surface	*Lubricant*	*Coefficient of friction at 20° C.*	*Breakdown temperature, °C.*
Platinum	Smear of cadmium mercapto-palmitate	0·1 smooth	140
,,	Smear of cadmium mercapto-palmitate covered with paraffin oil	,,	50
,,	Smear of cupric mercapto-palmitate	,,	180
,,	Smear of cupric mercapto-palmitate covered with paraffin oil	,,	50
,,	Smear of cupric laurate	,,	100
,,	Cupric laurate in paraffin oil	,,	60

(*f*) *The physical state of the soap film.* Some interesting experiments were carried out on relatively thick films of sodium stearate deposited on steel surfaces. If the soap film was applied from an aqueous solution as a wet smear breakdown occurred around 100° C. This is apparently due to the ebullition of the excess water which disrupts the lubricating layer. If, however, the soap is applied from an ethereal solution which is allowed to dry by evaporation it retains its lubricating properties up to 280° C., which corresponds closely to the temperature at which sodium stearate softens markedly (Peart and Tabor, 1944).

(*g*) *Soap films formed by chemical attack and by deposition.* As we have seen, metallic soap films, when deposited on solid surfaces, can provide effective lubrication even on non-reactive metals such as platinum (or glass). They give smooth sliding and low friction, and under suitable conditions the breakdown occurs at the bulk softening-point of the soap. In some cases even a monolayer of a metallic soap will provide effective lubrication. For example a monolayer of cadmium stearate will lubricate glass and steel surfaces.

There is, however, a marked difference between soap films produced on the surface by reaction between the fatty acid and the metal, and soap films deposited on to the metal surfaces in any other way. The soap film formed *in situ* by chemical reaction is, as a rule, firmly linked to the metal surface. Even if it is covered with excess paraffin oil, it will in general lubricate up to its softening-point. With soap films deposited on metal surfaces, however, in the presence of excess paraffin oil, breakdown may occur at a lower temperature due to the increased solubility of the soap in the superincumbent liquid and its weak attachment to the surface. Furthermore, as we have seen, the physical texture of the soap film may have a profound effect on its lubricating properties.

These results show that there is a general similarity between the lubricating properties of long-chain hydrocarbons, alcohols, fatty acids, and metal-soap films and the frictional behaviour of thin films of soft metals deposited on hard metal substrates. Provided the lubricant film is sufficiently thick to prevent appreciable contact between the rubbing surfaces, the friction is low and the lubricating properties are maintained until the melting-point of the lubricant film is reached. At this

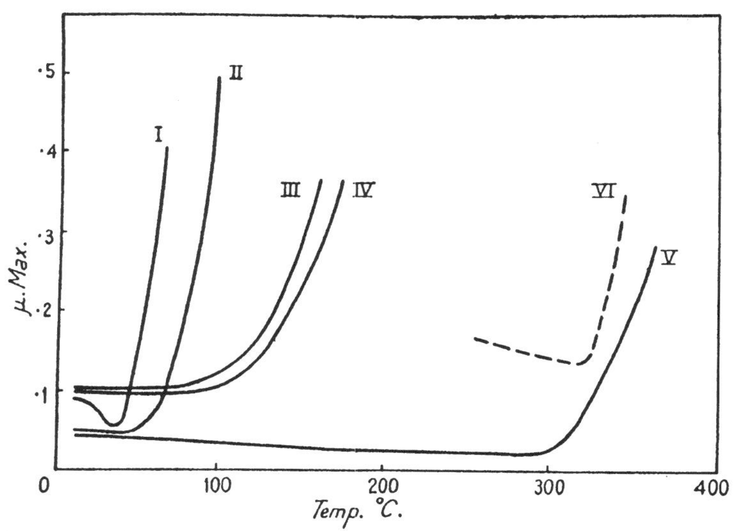

FIG. 78. Effect of temperature on friction of lubricated surfaces. I. Solid docosane (m.p. 44° C.) on platinum surfaces. II. Solid stearic acid (m.p. 69° C.) on platinum surfaces. III. Solid copper laurate (softening-point 110° C.) on platinum surfaces. IV. One per cent. lauric acid in paraffin oil on copper surfaces. V. 'Dry' film of sodium stearate (softening-point 280° C.) on steel surfaces. VI. Thin film of lead (m.p. 327° C.) on hard steel surfaces. In all cases lubrication ceases to be effective at the melting- or softening-point of the lubricant film.

point breakdown occurs. These similarities are shown in Fig. 78, where it is seen that docosane (m.p. 44° C.) and stearic acid (m.p. 69° C.) lubricate platinum surfaces until they melt; a thin film of lead (m.p. 327° C.) lubricates steel until the melting-point of the lead film is reached; sodium stearate lubricates steel until the stearate softens and melts (about 300° C.). Finally, lauric acid in paraffin lubricates copper until the softening-point of copper laurate is reached (about 100° C.).

STRUCTURE OF THE LUBRICATING LAYER: ELECTRON DIFFRACTION EXPERIMENTS [A]

The study of the structure of lubricant layers provides general support for the frictional results described above. Early work by Bragg (1925), Müller (1923), and Trillat (1925) using X-rays revealed the structure of thick films, providing information on the basic dimensions of the hydrocarbon molecules and the crystalline forms in which they appear. Later work using electron diffraction has provided a much more complete picture of the structure and orientation of hydrocarbon films deposited on solid substrates. The great advantage of electron diffraction methods over X-ray methods is the comparatively small penetrating power of the former. Thus in reflection experiments where a beam of electrons, accelerated by a potential of the order of 40 kV., strikes the surface at a very small glancing angle, the depth of penetration normal to the surface is only of the order of 20 A. This means that the diffraction patterns obtained reveal the structure of the uppermost layers only. In some cases this may be a disadvantage, particularly if we are interested in the general structure of thick films, since the electron beam will be entirely absorbed by the first few molecular layers. However, for films which are only a few molecules thick the diffraction of electrons at glancing incidence is probably the most suitable known method of examining molecular structure and orientation. Hydrocarbon films may also be examined by the transmission technique which gives greater resolving power. In addition it yields pictures in which the pattern from the substrate may be more readily distinguished from that of the hydrocarbon film. By use of the transmission technique, specimens several hundred Ångströms in thickness may be investigated. This method has, however, considerable technical difficulties, since the solid surface on which the films are supported must be continuous and extremely thin. For an original account of electron diffraction see Thomson and Cochrane (1939) and Finch and Wilman (1937).

Experiments have shown that for paraffins exceeding a certain minimum chain length, the surface layers have an ortho-rhombic structure in which the carbon chains are normal to the solid surface. This structure remains essentially unchanged whatever the thickness of the film. Fatty acids, however, show the striking result that the orientation of the first monolayer is different from that of subsequent layers. Provided the acid chain contains more than about 12 carbon atoms, a diffraction pattern may be obtained showing that the first monolayer is oriented with the chains approximately normal to the surface. On reactive metal surfaces 8 carbon atoms are sufficient to give an oriented monolayer. The layers on top of the monolayer, however, usually crystallize into the standard crystalline form of the fatty acid (generally monoclinic) and the hydrocarbon chains are inclined at an appreciable angle to the surface normal. The structure of these upper layers can be modified by rubbing with a degreased cloth, whereas the first monolayer remains unaffected. This suggests that the first layer is more firmly attached to the substrate than subsequent layers.

It is interesting to note that a monolayer of a soap such as barium stearate is more completely oriented than stearic acid itself on a non-reactive surface. Some workers have indeed suggested that when the orientation of a fatty acid on a metal substrate is very well defined it means that chemical action has occurred at the surface with the formation of a soap.

More recently, several workers, for example Tanaka (1941), Cowley (1948), Brummage (1947), have examined the effect of temperature on the orientation of hydrocarbon films deposited on solid surfaces. As the temperature is raised, there is increased thermal agitation of the molecules and the diffraction pattern grows more diffuse until at a critical transition temperature when the molecules become almost completely disoriented it fades out altogether. Dr. Menter (1949) and Mr. Sanders (unpublished) have investigated a range of acids from caprylic to octacosanoic on a number of different metals. The layers were deposited by melting the acid on to the clean polished surface and rubbing off the excess with filter-paper. They find with platinum that disorientation occurs about 10° C. below the melting-point of the acid. Typical diffraction patterns of stearic acid (m.p. 69° C.) on platinum are shown in Plate XXII. *a*, *b*. It is seen that a few degrees below the melting-point the oriented structure has completely disappeared. Moreover, after the disorientation has occurred the diffraction pattern from the platinum substrate becomes progressively clearer. If the specimen is heated to about 40° C. above the transition temperature it is found that, after removal from the camera, it can be wetted by water. This indicates that after disorientation the molecules are very poorly adsorbed on to the surface and can easily be removed by evaporation when the temperature is raised. On the other hand, with zinc, cadmium, and mild steel, disorientation does not occur until the temperature is well above the melting-point of the acid. For example, with stearic acid (m.p. 69° C.) on cadmium the transition does not occur until about 135° C., as is shown in the diffraction patterns in Plate XXII. *c*, *d*, *e*, and *f*. The results for other fatty acids are shown graphically in Fig. 79, where the transition temperature on various substrates is plotted against the chain length of the acid. *The temperature at which the films on zinc, cadmium, and mild steel lose their orientation corresponds approximately to the melting-point of the metallic soap formed by chemical reaction with the metal surface.* This temperature is close to that at which lubrication ceases to be effective, as may be seen in Fig. 77. It is also interesting to note that with these metals the pattern of the metal substrate is not clear immediately after disorientation of the surface film has occurred. It remains diffuse for a considerable further rise of temperature. This suggests that even after disorientation, the soap film remains well

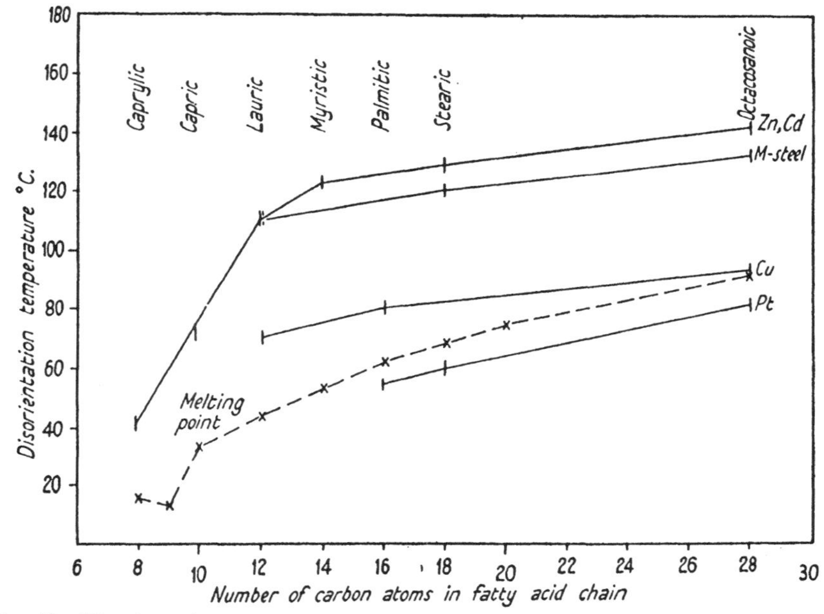

FIG. 79. Disorientation temperature of fatty acid films on substrates of Zn, Cd, mild steel, Cu, and Pt, as function of chain length. The broken line represents the bulk melting-point of the fatty acid. It is seen that on reactive metals the disorientation temperature is considerably higher than the melting-point of the acid and corresponds approximately to the softening temperature of the metallic soap formed by chemical reaction on the surface.

adsorbed on to the surface. This is confirmed by the observation that the surface is still hydrophobic on removal from the camera, even after this drastic heating.

The results with copper suggest that soap formation occurs readily only with the shorter chain fatty acids. For example, with lauric acid (m.p. 44° C.) disorientation is observed at about 80° C. With octacosanoic acid (m.p. 90° C.) disorientation occurs at the melting-point of the acid itself. It is possible that under different surface conditions, reaction with the longer chain acids could also occur. As we shall see, the lubrication of copper by fatty acids often depends on the way in which the copper surface is prepared.

It is clear from these results that the electron diffraction studies are in reasonably close agreement with the friction results. The first monolayer is strongly adsorbed on the solid surface and subsequent layers are well oriented, in their characteristic crystalline form, on top of the monolayer. As the temperature is raised, the lateral adhesion between the lubricant molecules is increasingly overcome by the thermal motion and the diffraction pattern grows less distinct. The transition temperature at which the pattern disappears corresponds closely to the bulk melting-point of the surface film and is close to the temperature at which the film ceases to be effective as a boundary lubricant. There is indeed close agreement between the friction results plotted in Fig. 77 and the electron diffraction results plotted in Fig. 79 for mild steel surfaces. The importance of soap formation is borne out in both types of measurement.

Mechanism of Soap Formation: Influence of Water [A]

We have already seen that with many metals the lubrication provided by fatty acids is really due to the formation of a metal soap as a result of chemical reaction with the metal surface. We may now ask, What is the mechanism of this chemical reaction? Experiments by Dubrisay (1940) and Prutton (1945) on the corrosion of metals by solutions of fatty acids in hydrocarbons show that for copper, cadmium, zinc, and other metals reaction takes place via the oxide film. On this view, therefore, fatty acids should not react with an electropositive metal such as copper or lead if metal oxide and oxygen are completely excluded. Under such conditions, therefore, fatty acids should prove incapable of providing effective lubrication.

Some experiments were recently carried out to test this view by Dr. Tingle (1947). Measurements were made of the lubrication obtained with a 1 per cent. solution of lauric acid in paraffin oil on metal surfaces which were freed from oxide and surface films by cutting a fresh surface under the applied lubricant. The large friction apparatus described earlier was used. By means of a specially designed cutting-tool, a smooth shallow track was laid bare on the lower metal surface immediately before its passage under the loaded hemispherical slider used to measure the force of friction; this is illustrated in Fig. 80 (*a*). The results show at once that *a 1 per cent. solution of lauric acid in paraffin oil is no more effective in lubricating the freshly cut metal surface than the paraffin oil itself*. Further, if the lubricant is left on the surface for several hours, the lubrication remains poor, and this is true even if the temperature is raised to 100° C.

In other experiments newly cut tracks were exposed to air for varying periods up to 24 hours; attempts were then made to lubricate them with the fatty acid solution. In the case of copper, lubrication was still ineffective even after 24 hours' exposure. Magnesium and cadmium showed

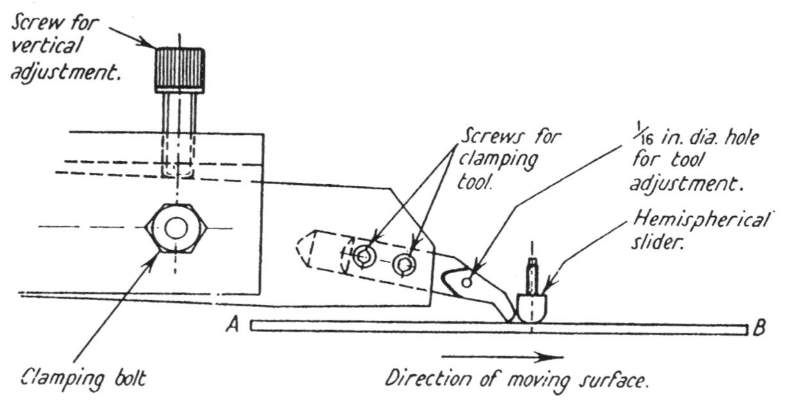

FIG. 80 (*a*). Experimental arrangement for laying bare a shallow track of 'clean' metal in the lower surface, before it passes under the slider.

the same effect, although to a lesser degree. In all cases, however, if the newly cut track was made wet with distilled water and dried before application of the lubricant, or was exposed to an atmosphere saturated with water vapour, good lubrication was obtained after a relatively short period. For effective lubrication it is apparently necessary for water to have been present as well as oxygen.

The effect of surface treatment on the lubrication of various metals is shown diagrammatically in Fig. 80 (*b*). The friction was measured when a 1 per cent. solution of lauric acid in paraffin oil was applied to the metal surfaces prepared in the following three ways:

A. Surfaces abraded on fine emery paper under cold water and then washed with warm water.

B. Surfaces from which surface oxide films had been removed under a pool of lubricant.

C. Surfaces cut in air to remove surface oxide films: cut tracks then washed with cold, warm, or boiling distilled water for various periods, the detailed treatment depending upon the metal.

The results show at once that for the metals Mg, Cd, Zn, Cu, and Fe the friction on the surfaces freed of oxide films (B) is very high and examination of the track shows that there is considerable plucking and tearing of the surfaces. The behaviour is, in fact, essentially the same as that observed with the paraffin oil alone. With the surfaces abraded under water (A) and with the cut track exposed to water (C) the friction

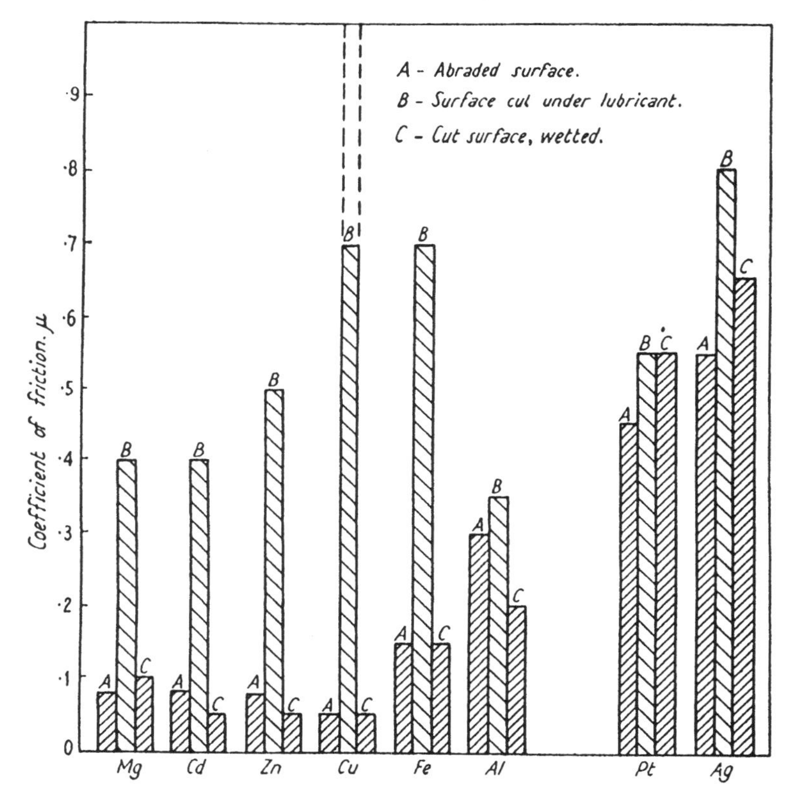

FIG. 80 (*b*). Friction of metals lubricated by 1 per cent. solution of lauric acid in paraffin oil. *A*, abraded surface; *B*, surface cut under pool of lubricant; *C*, cut surface, wetted. Lubrication of the cut surface *B* is no better than that observed with the paraffin oil alone. On abraded or wetted surfaces effective lubrication is obtained with the reactive metals. The non-reactive metals (Pt and Ag) are not lubricated whatever the surface treatment.

is low and the lubrication is effective. Aluminium shows a similar effect, though it is less marked. It is clear that with these metals the surface treatment plays an important part in determining whether chemical reaction will occur and consequently whether effective lubrication will be obtained. On the other hand, as we should expect, with platinum and silver the friction is high and has approximately the same value on surfaces prepared in the three different ways. It is evident that with the noble metals, chemical reaction does not occur whatever the way in which the surface is prepared.

These results provide strong additional evidence for the view that chemical reaction, with the formation of a soap film, is necessary for the effective lubrication of metal surfaces by solutions of fatty acids in paraffin oil. It is also evident that with dilute solutions of fatty acid the chemical attack is favoured by the presence of water as well as oxygen. The water may lead to the formation of a loose hydroxide or hydrated oxide layer of a character suitable for penetration and reaction by the acid. Alternatively, water may facilitate a local ionization of the acid molecules at the seat of reaction.

With more concentrated acid solutions, or with pure fatty acids, chemical attack and subsequent lubrication may occur in the absence of water, as Dr. Shooter (1951) has shown. On general grounds, however, we should expect that even with the pure fatty acid some oxide must be present on electropositive metals for this to occur.

The fact that metal surfaces which are freshly cut or exposed under the oil are difficult to lubricate effectively by fatty acid solutions is of some interest. Under practical conditions, where metal surfaces slide on one another in the presence of a lubricant the abrasion of the surfaces through the lubricant film may expose portions of metal free from surface oxide. These portions of the surface may not readily be lubricated by fatty acids in the oil unless oxygen (and water) is present to enable chemical attack to occur.

Investigation of Surface Adsorption by Radio-active Methods [*A*]

The use of artificially radio-active metals provides a sensitive method for investigating the nature of the boundary film present on the surface. Consider, for example, a monolayer of a fatty acid which is adsorbed on to the surface of a radio-active metal and is then removed. If it has merely been adsorbed on the surface by physical van der Waals forces it will, when desorbed and examined, show no radio-activity. If, however, it was chemisorbed and has reacted with the metal to form the soap, the removal of the film will bring the metal atom with it, so that the film when desorbed should be radio-active. A series of experiments has been carried out by Mr. A. C. Moore to investigate the nature of the boundary layers of fatty acids of alcohols and of esters when they are adsorbed from solution on to a number of metal surfaces.

The general method involves the use of thin spectroscopically pure ts of various metals which were made artificially radio-active in E.E.P. The cleaned metal sheet was then immersed in a dilute ene solution of the alcohol, acid, or ester so that adsorption could r. The sheet was removed from the solution, washed in cold ben- , and then the adsorbed film of hydrocarbon present on its surface removed by refluxing with hot benzene in a soxhlet. The radio- ity of the solution was then estimated (Bowden and Moore, 1949).

y acids

he non-reactive metals such as platinum and gold gave no evidence ll of chemical reaction or soap formation with stearic acid. The rbed film showed no radio-activity. The method is a sensitive one in the case of gold, metal equivalent to about one-thousandth of a nolayer of soap could have been detected. It was clear that the acid s physically adsorbed on the surface. With reactive metals such as zinc and cadmium, however, chemical attack occurred. The film, when removed from the surface, was highly radio-active and corresponded to the formation of soap films several molecular layers thick. The thickness of the surface film increased with time and in addition there was a slow but continuous dissolution of the metal into the benzene solution of the fatty acid.

Alcohols

Long-chain alcohols formed an adsorbed film (probably a monolayer) on the surface, but when this was removed it showed no radio-activity. The behaviour was the same on base metals such as zinc and cadmium as it was on the noble metals, showing that in all cases the adsorption was physical.

Esters

The behaviour of an ester was interesting. Experiments were carried out with ethyl stearate. Again this was adsorbed and on the noble metals there was no reaction. On the base metals, however, there was evidence for a slight but definite chemical attack. It is probable that the reaction with the ester is due to hydrolysis. Although the benzene solution was dried by refluxing and redistillation over sodium, it is most difficult to remove the last trace of water. The hydrolysis of the ester would lead to the formation of a very small amount of the fatty acid which when adsorbed could attack the metal to form the corresponding soap film. This offers an explanation of the previously rather puzzling observation (Frewing, 1944) that esters can lubricate steel surfaces at temperatures above the melting-point of the ester.

Adsorption of Fatty Acids, Alcohols, and Esters on Metals [A]

We have assumed that when a metal is placed in contact with a dilute solution of a fatty acid, alcohol, or ester a complete monolayer is rapidly adsorbed on the surface. Zisman's (1946) experiments on oleophobic films lend strong support to this view. Some experiments have

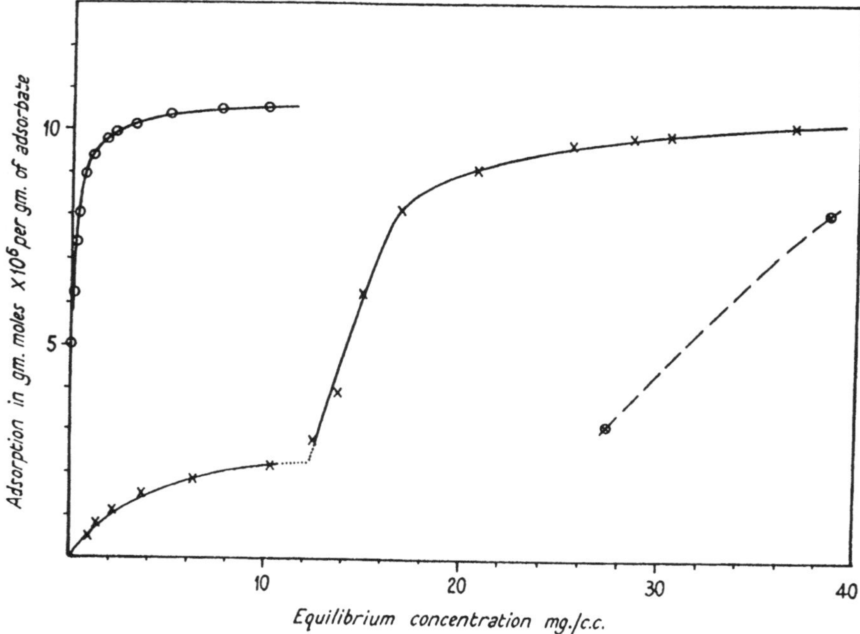

Fig. 81 (a). Adsorption of stearic acid ○, octadecyl alcohol ×, and ethyl stearate ⊗ on nickel powder at 23° C. The kink in the curve for the alcohol probably corresponds to a phase change in the adsorbed layer.

recently been carried out by Dr. Greenhill (1949) and Dr. Daniel to determine by direct experiment the amount adsorbed as a function of the concentration of the solution and of time (Daniel, 1949).

The method employed by Akamatsu (1942) and more recently in England by Hutchinson (1947) was used. Briefly, the method consists of shaking a solution (in a volatile solvent) of the compound under investigation with a weighed quantity of metal powder. The concentration of the solution before and after adsorption is determined by spreading a suitable volume (if necessary, after dilution) by means of a micropipette on a Langmuir trough and compressing the monolayer so formed to a predetermined surface pressure. The concentrations are then proportional to the area covered by the monolayer. From a knowledge of the force–area curve of the adsorbate (Adam, 1941), it is then a simple matter to calculate the adsorption which has taken place. Some typical isotherms determined in this manner are shown in Figs. 81 (*a*), (*b*), and (*c*).

Experiments showed that about 90 per cent. of the adsorption occurs during the first 5 minutes, but final equilibrium does not appear to be

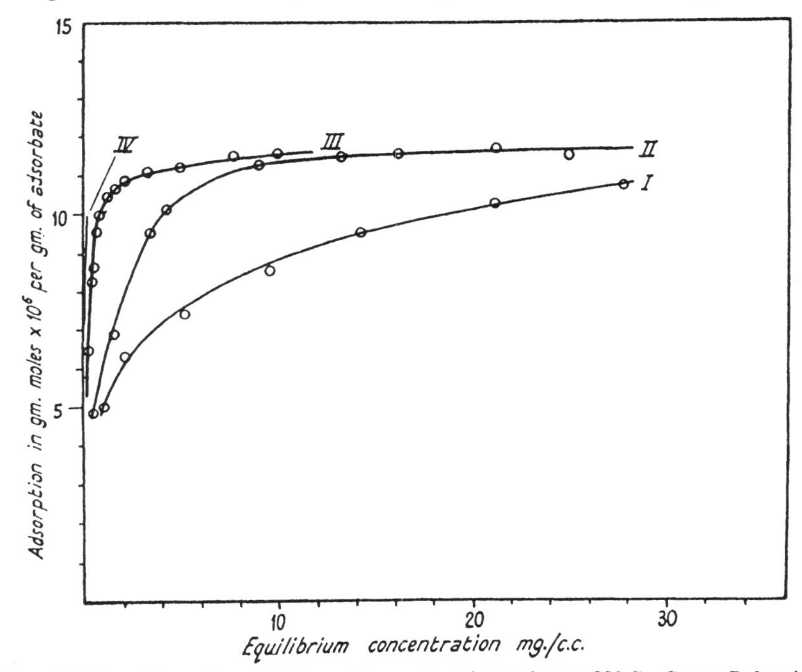

FIG. 81 (*b*). Adsorption of fatty acids on nickel powder at 23° C. Curve I, lauric acid; curve II, palmitic acid; curve III, stearic acid; curve IV, octacosanoic acid.

established for a considerable period of time. So far only silver, nickel, platinum, iron, and copper powders have been examined and in all cases the general behaviour is the same. With all metals, rapid adsorption occurs and the adsorption of the alcohol appears to be complete after about 1–4 hours, a further increase of a few per cent. (barely outside experimental error) occurring in the following 24 hours. With fatty acids, the same holds true for nickel, platinum, silver, and iron. With copper, however, the adsorption increases steadily with time (see below). It is apparent that a slow but continuous chemical reaction is occurring between the acid and the copper.

In the isotherms given in Figs. 81 (*a*) and (*b*) the concentration of solution in contact with the powder after 4 hours has been termed the equilibrium concentration. It is seen (Fig. 81 (*a*)) that for the three compounds containing the C_{18} chain, the fatty acid is the most strongly adsorbed, 90 per cent. of the maximum uptake being reached at a concentration of about 1 mgm./c.c. The same uptake is reached with the alcohol at about 20 mgm./c.c., while the ester requires about 40 mgm./c.c. (The inflexion in the curve for the alcohol is probably due to a phase-

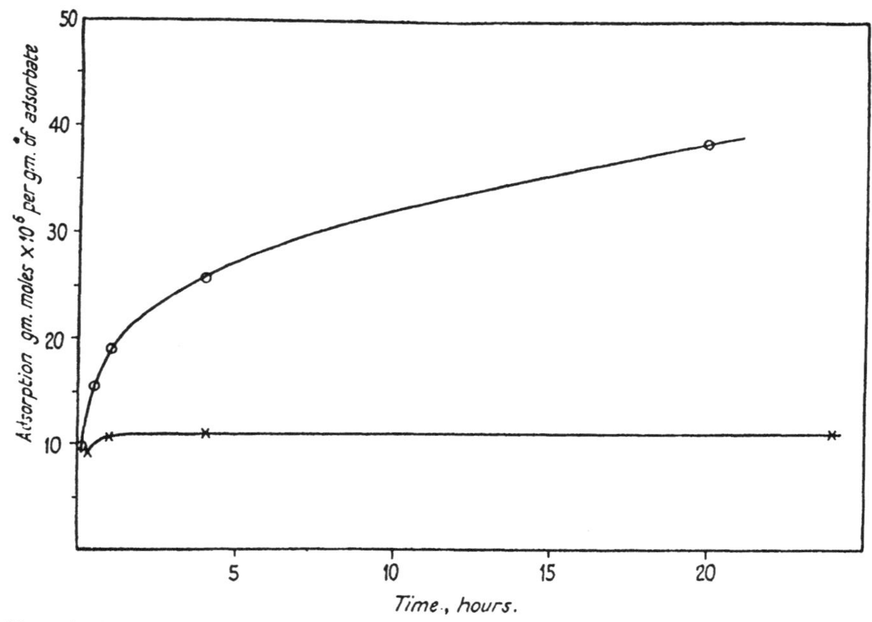

FIG. 81 (*c*). Adsorption of stearic acid ○ and octadecyl alcohol × on copper powder at 23° C. The steady increase of adsorption of the fatty acid indicates continuous reaction between the acid and the copper.

change in the adsorbed layer.) There is considerable evidence that the fatty acid is adsorbed as a unimolecular layer; in fact, by assuming that each molecule occupies an area of 20 A^2 several workers (Harkins and Gans, 1931; Smith and Fuzek, 1946) have used the maximum uptake to obtain a value for the surface area of a powder. It is evident from Fig. 81 (*a*), therefore, that at maximum uptake, the acid, the alcohol, and the ester are adsorbed as close-packed monolayers.

Considering the fatty acids, Fig. 81 (*b*), it is seen that the longer the CH_2 chain, the more strongly is the acid adsorbed at a given concentration. Thus, with stearic acid the monolayer is practically complete at 5 mgm./c.c., palmitic at 12 mgm./c.c., while lauric acid is only about 90 per cent. complete even at a concentration of 30 mgm./c.c. This is undoubtedly closely related with the solubility of the acid in the solvent. In this connexion it is interesting and somewhat surprising to find that octacosanoic acid does not form a complete monolayer from benzene solution. This must presumably be due to the extremely small solubility of octacosanoic acid in benzene. If, in fact, the concentrations in Fig. 81 (*b*) are plotted as fractional concentrations of the solubility, we should expect the isotherms for all the acids to lie on the same curve.

The adsorption of some of these compounds on copper has been examined in some detail (see Fig. 81 (*c*)). The alcohol gives an isotherm very similar to that observed on nickel. Stearic acid, however, reacts slowly with the copper; solutions in contact with it develop a pale blue colour after some hours. In addition, assuming that at maximum uptake the alcohol forms a close-packed monolayer, then Fig. 81 (*c*) shows that more than one layer is adsorbed on the copper. It is also evident that the adsorbed layer gradually increases in thickness with time as the reaction proceeds.

The solvent employed does not, within the narrow limits examined, have any appreciable influence on the maximum uptake. For example, a 1 per cent. solution of stearic acid in benzene, cyclohexane, octane, and hexane gave adsorptions of 3·32, 3·43, 3·38, and 3·25 mgm./gm. on the same sample of nickel powder. However, it is possible that more polar solvents may influence the maximum adsorption. On more reactive metals, such as zinc and cadmium, we should again expect to find a continuous uptake of the acid due to chemical attack.

Some experiments have also been carried out to measure the heats of adsorption of these compounds on metal surfaces. The difference in the heat of physical and chemical adsorption for these compounds is small. However, recent results indicate that with non-reactive metals (silver, nickel) the adsorption is physical whilst with reactive metals (iron, copper) the adsorption involves chemical reaction (Daniel, 1949).

Mechanism of Boundary Lubrication [*A*]

According to the earliest workers, the friction of clean solid surfaces was due to the interlocking of surface asperities and the frictional work was dissipated in raising one set of surface roughnesses over the other. On this view, the effect of a lubricant was to form a film so thick that the roughnesses of the surfaces could not come into contact with one another, interlock, and so cause friction.

The more modern theories of boundary lubrication, however, assume that the resistance to motion is due to intermolecular forces at the points of contact. The most systematic formulation of this theory is due to the work of Hardy, who showed that the friction was not only influenced by the chemical nature of the lubricant but also by the nature of the underlying surface. Working with homologous series of paraffins, alcohols, and fatty acids on various surfaces, Hardy found that the static friction was a function of separate contributions by the solid surfaces, the chemical series to which the lubricant belonged, and the number of carbon atoms in its chain. To interpret these data Hardy assumed that the friction between unlubricated surfaces is due to the surface fields of force. When the lubricant is added, the lubricant molecules are physically adsorbed and orientate themselves at each of the solid surfaces to form a unimolecular film. The solids sink through the lubricant layer until they are separated by only the unimolecular adsorbed film. Since the polar groups adhere to the metal surface, contact takes place, not between the metal surfaces themselves, but between the non-polar groups at the other end of the lubricant molecules. Slip then takes place between these non-polar molecular sheets, and the efficiency of a boundary lubricant is measured by the extent to which these films can mask the field of force of the underlying surfaces. It is apparent that this effect will depend on the polarity of the lubricant molecule and on its chain length. Along these lines Hardy was able to explain the linear relationship observed between the friction and the molecular weight for different members of a homologous series. This theory received indirect confirmation from the X-ray experiments of Trillat, Bragg, and Müller, who showed the existence of orientated films at the surface of a metal. It also received some support from the electron diffraction investigations described above on the structure of surface films.

Nevertheless, the results quoted in the preceding chapters indicate that Hardy's theory of boundary lubrication is an over-simplification. Even with the best boundary lubricants and with very light loads there is some wear of the underlying surface once sliding has commenced. A detailed microscopic examination of the surfaces such as that described in Chapter IV shows, in fact, that the metal is torn to a depth which is large compared with the dimensions of a molecule. Even more striking are the results obtained using electrographic surface analysis, and radio-active tracer techniques where the adhesion and smearing of one metal surface on to the other, through the lubricant film, are clearly demonstrated. With some metals the same effects are observed (using the electron microscope) at loads much less than a gramme. These results mean that the friction of lubricated metals cannot, in general, be due simply to the sliding of one lubricant monolayer over the other, nor indeed can it be merely a function of the surface forces as Hardy supposed. It must be greatly influenced by the bulk properties of the metals concerned.

Further, as we have seen, physically adsorbed monolayers of long-chain polar compounds may not give effective lubrication. For example, liquid alcohols which are highly polar and are presumably well adsorbed at the metal surface are poor boundary lubricants and scarcely better than liquid paraffins. Again, on unreactive surfaces, liquid fatty acids are almost as ineffective as liquid paraffins and alcohols. Finally, though the friction decreases with increasing chain length the longest chain molecules do not in fact give zero friction as Hardy suggested; the friction reaches a low but finite value.

As has been pointed out with unlubricated surfaces, the processes occurring during sliding are complex and it is difficult at this stage to give a really satisfactory quantitative theory of boundary lubrication which is applicable to ordinary conditions of sliding. It is clear that any such theory must take into account the metallic adhesion which occurs through the lubricant film. We may do this by modifying the theory of unlubricated surfaces to allow for the protection provided by the lubricant film. If the ploughing term is small, the frictional resistance of unlubricated surfaces may be written as $F = As$, where A is the real area of contact between the metals and s is the shear strength of the metallic junctions formed at the points of contact. With clean metals the welding is strong and the break frequently occurs within the metal itself. In this case s will be equal to the shear strength of the metal. With lubricated surfaces there is considerably less tearing of the metal surface and the amount of intimate metallic contact is very greatly reduced.

When lubricated metals are placed in contact under an applied load, plastic flow occurs until the area is large enough to support the applied load. As a result of this deformation a film of lubricant will be trapped between the two metal surfaces and there subjected to very high pressures. The pressure, however, will not be uniform over the whole region of contact. In the regions where the pressure is highest a local breakdown of the lubricant film may occur, resulting in metallic adhesion. The extent of this breakdown will naturally depend on the nature of the lubricant film. Further, if the speed of sliding is appreciable, it will be aided by local high temperatures developed during sliding. As a result of the partial breakdown of the lubricant film, metallic junctions which are large in comparison with the size of a molecule are formed between the surfaces.

The resistance to motion consists partly of the force necessary to break these junctions. There will also be some resistance to sliding by the lubricant film itself, and we may write:

$$F = A\{\alpha s_m + (1-\alpha)s_l\}, \tag{1}$$

where A = the area which supports the applied load,

α = the fraction of this area over which breakdown of the film has occurred,

s_m = the shear strength of the junctions at the metal-metal contact,

s_l = the shear strength of the lubricating film.

Thus if α remains essentially constant for a given lubricant and surface, the frictional force will be proportional to A, so that Amontons's law will hold as for unlubricated surfaces. This is observed over a very large range of loads. For very small loads, as we saw in Fig. 74, Amontons's law may no longer be valid and the coefficient of friction may reach very high values. It is probable that at these loads the deformation of the underlying metal is too small to produce appreciable breakdown of the lubricant film. The frictional force is largely due to the interaction of the hydrocarbon film itself and the area over which shearing occurs is no longer proportional to the applied load. The behaviour is not unlike that observed with thin metallic films where at small loads the coefficient of friction may be very high (see Fig. 48). This would explain the surprising observation that at very small loads the friction of a lubricated surface may be higher than that observed with unlubricated surfaces.

In general, however, at loads above a few grammes, there is some breakdown of the lubricant film and the load is supported by the underlying surfaces so that Amontons's law is obeyed. With a good lubricant the area over which the metallic junctions are formed may be very small indeed. Nevertheless, the shear strength of these junctions may be so high compared with that of the lubricant that they may be responsible for some part of the resistance to motion.

The main difference between this view of the mechanism of boundary lubrication and that suggested by Hardy is shown diagrammatically in Fig. 82. Fig. 82 (*a*) depicts the mechanism proposed by Hardy where interaction between the outer surfaces of the adsorbed monolayers is envisaged, but where no metallic contact occurs. Fig. 82 (*b*) depicts the mechanism described above which involves a breakdown of the lubricant film at small localized regions. The metallic junctions so formed are partly responsible for the friction and almost entirely responsible for the wear and surface damage involved.

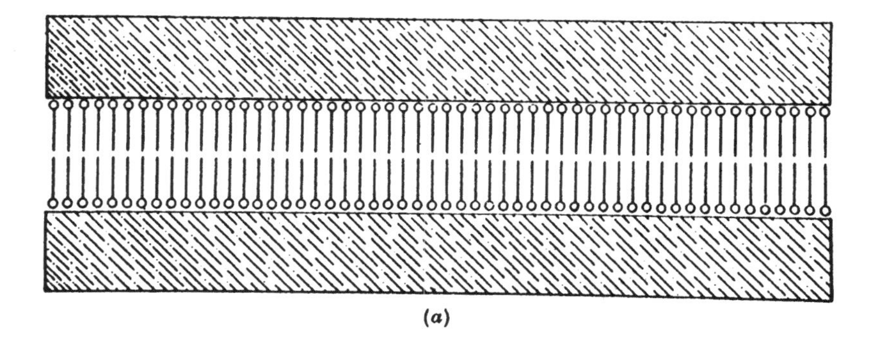

(a)

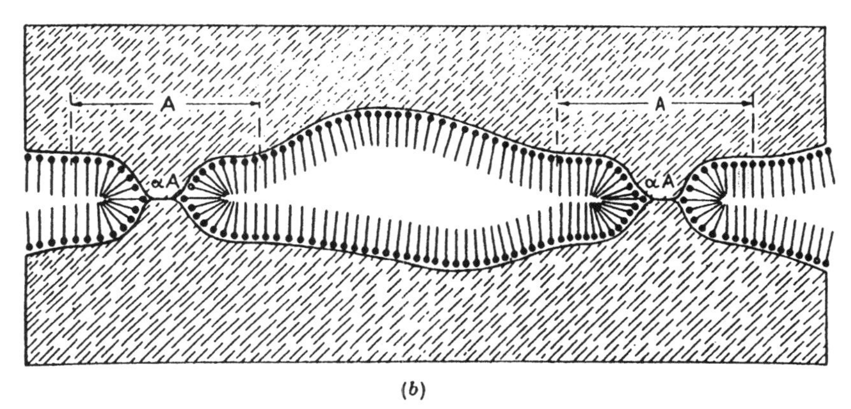

(b)

FIG. 82. The sliding of solid surfaces under conditions of boundary lubrication. (a) Hardy's view which envisages physical adsorption of the hydrocarbon, alcohol, or fatty acid. The frictional resistance is due to interaction between the outer surfaces of the adsorbed monolayers without any metallic contact occurring. (b) Mechanism involving breakdown of the lubricant film at small localized regions. The load is supported over an area A, whilst metallic junctions are formed through the lubricant film over a much smaller area αA. The force to shear these metallic junctions ($\alpha A s_m$) in general constitutes part of the friction observed. The remainder of the frictional force is involved in shearing the lubricant film $\{(1-\alpha)A s_l\}$. Further, if the lubricant is a fatty acid it may react with the metal to form a metallic soap.

On this view, therefore, the main function of the lubricant film is to reduce the amount of intimate metallic contact between the surfaces by interposing a layer that is not easily penetrated and that possesses a relatively low shear strength. In order to prevent appreciable contact the lubricating layer must, in addition to being firmly attached to the surface, possess strong lateral adhesion between the hydrocarbon chains. The results described in this and the previous chapter suggest that the lateral adhesion between the long chains of the lubricant molecules are at least as important as the strength of attachment to the surface. For example, liquid long-chain alcohols (or liquid fatty acids on unreactive surfaces) which are highly polar and well adsorbed at the surface are worse boundary lubricants than solid paraffins which are only weakly adsorbed at the metal surfaces. When these solid paraffins are melted, however, the lateral adhesion falls to a low value and metallic welding occurs with a corresponding increase in friction and wear.

In the liquid state, paraffins, alcohols, esters, etc., are not good boundary lubricants. They give a relatively high coefficient of friction and there is appreciable seizure and wear of the surfaces. Nevertheless, they do reduce the friction and wear considerably. For example, clean copper surfaces have a coefficient of friction of $\mu = 1{\cdot}4$; with liquid paraffin oil, the friction is reduced to about $\mu \approx 0{\cdot}8$ and there is a corresponding reduction in wear. It is clear that the liquid paraffin molecules have been able to reduce the amount of metallic contact even though they possess little lateral adhesion and relatively weak attachment to the surface. Further, with the pure liquid paraffins, alcohols, and acids there is a steady decrease of friction with increasing chain length. This effect is seen in Hardy's results and in the results quoted in the previous chapter. Apparently the longer chain length gives a greater measure of lateral adhesion between the lubricant molecules or provides a greater separation between the surfaces so that the lubricant film is better able to prevent breakdown and metallic seizure. The decrease in friction continues with increasing chain length until the stage is reached at which the molecule is solid at room temperature. The friction now reaches its minimum value, which does not fall below about $\mu = 0{\cdot}05$–$0{\cdot}1$. Even at this stage the solid film of paraffin or alcohol is usually unable to prevent some metallic contact occurring between the metal surfaces.

The importance of strong lateral adhesion between the lubricant molecules is again shown by the lubricating properties of fatty acids. At the loads and speeds used in the experiments described in this chapter, fatty acids on non-reactive metals prove to be hardly more effective as boundary lubricants than aliphatic hydrocarbons. Even when they form well-adsorbed and oriented films on platinum and nickel surfaces, the fatty acids do not lubricate effectively above their bulk melting-points. That is to say that, though they produce a marked reduction in friction and wear, their behaviour is similar to that of liquid aliphatic hydrocarbons. If, however, they are applied to surfaces with which they can react they readily form a film of metallic soap. These soap films possess strong lateral cohesion and their softening-point is usually very much higher than the melting-point of the pure acid. As a result the lubricating film so formed is capable of providing effective lubrication to a higher temperature. In addition the metal soap is formed *in situ* and will, in general, be well attached to the surface. On account of their high lateral cohesion and the fact that they can withstand appreciable deformation without rupture, these soap films are usually very protective and with some metals even a monolayer is sufficient to reduce the penetration and metallic contact to a very low value. With other metals, however, considerably thicker films may be necessary. These films maintain their lubricating properties with increasing temperature until the softening-point of the soap is reached. At this stage the film softens or melts and increased metallic contact occurs through it, with a corresponding increase in friction and wear. As the soap film loses its lateral cohesion its linkage to the metal surface also grows weaker and the breakdown of the lubricant film is often accompanied by its dissolution into the superincumbent oil or excess acid.

We see, therefore, that fatty acids are 'good' boundary lubricants because of the special properties possessed by the metallic soap. There is, however, another way in which the formation of a metallic soap may affect the frictional behaviour of sliding surfaces. At higher sliding speeds the formation of a viscous soap film of appreciable thickness may lead to a relatively large separation of the surfaces even at speeds well below those at which hydrodynamic lubrication might be expected to begin. As a result, the conditions are no longer those of true boundary lubrication and the friction and wear may fall to very low values. This effect has been described in an interesting paper on quasi-hydrodynamic lubrication by Beeck, Givens, and Smith (1940).

Because of their remarkable protective properties thin films of metallic soaps may be applied directly to metal surfaces where they will often function as extremely effective boundary lubricants. With these films, however, the physical texture and the mode of deposition may be very important. The best frictional properties are obtained when the soap film has a close coherent texture, is evenly deposited over the surfaces, and is firmly adsorbed. These films will lubricate until the softening-point of the soap occurs. If, however, the attachment of the soap to the surface is weak and there is a superincumbent layer of oil present, it may dissolve in the excess oil at a temperature lower than its softening-point.

Again, if there are solvents present in appreciable quantities in the soap film, violent ebullition may cause them to disrupt the lubricant film. Under these conditions, the breakdown of the lubricant film will occur at temperatures below the softening-point of the metallic soap.

The softening-point of a soap is not clearly defined and may cover a wide temperature range. The actual temperature at which the weakening of the soap film will allow sufficient penetration to cause an appreciable increase in metallic adhesion (and hence in the friction and wear) may depend on the physical conditions of the experiment. With any given metal and lubricant the transition temperature is in fact dependent upon the load, speed, and shape of the sliding surfaces.

A comparison has already been made between the lubricating properties of thin metallic films and of boundary lubricant films. As we saw, both types of film can produce a substantial reduction in friction, and their behaviour with respect to wear and temperature is similar. There are, however, two marked differences. Firstly the melting-point of metallic films and films of pure paraffins, alcohols, etc., is clearly defined and the breakdown occurs sharply at the melting-point. With metal soaps, however, the softening-point is not at all clearly defined and the breakdown occurs at temperatures depending on the physical constants of the apparatus. The second and most striking difference between thin metallic films and soap films is that even on rough surfaces a single molecular layer of soap may provide effective lubrication, whereas metal films must be appreciably thicker to be effective (*c.* 10^{-5} cm.). Solid hydrocarbons and alcohols also need to be relatively thick to provide effective lubrication. The extremely high tenacity of the soap monolayer and its ability to prevent metallic seizure is, indeed, remarkable.

REFERENCES

N. K. Adam (1941), *The Physics and Chemistry of Surfaces.* Oxford Univ. Press.
H. Akamatsu (1942), *Bull. Chem. Soc. Japan,* **17**, 161.
O. Beeck, J. W. Givens, and A. E. Smith (1940), *Proc. Roy. Soc.* **A 177**, 90.
W. C. Bigelow, D. L. Pickett, and W. A. Zisman (1946), *J. Colloid Science,* **1**, 513.
W. C. Bigelow, E. Glass, and W. A. Zisman (1947), *J. Colloid Science,* **2**, 563.
F. P. Bowden and A. C. Moore (1949), *Research,* **2**, 585.
F. P. Bowden, J. N. Gregory, and D. Tabor (1945), *Nature,* **156**, 97.
W. L. Bragg (1925), ibid. **115**, 269; *Proc. Roy. Inst.* **24**, 481.
K. G. Brummage (1947), *Proc. Roy. Soc.* **A 188**, 414; **A 191**, 243.
J. M. Cowley (1948), *Trans. Faraday Soc.* **44**, 60.
S. G. Daniel (1949), Ph.D. Dissertation, Cambridge.
R. Dubrisay (1940), *Compt. rend.* **210**, 533.
G. I. Finch and H. Wilman (1937), *Ergebn. exakt. Naturwiss.* **16**, 353.
J. J. Frewing (1944), *Proc. Roy. Soc.* **A 182**, 270.
E. B. Greenhill (1949), *Trans. Faraday Soc.* **45**, 625.
J. N. Gregory (1943), C.S.I.R. (Australia) Tribophysics Division, Report A 74.

SIR. W. B. HARDY (1936), *Collected Works.* Cambridge Univ. Press.
W. D. HARKINS and D. M. GANS (1931), *J.A.C.S.* **53**, 2804.
T. P. HUGHES and G. WHITTINGHAM (1942), *Trans. Faraday Soc.* **38**, 9.
E. HUTCHINSON (1947), *Trans. Faraday Soc.* **63**, 439.
J. W. MENTER (1949), Ph.D. Dissertation, Cambridge.
A. MÜLLER (1923), *J. Chem. Soc.* **123**, 2043; (1925) *Proc. Roy. Soc.* **A 114**, 542.
J. PEART and D. TABOR (1944), C.S.I.R. (Australia) Tribophysics Division, Report A 99.
C. F. PRUTTON *et al.* (1945), *Ind. Eng. Chem.* **37**, 90.
J. SAMESHIMA *et al.* (1940), *Rev. Physical Chem. Japan*, **14**, 55.
K. V. SHOOTER (1951), Ph.D. Dissertation, Cambridge.
H. A. SMITH and J. F. FUZEK (1946), *J.A.C.S.* **68**, 229.
D. TABOR (1941), *Nature*, **147**, 609.
K. TANAKA (1941), *Mem. Coll. Sci. Kyoto*, **A 23**, 195.
G. P. THOMSON and W. COCHRANE (1939), *Theory and Practice of Electron Diffraction*, Macmillan.
E. D. TINGLE (1947), *Nature*, **160**, 710; (1950), *Trans. Faraday Soc.* **46**, 93.
J. J. TRILLAT (1925), *Compt. rend.* **153**, 280.

For an earlier account of the structure and orientation of Langmuir–Blodgett layers, as determined by electron diffraction, see the paper by L. H. GERMER and K. H. STORKS (1938), *J. Chem. Phys.* **6**, 280.

XI

ACTION OF EXTREME PRESSURE LUBRICANTS

In the previous chapters we saw that the main function of a lubricant film is to reduce the amount of intimate metallic contact between sliding metal surfaces. Under extreme conditions of load and speed, even the best boundary lubricants of the fatty acid or metallic soap type may break down with a consequent increase in the friction and amount of surface damage. This occurs, for example, with hypoid gears where the rubbing surfaces are made of very hard steel and the sliding speeds are relatively high. Here it is customary to use lubricants to which certain active radicals or groups have been added. These additives are referred to as 'extreme pressure' additives, and the resultant lubricants are termed 'extreme pressure' or 'E.P.' lubricants, since they enable the rubbing surfaces to run satisfactorily under conditions where ordinary mineral or vegetable oils would prove inadequate. The term 'extreme pressure' lubricant is somewhat misleading. It is true that with hard surfaces the area of contact for a given load will be smaller so that the pressure on the lubricant film will be greater. But it is *the high temperatures* developed between the rubbing surfaces under these severe conditions of sliding that are primarily responsible for the breakdown of the lubricant film, and this temperature effect will be more marked if the surfaces are hard. In addition, as we shall see, it is the high surface temperatures which play an important part in rendering these 'extreme pressure' lubricants effective.

A review of the patent literature over the last fifteen years indicates that almost every element in the periodic table has, in one form or another, been claimed to possess 'extreme pressure' properties. A more systematic survey shows, however, that the most widely used substances in 'extreme pressure' additives are sulphur, chlorine, and phosphorus. The general view of the action of these substances is that, owing to the thermal decomposition of the additive during running, a compound is formed with the metal surface which reduces the friction or the amount of surface damage. Thus Beeck, Givens, and Williams (1940) have shown that materials such as tricresyl phosphate react with the metal surfaces to form phosphides which have a very specific effect on the frictional properties of the surfaces. Davey (1945), working with additives containing chlorine, has shown that they function by the liberation of free chlorine and the formation of metal chlorides, while sulphurized additives have long been known to form sulphide films. In the first part of this chapter we shall describe an investigation carried out mainly by Mr. Gregory on the lubricating properties of thin films of metallic chlorides and on the way in which their behaviour compares with that observed when surfaces are lubricated with certain organic compounds containing chlorine. In this investigation a study has been made of the frictional properties of these compounds over a wide temperature range. In the second part we shall describe a similar investigation carried out mainly by Dr. Greenhill (1948) on sulphide films and compounds containing sulphur. In the third part we shall discuss an investigation on the use of typical extreme pressure lubricants in certain machining operations.

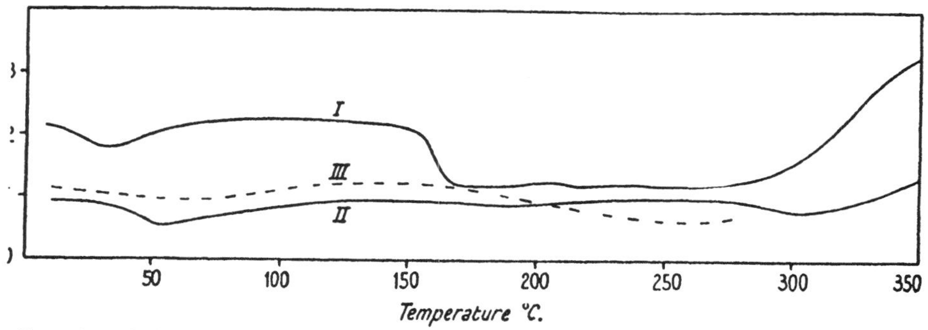

FIG. 83. Friction of chloride films on steel as a function of temperature. I, Chloride film alone. II, Chloride film covered with paraffin oil. III, Ferric chloride film.

LUBRICATION OF METALS BY COMPOUNDS CONTAINING CHLORINE

Chloride films

Experiments were carried out on chloride films produced by exposing steel, copper, and cadmium surfaces to dry chlorine gas for a definite period. The frictional measurements show that if the films are thick enough to produce interference colours, i.e. about 1,000 A thick, they can cause a very large reduction in friction and surface damage up to elevated temperatures. If the chloride film is covered with paraffin oil the friction is reduced even further. The effect of chloride films is particularly marked with steel surfaces, where the coefficient of friction does not exceed $\mu = 0{\cdot}2$ even at a temperature of 300° C., as is seen in Fig. 83 (curves I and II). That this is due to the chloride film is supported by the fact that ferric chloride deposited on steel from an ethereal solution gives similar results (Fig. 83, curve III).

With copper and cadmium surfaces the reduction in friction is not so marked [μ lies between 0·3 and 0·4], but the low coefficient of friction is maintained up to temperatures exceeding 300° C. In all cases the chloride films are readily hydrolysed by the presence of moisture in the air and rapidly lose their lubricating properties. This is especially marked with steel surfaces, and special precautions were taken to form the chloride film in a desiccator and carry out the frictional measurements immediately after removal from the desiccator.

Compounds containing chlorine

Long-chain paraffinic halides. Frictional experiments were carried out using octadecyl chloride, cetyl bromide, and cetyl iodide (m.p. *ca.* 20° C.).

CH_3	CH_3	CH_3
$(CH_2)_{16}$	$(CH_2)_{14}$	$(CH_2)_{14}$
HCH	HCH	HCH
Cl	Br	I
Octadecyl chloride	Cetyl bromide	Cetyl iodide

In all cases these compounds gave a low friction and slight surface damage when solid, but increased friction and surface damage above their melting-points, *ca.* 20° C. This behaviour is similar to that of straight-chain paraffins except that the coefficient of friction is somewhat lower. It would appear, therefore, that under the experimental conditions involved there is no reaction between the compounds and the surface to form a metallic chloride film. Even at elevated temperatures the friction remains high, showing that the compounds do not react.

Compounds containing the $>SeCl_2$ *group.* Friction measurements show that a number of selenium compounds containing the $>SeCl_2$ group are very effective as boundary lubricants. A typical compound investigated was $\beta\beta'$-dichlor dicetyl selenium dichloride, prepared by the reaction of cetene with selenium tetrachloride according to the following equation:

$$2CH_3(CH_2)_{13}{-}CH{=}CH_2 + SeCl_4 \rightarrow \begin{matrix} CH_3(CH_2)_{13}{-}CHCl{-}CH_2 \\ CH_3(CH_2)_{13}{-}CHCl{-}CH_2 \end{matrix}\Big> SeCl_2.$$

Results obtained on steel surfaces with this compound are shown in Fig. 84. It is seen that at a temperature between 150° and 180° C. a large reduction in friction occurs (curve I) and this effect is more marked at higher concentrations (curve II). This is due to the reaction of the compound at this temperature with the steel surface to form the metal chloride. Since stearic acid lubricates steel surfaces up to about

160° C., effective lubrication up to the temperature of reaction may be expected if we add a small quantity of stearic acid to the lubricant. Fig. 84, curve III, shows that, with a solution of paraffin oil containing 1 per cent. of the chloride and 1 per cent. stearic acid, this does in fact occur. A low coefficient of friction is obtained from room temperature to over 300° C.

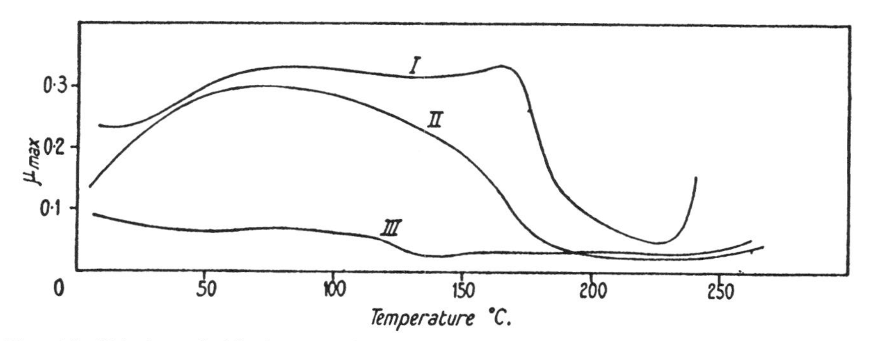

Fig. 84. Friction of chlorine-containing compounds on steel surfaces as a function of temperature. I, 0·1 per cent. solution of A in paraffin oil. II, 1·0 per cent. solution of A in paraffin oil. III, 1 per cent. solution of $A + 1$ per cent. solution stearic acid in paraffin oil. A is $\beta\beta'$-dichlor dicetyl selenium dichloride.

With copper and cadmium surfaces a solution of the compound in paraffin gave better results than the paraffin oil alone, but the behaviour in no way resembled the extremely good lubrication obtained with steel surfaces.

The effectiveness of this compound and of other compounds containing the $>SeCl_2$ group is not due primarily to the presence of the selenium atom, but to the presence of the chlorine attached to the selenium. If the chlorine is removed by hydrolysis (so producing the dihydroxide) or removed with zinc dust (to give the selenide), the resulting compounds are poor boundary lubricants and show no sudden decrease in μ at a critical temperature. Further, in the lubrication of steel surfaces with $>SeCl_2$ compounds, there was direct evidence that the lubricating film formed at the higher temperatures was the metal chloride. This was shown by refluxing iron filings with an alcoholic solution of a typical selenium dichloride compound. After reaction it was found that iron chloride had been formed, and titration in aqueous solution showed that about 80 per cent. was ferrous chloride, the remainder being ferric.

Long-chain acid chlorides. A few experiments were carried out using dilute solutions of stearyl chloride in paraffin oil. Stearyl chloride has the formula $CH_3(CH_2)_{16}COCl$, and it differs from a long-chain paraffinic halide (e.g. octadecyl chloride) in that the end group is COCl, in which the chlorine is reactive, instead of CH_2Cl. With steel surfaces a solution containing as little as 0·5 per cent. stearyl chloride gave smooth sliding and low friction up to 300° C. With copper surfaces, however, solutions containing up to 5 per cent. stearyl chloride did not provide effective lubrication.

Importance of chloride formation

As we have seen, chloride films themselves can be very efficient boundary lubricants on steel surfaces, both when dry and when in the presence of paraffin oil. They are less effective on copper or on cadmium surfaces. On steel surfaces a thin film of iron chloride causes a great reduction both in the frictional force and in the amount of wear and seizure. This means, in terms of equation (1), Chapter X, that the chloride film formed on steel gives a low value to both s_t and α. These films have a lamellar structure and are sheared easily. They are, however, only effective when they are anhydrous. Hydrolysis due to moisture causes a rapid deterioration in their lubricating properties. In addition the formation of hydrochloric acid may lead to excessive corrosion of the surfaces.

The lubricating behaviour of organic compounds containing chlorine varies with the stability of the compounds and with the nature of the metal surfaces. Long-chain halides, such as octadecyl chloride, are stable even at 300° C., so that no chloride film is formed; these compounds are relatively poor lubricants. On the other hand, compounds containing labile chlorine atoms which may become unstable at moderate temperatures are efficient lubricants, particularly on steel surfaces. The properties of the compounds containing the $\rangle SeCl_2$ group seem to be mainly dependent on the chlorine atoms and not on the selenium. This is shown by hydrolysis or dechlorination. If the chlorine is removed by either method, the lubricating properties disappear. Further, the formation of the chloride film may be observed visually at a temperature characteristic of the given compound, and the drop in friction occurs just at this temperature.

Although it is clear that the lubrication produced on steel surfaces is due to the formation of the metal chloride, it is not clear whether the film is ferric, ferrous, or both. Chemical tests indicate that the anhydrous chloride produced from the $\rangle SeCl_2$ compounds is predominantly ferrous. On the other hand, the film formed by exposing steel to chlorine is ferric chloride, but it is possible that at higher temperatures the reaction $2FeCl_3 + Fe \rightarrow 3FeCl_2$ may occur. On the other hand, electron diffraction work by Brummage (1947) shows that at high temperatures the film may be an oxychloride. The composition of the lubricating film, therefore, is not exactly known, nor is it clear whether the composition is the same at high temperatures as at room temperature.

These experiments show that compounds containing a labile chlorine atom may provide very effective lubrication of steel surfaces by the formation of a surface film of chloride. With copper and cadmium, however, although similar surface films may be formed the lubrication is far less effective, and this is presumably due to the fact that these chloride films do not possess the appropriate shear properties.

Lubrication of Metals by Compounds containing Sulphur

Sulphide films

Thin films of metal sulphides were formed on steel, copper, silver, and cadmium surfaces by immersing the metals in an ammonium polysulphide solution or in a dilute solution of sodium sulphide. Friction experiments show that these films produce their maximum reduction in friction when their thickness exceeds about 1,500 A. This is shown in Fig. 85 for copper surfaces where the thickness of the film has been estimated from an optical interference method due to Constable (1929). It is seen that the coefficient of friction does not fall below about 0·5, so that the reduction in friction is not as marked as with chloride films. However, these films are not sensitive to the presence of moisture and they retain their properties up to very high temperatures. Further, taper sections of the tracks formed show that they produce a very marked reduction in the amount of seizure and surface damage. Typical results for copper surfaces are shown in Plate XXIII. *a* and *b*.

If the sulphide films are covered with paraffin oil the frictional behaviour is better than that obtained with either paraffin oil or a sulphide film alone. The best results, however, are obtained if the paraffin oil contains a small percentage of fatty acid. Here the fatty acid provides a low coefficient of friction up to about 160° C., whilst above this temperature, when the acid or soap film breaks down, the sulphide film prevents heavy seizure. Fig. 86 shows a comparison of these friction results on steel. Photomicrographs of the tracks formed show corresponding reductions in the amount of surface damage.

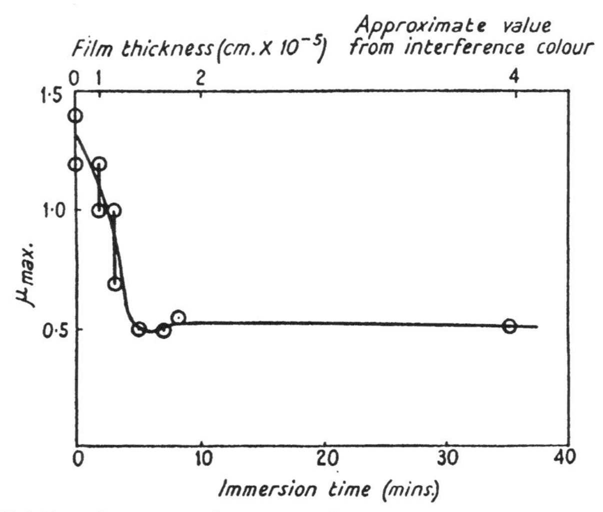

Fig. 85. Friction of copper surfaces covered with sulphide films of various thicknesses. The sulphide film is formed by immersing the metal in an ammonium polysulphide solution for various times, and its thickness is estimated from interference colours. There is a marked reduction in friction when the film thickness exceeds about 2×10^{-5} cm.

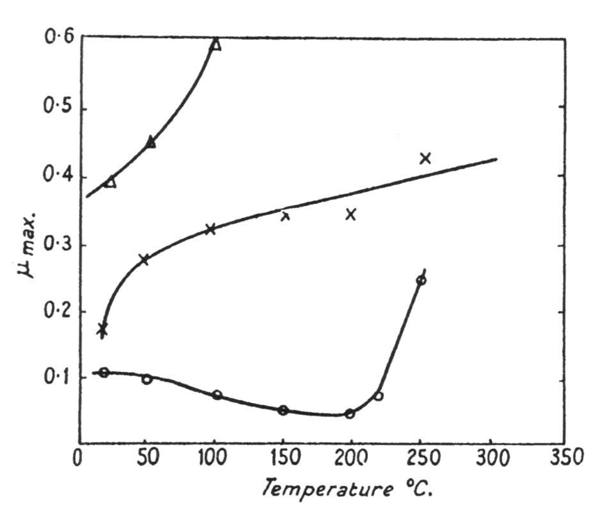

Fig. 86. Friction of sulphur films on steel as a function of temperature. △ Paraffin on steel. × Paraffin on sulphided steel. ○ 1 per cent. fatty acid in paraffin oil on sulphided steel surface.

Sulphurized compounds

Sulphurized oleic acid and sulphurized cetene. Some friction measurements were made with two sulphurized compounds which approximate to the conventional sulphurized lubricants used in machining operations and other extreme pressure processes. The first was sulphurized oleic acid and was made by heating oleic acid with the calculated amount of sulphur. The second was sulphurized cetene and was prepared in a similar manner. The formulae of these substances are doubtful, but the sulphur is probably attached across the double bond and is therefore loosely bound to the molecule. Both lubricants gave similar results on steel and silver. The friction was low ($\mu \approx 0{\cdot}1$) and the surface damage slight over the whole temperature range from 20° to over 300° C. A visible sulphide film was formed in all cases and it was evident that the behaviour was essentially the same as that of sulphided surfaces covered with a layer of oleic acid or of cetene respectively.

Pure long-chain sulphur compounds. Frictional experiments were carried out using long-chain aliphatic compounds containing sulphur. These compounds, which are described below, are highly polar but are readily soluble in paraffin oil and were usually tested as a 1 per cent. solution. Not one of the compounds examined had any appreciable effect on platinum or silver surfaces. The friction remained as high as when paraffin oil itself was used. On steel, copper, and cadmium surfaces, however, it was found that the compounds could be divided into three main classes. The first class consists of long-chain sulphides, disulphides, and thiocyanates such as:

Cetyl methyl sulphide	Dicetyl sulphide	ββ'-dichlor dicetyl sulphide	Cetyl thiocyanate
CH_3	CH_3 CH_3	CH_3 CH_3	CH_3
$(CH_2)_{14}$	$(CH_2)_{14}$ $(CH_2)_{14}$	$(CH_2)_{13}$ $(CH_2)_{13}$	$(CH_2)_{14}$
CH_2	CH_2 CH_2	$CHCl$ $CHCl$	CH_2
S	S	CH_2 CH_2	$S—CN$
CH_3		S	

None of these compounds was effective, and it was evident that under the conditions of the experiment there was no chemical reaction with the surfaces. Similar results were obtained with two aromatic sulphur compounds, though there was evidence that at very much higher temperatures they broke down and attacked the surface to form the metal sulphide.

The second class consists of sulphur compounds possessing a replaceable hydrogen atom. The compounds investigated included:

CH_3	CH_3	CH_3	CH_3
$(CH_2)_{14}$	$(CH_2)_{14}$	$(CH_2)_{10}$	$(CH_2)_{13}$
CH_2	CH_2	CH_2	H—C—SH
SH	O=S=O	C—SH	C=O
	OH	‖ S	OH
Cetyl mercaptan	Cetyl sulphonic acid	Di-thio tridecylic acid	α-mercapto palmitic acid

These substances showed little sign of reacting with the metal surface to form a sulphide film. Further, although they provided effective lubrication on steel, copper, and cadmium surfaces they were no more effective than paraffin oil alone on silver or platinum. The frictional behaviour of these compounds is thus very similar to that of fatty acids and the results suggest that their effectiveness as boundary lubricants is due to their reaction with the metal to form a compound analogous to a metallic soap. This view is confirmed by experiments carried out on the lubricating properties of the copper and cadmium derivatives of cetyl mercaptan and α-mercapto-palmitic acid. Their frictional behaviour on any metal surface is similar to that observed when the original compounds are applied to copper or cadmium surfaces. This similarity has already been discussed in Chapter X.

The third class consists of active compounds not containing a replaceable hydrogen atom, such as dithiocyanates. Dithiocyano-stearic acid prepared from oleic acid and free thiocyanogen $(SCN)_2$ provides very good lubrication on steel, the friction remaining low up to 300° C. (de Kadt, unpublished). This compound is of uncertain composition but the CNS group is present. It is not, however, effective on copper or cadmium. Similar results are obtained with a compound prepared from cetene and thiocyanogen. The behaviour of these compounds is obviously not due to sulphide film formation, and the evidence points to the formation of a film of ferric thiocyanate. This view is confirmed by experiments on the lubricating properties of thin films of ferric thiocyanate deposited on steel surfaces from an ethereal solution and by direct exposure of the steel to ethereal solutions of thiocyanogen. If these films are kept dry by covering with a thin layer of paraffin oil, they give smooth sliding and a low friction up to temperatures well over 200° C. In the presence of moisture, however, these films hydrolyse and cease to provide effective lubrication.

Importance of chemical attack and of nature of surface film

We see that sulphide films, provided they are above a critical thickness, are able to reduce the friction and surface damage between sliding metals. The reduction in friction, however, is not as marked as that observed with chloride films. In terms of equation (1), Chapter X, we may say that sulphide films produce a great reduction in α, so that the wear and surface damage are reduced, but s_l is still fairly high. The coefficient of friction for sulphide films in fact never falls below about $\mu = 0{\cdot}50$. In contrast, chloride films give a low value for both s and α, so that both the friction and the wear are greatly diminished. The low friction of the chloride films on iron is probably bound up with the physical structure of iron chloride, which has a lamellar structure and is readily sheared. The measurements show, in fact, that its frictional properties are similar to those obtained with fatty acid lubrication. On other metals the chloride films do not give so low a friction, probably because of the different structure of these films. Sulphide films on iron and other metals are comparatively hard, and although they are effective in preventing metallic contact they do not give a particularly low friction. They are, however, very stable, are unaffected by the presence of moisture, and retain their anti-seizing properties up to very high temperatures.

When paraffin oil containing a small proportion of fatty acid is added to the sulphide film, the fatty acid reacts to form a metallic soap on top of the sulphide layer. This produces a very low coefficient of friction up to the temperature at which the metallic soap softens or melts. Although the friction increases rapidly at this stage, the presence of the sulphide film prevents excessive wear and seizure up to very much higher temperatures. As the experiments with sulphurized oleic acid and sulphurized cetene show, this mechanism probably accounts for the behaviour of the majority of extreme pressure lubricants containing sulphur. These results therefore, confirm the accepted view that for a very large class of sulphurized lubricants the essential action of the lubricant is to react with the sliding surfaces to form a sulphide film. If the compounds are so stable that they do not react with the surface they are no more effective in paraffin oil solutions than the paraffin oil itself. Thus long-chain sulphides and disulphides which are non-reactive under the conditions of the experiments do not lubricate any metal.

The experiments show, however, that some sulphur compounds which do not form sulphide films may yet provide effective boundary lubrication. This occurs, for example, with sulphur compounds which are acidic in nature. They function by reacting with the surface to form metallic compounds analogous to metal soaps, so that the sulphur really plays no part in their operation. These 'soaps' may serve as very effective lubricating films if their softening-points are high. At sufficiently elevated temperatures, however, they probably decompose to form sulphide films.

Another type of sulphur compound which is effective as a boundary lubricant but which does not form a sulphide film is the group containing the thiocyanate radical CNS. These substances decompose and the thiocyanate group attacks the surface, forming a thiocyanate film. Such compounds only appear to be effective on steel surfaces and therefore resemble the chlorine extreme pressure additives very closely. Indeed, anhydrous ferric thiocyanate $Fe(CNS)_3$ and ferric chloride $FeCl_3$ are both efficient boundary lubricants up to high temperatures.

It is not suggested that the compounds described in this and the preceding part are necessarily good practical 'extreme pressure' additives, but that the action of these compounds is similar to that of conventional 'E.P.' additives. The results show, as with chlorine compounds, that many additives containing sulphur are specific for a given metal or group of metals. The problem of choosing an extreme pressure additive depends, therefore, not only on the conditions such as temperature and pressure, but also on the type of metal to be lubricated.

Phosphorus Additives

Beeck, Givens, and Williams (1940) have shown that some phosphorus additives function by reacting with the metal surface to form metal phosphides. These workers suggest that the phosphide film produced at the high spots forms a low-melting eutectic with the underlying metal and this material is wiped or polished away by the sliding process to leave a smooth surface. On this view the 'extreme pressure' properties (or load-carrying capacity) of phosphorus additives are essentially due to a process of 'chemical polishing'. In some cases it would seem that chemical polishing is not the main factor, but more work needs to be done on this. It is probable that the phosphorus compounds often act in a manner similar to that of compounds containing sulphur and chlorine in that surface films are formed which are protective and which, with some metals, may possess a low shear strength.

REACTIVITY OF EXTREME PRESSURE ADDITIVES [A]

With many types of extreme pressure additives, reaction with the metal surface does not take place very rapidly at room temperature. Consequently until the temperature of reaction is reached, these substances may prove relatively ineffective as boundary lubricants. For this reason it is often advantageous to include in the lubricant a small

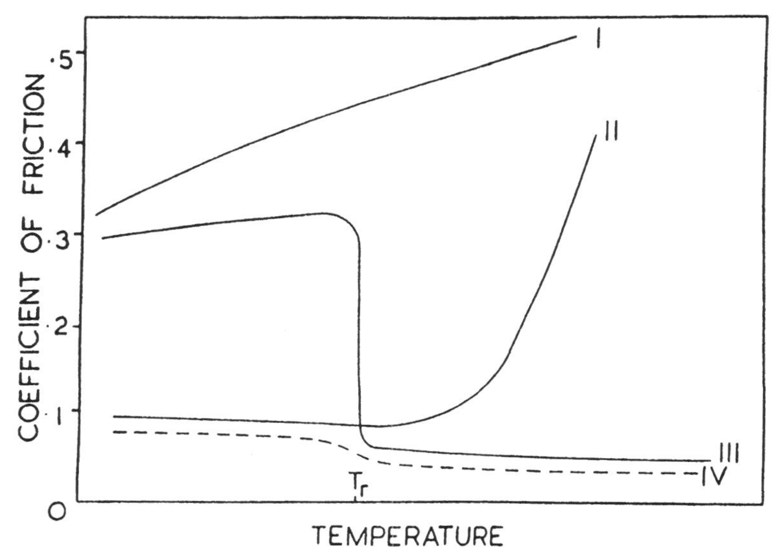

FIG. 87. Frictional behaviour (schematic) of various lubricants, as a function of temperature. I, paraffin oil; II, fatty acid; III, E.P. lubricant which reacts with the surfaces at temperature T_r; IV, mixture of E.P. lubricant and fatty acid. The fatty acid provides effective lubrication up to the temperature at which the E.P. additive reacts with the surface.

quantity of fatty acid which may provide effective lubrication at temperatures below the reaction temperature of the additive. A stylized description of this behaviour is shown in Fig. 87 where the coefficient of friction is plotted against the temperature. Curve I is for paraffin oil and shows that the friction is initially high and increases as the temperature is raised. Curve II is for a fatty acid which reacts with the surface to form a metallic soap; this provides good lubrication from room temperature up to the temperature at which the soap begins to soften. Curve III is for a typical extreme pressure additive which reacts very slowly below the temperature T_r so that in this range the lubrication is poor: above T_r the protective film is formed and effective lubrication is provided up to a very high temperature. Curve IV is the result obtained when some fatty acid is added to the additive. Good lubrication is provided by the fatty acid below T_r, whilst above this temperature the greater part of the lubrication is due to the additive. At still higher temperatures, not shown on this figure, a deterioration in lubricating properties will also occur for both curves III and IV.

The reactivity of the additive will, of course, depend largely on the type of operation for which it is being used. It is clear that for such processes as cutting and drawing a relatively unstable compound may be used, since corrosion is not of very great importance. In a running machine, however, an additive which is too unstable under running conditions may do more harm than good if the high reactivity of the additive results in excessive chemical corrosion. In this type of machine the additive should be one which will give rise to chemical reaction only at temperatures or pressures in the region of danger, i.e. when welding and tearing become so great that high wear and subsequent seizure will otherwise occur. It follows from these considerations that extreme pressure lubrication is, in fact, largely a problem in controlled corrosion. The stability of the additive must be so chosen that under mild conditions no appreciable reaction occurs whilst under severe conditions there is adequate chemical reaction without excessive corrosion. When the reaction product is formed on the surface it should have the appropriate physical properties for reducing intermetallic contact at the operating temperature. It is an additional advantage if it possesses a low shear strength so that it also gives a low coefficient of friction.

Extreme Pressure Lubricants in the Cutting and Drawing of Metals [*A*]

When metals are machined or cut as in planing, turning, or grinding operations, the work required to separate the removed material from the parent material depends on the strength of the metal and on the friction between the nose of the cutter and the chip. As the removed material is continuously exposing clean metal to the nose of the tool, the frictional force involved may be very high indeed. The friction may, in fact, constitute an undesirably large part of the total work required in the cutting operation. This energy appears very largely as heat and, as Herbert (1926) showed many years ago, may produce temperatures of many hundreds of degrees at the rubbing surfaces. Further, the rubbing action between the work and the tool may produce very heavy wear of the tool nose. This in turn may lead to a deterioration in the surface finish of the machined material and to a marked reduction in the effective life of the tool. These factors and the general physical processes involved in the cutting operation have been described in an admirable booklet by Ernst and Merchant (1940).

For these reasons it is customary to use cutting fluids in all but the lightest cutting operations. The main function of the cutting fluid is to cool the work and to reduce the friction and wear between the rubbing surfaces. In spite of the cooling action of the fluid, the temperature between the metal and the nose of the tool may be very high. This is particularly marked if the metal is tough. For this reason a good cutting fluid should act as a boundary lubricant which can operate effectively at high temperatures and high pressures.

In a great number of machining operations the continuous rubbing of the metal on the tool rapidly leads to the formation of a very thin strongly-adhering layer of metal on the nose of the tool. As a result, most of the cutting takes place between this 'built-up' edge and the metal itself. If the lubrication is poor or completely absent, the built-up edge gradually grows in size until finally it is torn off together with a portion of the tool material. A damaged tool nose together with a typical wear 'crater' is shown in Plate XXIII. *c*. If, however, the lubrication is effective, the built-up edge remains small and protects the nose of the tool from excessive wear. It is clear, therefore, that a great part of the rubbing action may occur not between the tool surface and the metal but between the metal and the 'built-up' edge. That is to say in many machining operations lubrication is largely a problem of lubricating a metal sliding on itself at high temperatures.

It is interesting to compare the lubricating properties of an oil with its effectiveness as a cutting fluid. Investigations show that, under some conditions, there may be a close correlation between the two. Some experiments were carried out on the lubricating properties of a number of oils up to temperatures of 250° C. on surfaces consisting of a 3 per cent. Ni steel (Vickers hardness 200). The results were compared with the performance of the same oils in an actual machining operation (Brookman and Ham, 1941).

The frictional behaviour of some of the lubricants is shown in Fig. 88. It is seen that with some oils lubrication breaks down at a comparatively low temperature; the friction rises to a high value and intermittent motion occurs. This rise in friction is accompanied by a corresponding increase in surface damage. This is very marked for example with oils Nos. 1, 2, and 3, which must therefore be judged as possessing poor lubricating properties at high temperatures. With other lubricants, however, smooth sliding and low friction are maintained up to very high temperatures. A systematic survey of the effect of various ingredients showed that on the whole the best results were to be expected if the lubricant contained a few per cent. of fatty oils, between 2 and

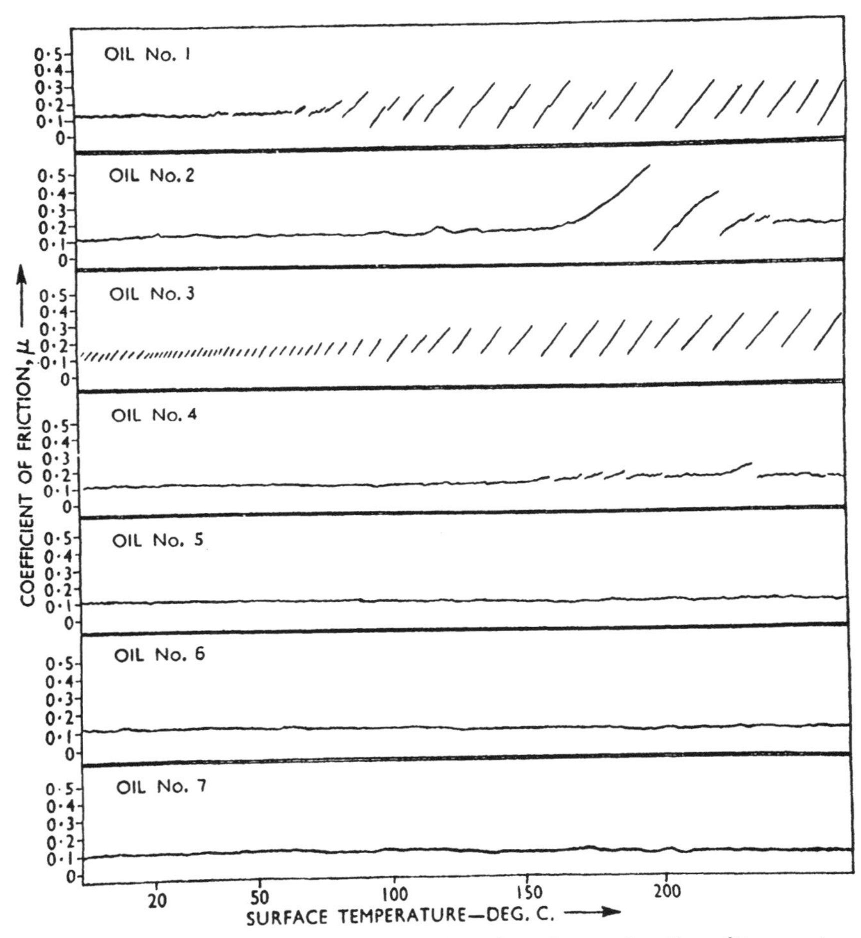

FIG. 88. Frictional behaviour of a number of cutting oils as a function of temperature. Steel surfaces. Oils 1 and 3 break down at temperatures below 100° C. Oils 5, 6, and 7 provide effective lubrication up to temperatures exceeding 200° C.

4 per cent. sulphur, and between 0·5 and 1 per cent. of chlorine. On the basis of these conclusions further samples of lubricants were prepared with this range of composition and their lubricating properties examined. The laboratory experiments showed that of all the oils tested, this group of lubricants possessed the best boundary lubricating

properties. The frictional behaviour of some of these products is shown in Fig. 88, the best results being given by Nos. 4, 5, 6, and 7 in increasing order of performance.

Some practical tests were then carried out on a cutting operation in the production shop. The machining performance was judged in terms of the tool life consistent with a certain quality of surface finish. The investigations showed that on the whole there was a good correlation between the lubricating properties of the oils at elevated temperatures as measured in the laboratory and the practical performance.

An interesting point, however, was the anomalous behaviour of oil 7. Oils 4, 5, and 6, which had increasingly good lubricating properties, gave an increasingly good performance in practice. Oil 7, however, which had the best lubricating properties, gave a worse performance; it gave a good finish but a smaller tool life. This result was interpreted as being due to the fact that as the lubrication was improved the built-up edge was reduced in size. If the lubrication was too good, however, as apparently occurred with oil 7, no built-up edge was formed. As a result the surface finish was good but the tool wear was greatly increased. These conclusions emphasize the important role of friction between the work and the tool in machining operations, a factor which has frequently been stressed by Ernst and Merchant and other workers.

It is of course clear that correlation between the frictional properties of a lubricant as measured in the laboratory and its performance as observed in a practical machining operation will not always be close. In some circumstances the actual conditions in the machining process may be very different from those applying in the laboratory experiments. The correlation between laboratory frictional measurements and practical performance may then be very poor indeed. Two very important factors, for example, are the sliding speeds and the accessibility of the work to the lubricant. As we have seen in Chapter IX, the frictional behaviour of lubricated surfaces may be markedly dependent on the speed of sliding. If the laboratory experiments are carried out at speeds very different from those at which the removed material slides over the nose of the tool, the frictional behaviour in the two cases may be significantly different. Again, in the laboratory experiments the lubricant is usually applied in a manner that ensures the presence of a lubricant film between the sliding surfaces. In a cutting operation, however, the lubricant must penetrate into the very narrow crevices ahead of the tool nose and it must react with the metal so rapidly that by the time the chip has reached the tool nose the lubricating film has been formed. It is clear, therefore, that the accessibility of the work to the lubricant and the rate at which the lubricating film can be formed on the metal are of the very greatest practical importance. These factors may under some conditions prove more significant than the intrinsic lubricating properties of the lubricant film itself.

Another metal-forming operation in which lubrication is of great importance is the drawing of metals. In this operation the metal is forced through a hard steel die and takes up the shape imposed by the geometry of the die and the plunger. The plastic deformation which takes place involves high pressures between the surface of the metal and the die and high surface temperatures. If lubrication is poor there is marked adhesion to the surface of the die. This causes bad scratching and tearing of the surface of the drawn article and the die must be resurfaced. Here, again, as with machining operations, drawing fluids are used to cool the work, to reduce the friction between the die and the work, and to diminish the amount of pick-up and seizure.

For lighter drawing operations a solution of soap, e.g. sodium stearate, is often satisfactory. For more extreme operations, however, where the amount of deformation of the metal is greatly increased, active ingredients, such as chlorine and sulphur, must be added to the lubricant. It is interesting to note that here again a close correlation exists between the lubricating properties of the drawing fluid and its performance in a drawing operation. This was shown in an examination of the lubricating properties of a number of lubricants over a temperature range of 20° to 250° C. The lower surface was of brass and the upper surface was a hemispherical slider of die steel which was covered, before each experiment, with a thin film of brass. The lubricant was applied to the surface in the form of a water emulsion (Greenhill, Ham, and Tabor, 1945).

The friction measurements showed that sulphur in the lubricant improved the lubricating properties at high temperatures, though if the 'free' sulphur content was too high it caused excessive staining of the brass surface. 'Combined' sulphur, that is, sulphur more firmly bound to the lubricant molecule, reduced the amount of staining but did not provide such effective lubrication. It was also found that the presence of reactive chlorine produced an additional improvement in the frictional behaviour. These results are in direct agreement with those described in the earlier portions of this chapter. On the basis of the frictional measurements a lubricant was prepared, as in the cutting experiments described above, which gave very effective lubrication up to a temperature of over 200° C.

Parallel with the laboratory investigation, a series of practical tests was carried out in the factory. In particular a comparison was made of the extent to which pick-up occurred on the die (necessitating re-surfacing of the die) when various types of drawing fluid were used. Results for the six chief types of lubricant tested are shown in Table XXXIII.

TABLE XXXIII

Product	*Main constituents*	*Frictional behaviour*	*Practical performance*
Commercial soap	Sodium stearate	Average. Marked breakdown above 150° C.	Satisfactory under favourable conditions. Little margin of safety.
EF 228	Sodium soap, potassium soap, olein.	Average. Marked breakdown above 150° C.	Slightly better than ordinary soap.
EF 221	Sulphurized olein, castor oil, ammonium soap.	Very good. Slight breakdown at 180° C.	Very satisfactory. Superior to ordinary soap. Staining.
EF 231	Sulphurized olein, castor oil, sodium soap.	Very good. Low friction. Slight breakdown at 200° C.	Extremely satisfactory. Far superior to ordinary soap. Staining.
EF 256	Chlorinated mixed acids, sodium soap.	Very low friction up to 120° C. Marked breakdown above 150° C.	Somewhat better than ordinary soap. Staining.
EF 268	Chlorinated mixed acids, sodium soap, sulphurized fatty oils.	Very low friction up to 200° C.	Extremely successful in drawing steel. Far superior to ordinary soap. Little staining.

These results show that there is, on the whole, a general correlation between the lubricating properties as measured in the laboratory and the performance as shown by practical drawing tests.

A further point of interest that developed in this investigation was the method of application of the lubricant emulsion. If the lubricating properties of a soap emulsion are examined in the laboratory it is found that even for chlorinated and sulphurized soaps there is a tendency for the lubricant to break down at 100° C. because of the ebullition of the water from the lubricant film. If, however, the lubricant film is deposited from a volatile solvent so that the water content of the lubricant film is extremely small, it lubricates effectively up to very much higher temperatures. This has already been described in Chapter X. This observation was found to have a direct application in practical drawing operations. If the lubricant emulsion is applied to the metal surface as a 'dry' film (by immersing it in a hot emulsion and taking it out so that most of the water in the film rapidly evaporates) a much better performance is obtained. The use of 'dry' films of this type in deep drawing has now become very widespread.

REFERENCES

O. Beeck, J. W. Givens, and E. C. Williams (1940), *Proc. Roy. Soc.* **A 177**, 103.
J. G. Brookman and R. B. Ham (1941), C.S.I.R. (Australia) Tribophysics Division Report A 37.
K. G. Brummage (1947), Private communication.
F. H. Constable (1929), *Proc. Roy. Soc.* **A 125**, 630.
W. Davey (1945), *J. Inst. Petrol.* **31**, 73.
H. Ernst and M. E. Merchant (1940), *Chip Formation, Friction and Surface Finish.* Cincinnati Milling Machine Co.
E. B. Greenhill (1948), *J. Inst. Petrol.* **34**, 659.
—— R. B. Ham, and D. Tabor (1945), C.S.I.R. (Australia) Tribophysics Division Report A 140.
J. N. Gregory (1948), *J. Inst. Petrol.* **34**, 670.
E. G. Herbert (1926), *Proc. Inst. Mech. Eng.* **2**, 289.
A useful survey of the various types of sulphurized E.P. lubricants has recently been given by H. Sellei (1949), *Petroleum Processing*, Sept., Oct.

XII

BREAKDOWN OF LUBRICANT FILMS [A]

As we have already seen, when surfaces slide under ideal hydrodynamic conditions there is no wear of the moving parts. In many mechanisms, however, it is not always possible to maintain completely hydrodynamic conditions and there may be a considerable amount of boundary lubrication during which wear of the surfaces may occur. For this reason it is important to be able to estimate the amount of non-hydrodynamic lubrication since this determines the life of the running parts. In practical tests the effectiveness of the lubrication in any particular mechanism is usually determined by measuring the amount of wear which occurs under given conditions, or the critical load and speed at which a marked increase of wear and seizure takes place. This method often provides useful empirical information, but it may shed little light on the detailed processes occurring during sliding.

A simple method of determining the extent to which fluid lubrication is occurring between any pair of sliding surfaces is to measure the electrical resistance between them. This method has already been described in Chapter V and we shall use the same method in Chapter XIII to investigate the mechanism of impact. If the surfaces are completely separated by a liquid hydrodynamic film the resistance will be large. For example the resistance of an oil film 10^{-5} cm. thick may be 10^6 to 10^{10} ohms/sq. cm. On the other hand, if metallic contact occurs across a metallic bridge only 10^{-6} sq. cm. in area the spreading resistance for a metal such as steel is of the order of $\frac{1}{100}$ ohm. Thus electrical resistance measurements provide a very sensitive means of discriminating between complete fluid lubrication and the smallest amount of metallic contact. The main disadvantage of this method is that it does not make a clear distinction between metallic contact and the contact between metal surfaces covered with boundary films. As we saw in Chapter I, if a lubricant film is placed between metal surfaces under loads sufficient to cause plastic flow of the metals at the regions of real contact, the electrical resistance between the surfaces is essentially the same as that observed with unlubricated surfaces. Similarly, when the surfaces are sliding, the electrical resistance for unlubricated metals is almost indistinguishable from that observed at the same load for the same metals lubricated with a boundary film, even though the coefficient of friction and the amount of surface damage may be very much lower.

It is clear, however, that the least separation of the surfaces beyond the region of boundary lubrication introduces resistance values ranging from several ohms to millions of ohms. Consequently a contact resistance exceeding, say, 100 ohms means that the surfaces are completely separated by a lubricant film, and no wear of the moving parts is possible. On the other hand, a contact resistance considerably less than 1 ohm means that there is either intimate metallic contact between the surfaces or at best a regime of boundary lubrication. If there is intimate metallic contact the wear may be high; if there is good boundary lubrication the wear may be slight, but it is clear that the wear cannot be less than that occurring under conditions of fluid lubrication. Consequently we may say that a high resistance means no metallic contact and no wear; whilst a very low resistance implies some wear, the extent of which depends on the effectiveness of the boundary film (Tabor, 1946). With these general observations in mind we may use resistance measurements to investigate the condition of lubrication between rubbing metal surfaces. In what follows we shall describe the application of this technique to an analysis of the lubrication between the piston rings and cylinder wall of a running engine, and between a journal and its bearing.

Lubrication between the Piston Rings and Cylinder Wall of a Running Engine

In this investigation, which was carried out by Mr. Courtney-Pratt and Mr. Tudor (1946), a small single-cylinder water-cooled engine was used. One of the piston rings was electrically insulated from the piston and leads were taken from the ring through a scissor link-motion to the bottom of the crankcase. The resistance of the oil film between the piston ring and the cylinder wall was measured by a current-potential method. Preliminary measurements showed that electrical breakdown of the oil film could be avoided if the potential drop across the film was less than 0·3 volt. In most of the measurements a considerably lower value was used. The resistance was recorded on a cathode-ray oscillograph the sweep of which could be synchronized with each cycle of the engine.

Effect of speed

The first and most striking result is that the surfaces of the piston ring and the cylinder wall are never separated by a continuous film of lubricant throughout the entire cycle. The resistance continuously fluctuates from 'infinity' to 'zero', indicating an intermittent breakdown of the oil film during the cycle. A typical result obtained with a given oil at three different speeds is shown in Plate XXIV. 1. When the trace is at the top of the figure it means that there is complete hydrodynamic lubrication. Any breakdown is shown by a drop in the resistance. It is seen that the amount of hydrodynamic lubrication is greatly increased with increasing speed. Nevertheless, even at the highest speeds there is a definite amount of breakdown of the hydrodynamic film.

This effect is shown in a different way in Plate XXIV. 2, where the resistance measurements are more sensitive and the cathode-ray trace has been synchronized with a single cycle of the engine. It is seen that the breakdown of the lubricant film is most marked near the ends of the stroke where the relative sliding velocity of the surfaces is the lowest during the cycle. It is clear that the slow speed of sliding accounts, at least in part, for the more rapid rate of wear observed in practice near the top and bottom dead centres.

Effect of viscosity and temperature

Two other results of considerable interest are the effect of viscosity and the effect of temperature. A series of oils was selected all of which were 'straight' mineral oils of varying viscosity but all distilled from the same crude. Their viscosities at the mean temperatures at which the runs were carried out are given below, and it is seen that the viscosities cover a range of 27:1.

Oil G 926 at 17° C. . . .	968 centistokes
,, G 917 at 18° C. . . .	264 ,,
,, G 915 at 20° C. . . .	108 ,,
,, G 913 at 20° C. . . .	35·4 ,,

The resistance traces are shown in Plate XXIV. 3. It is seen that the amount of hydrodynamic lubrication increases markedly for the oils of higher viscosity. This is in general agreement with practical observations. However, a reduction in the rate of wear cannot be effected simply by increasing the viscosity of the oil. In practice it is considerably more difficult to pump a thicker oil through the engine, so that there may be an appreciable delay at starting before the oil reaches the rubbing surfaces. Further, a thicker oil involves increased viscous-frictional resistance in the engine. It should also be noted that although the lubrication is improved by using an oil of high viscosity, hydrodynamic lubrication is never completely maintained throughout the entire cycle.

Similar experiments show that the temperature at which the oil is circulated through the engine has a marked effect on the lubrication. A typical result is shown in Plate XXIV. 4, where it is seen that a small temperature increase from 28° to 54° C. with a corresponding decrease in viscosity from 330 to 64 centistokes causes a very marked deterioration in the lubrication. This effect is reversible on cooling and is not due solely to the decrease in viscosity. This was shown very clearly by a second set of experiments carried out at 20° C. with oils which had the same viscosities as the hot oils used in Plate XXIV. 4. In every case the lubrication with the hot oil was worse than that obtained with the cool oil of equal viscosity. Thus apart from considerations of viscosity there is a marked deterioration in lubrication with increasing oil temperature, an effect that appears to be directly connected with the breakdown of boundary lubricants as the temperature is raised (see Chap. IX). These results suggest that, from the point of view of wear-reduction, the engine should be run as cool as possible; in practice, however, as the researches of C. G. Williams (1940) show, the lower limit is determined by the temperature at which the products of combustion condense on the cylinder and so produce increased corrosive wear. A good general account of the part played by *corrosion* in the wear of an internal combustion engine has been given by Williams (1940).

Lubrication between a Journal and Bearing

Similar experiments have been carried out by Mr. Tudor (1947) on the lubrication of a steel shaft running in a white-metal sleeve bearing. The bearing, which was of conventional design, was $1\frac{3}{4}$ in. diameter and $1\frac{3}{4}$ in. long and the nominal loading could be varied from about 200 to 600 lb./sq. in. By a technique similar to that described above the electrical resistance of the oil film was recorded on a cathode-ray oscillograph. No measurements were made of the frictional torque. Under ideal conditions the rotating shaft drags a converging wedge of oil between it and the bearing and the hydrodynamic pressures developed in the lubricant film are sufficient to keep the surfaces completely separate (Reynolds, 1886). The pressure distribution in the oil film is shown in Fig. 89. There is no wear of the rotating parts and the electrical resistance of the oil film is very high. Under these conditions the coefficient of friction is directly proportional to the dimensionless parameter ZN/P, where Z is the viscosity of the oil, N is the speed of revolution of the shaft, and P is the nominal pressure on the bearing.

Earlier experiments in which the frictional torque was measured show that this is generally true for larger values of ZN/P; below a critical

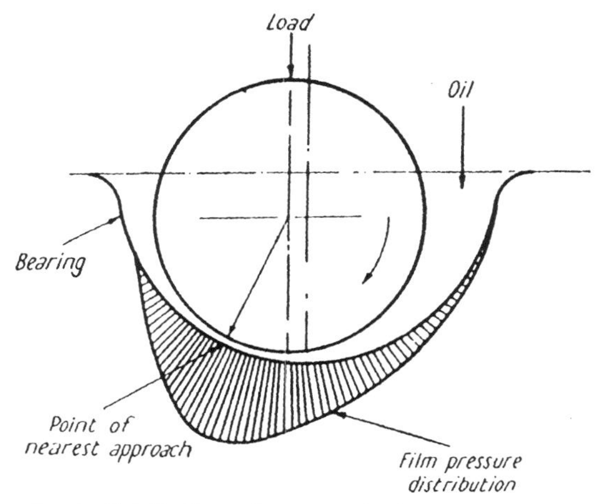

FIG. 89. Approximate distribution of pressure in the oil wedge between a journal and a 180°-bearing, under conditions of hydrodynamic lubrication.

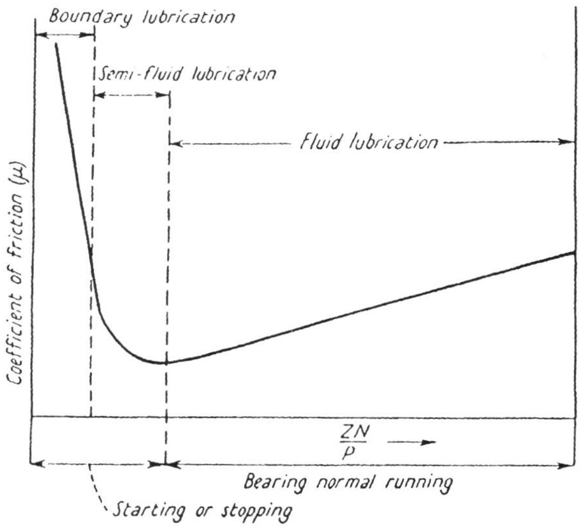

FIG. 90. Frictional behaviour of a journal and bearing as the operating conditions are varied. μ = coefficient of friction, Z = viscosity of lubricant, N = number of revs. per min., P = nominal pressure. Under purely hydrodynamic conditions μ is low and is proportional to ZN/P. Under less favourable conditions the lubricant film breaks down and there is a marked rise in friction.

value of ZN/P, however, there is a rise in friction which may be very steep (Fig. 90). This transition marks the stage at which there is no longer complete fluid lubrication (see, for example, Hersey, 1938). Consequently in this range the electrical resistance is low whilst the amount of surface damage and seizure may be high. In order to reduce the amount of wear every effort is made in engineering practice to keep the journal running on the straight portion of the curve where conditions are those of purely hydrodynamic lubrication. As we shall see below, however, it is very difficult in practice to attain complete separation of the rotating surfaces.

Effect of load, speed, viscosity, and temperature

The effect of load, speed, and viscosity on the electrical resistance between the shaft and the bearing is shown in Plate XXV. In these figures each trace represents approximately one complete revolution of the shaft. It is seen that the smaller the load, the less is the amount of breakdown of the lubricant film (Plate XXV. 1). A similar effect is observed when the speed of rotation (Plate XXV. 2) or the viscosity of the oil (Plate XXV. 3) is increased. These results are in direct agreement with the ZN/P characteristics discussed above. The traces show, however, that even under the most favourable conditions of low load, high speed, and high viscosity, there is a definite amount of breakdown of the hydrodynamic film.

The effect of the temperature of the circulating oil is shown in Plate XXVI. 1. It is seen that a temperature rise of only 11° C. (from 19° to 30°) is sufficient to change the resistance trace from one that is predominantly hydrodynamic (trace (1 *a*)) to one that is almost completely non-hydrodynamic (trace (1 *c*)). This effect is mainly due to the decrease in viscosity of the oil with increasing temperature.

There are two other effects of particular interest that the investigation reveals. The first is the influence of 'running-in'. Plate XXVI. 2 shows the results obtained at various stages of an experiment, each section representing approximately a single revolution of the shaft, commencing at approximately the same relative position of the shaft and bearing. It is seen that 5 minutes after starting there is a great deal of non-hydrodynamic lubrication. As the time of running increases the amount of breakdown decreases and after about 2 hours a considerable portion of the revolution is operating under hydrodynamic conditions. Because of the running-in effect the comparative traces in Plates XXV and XXVI. 1 were carried out after equal periods of running from standstill.

The running-in process presumably consists in removing surface asperities which normally penetrate the lubricant film, and in slightly modifying the contour of the bearing to provide a better 'fit' for the journal. These conclusions are in complete agreement with experimental measurements of the frictional torque. Fig. 91 shows some friction-ZN/P curves obtained by McKee (1927) in his classical work on the lubrication of a bearing. It is seen that the point at which the

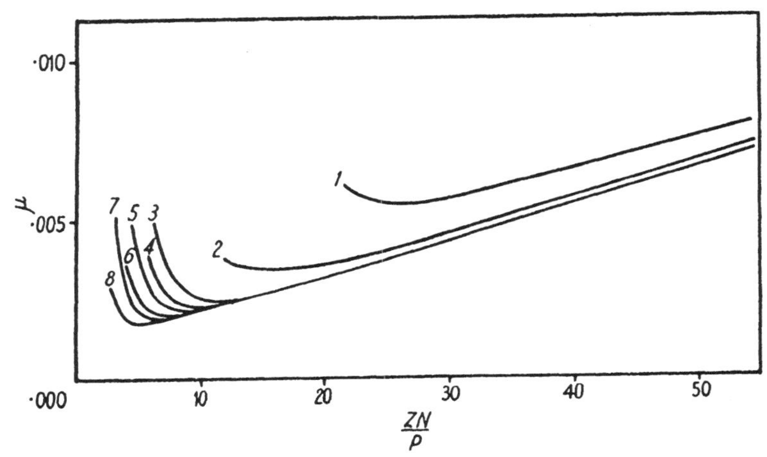

FIG. 91. Coefficient of friction of a journal and bearing as running-in proceeds. From McKee's paper (1927). μ = coefficient of friction, Z = viscosity, N = revs. per min., P = nominal pressure. As running-in proceeds through eight consecutive runs the range over which hydrodynamic lubrication is obtained is extended into the more unfavourable region of ZN/P. Compare, for example, run No. 8 after prolonged running-in, with run No. 1 at the commencement of the experiment.

minimum value of the friction occurs moves down the ZN/P axis as the time of running proceeds. This means that hydrodynamic lubrication persists into a more extreme range of loads, speeds, and viscosities as the bearing is run in.

Similar experiments show that the 'running-in' process may be greatly facilitated if the bearing is run at a heavier load for a few hours. Nevertheless the resistance measurements show that although the 'running-in' of the bearing may greatly reduce the extent to which asperities break through the hydrodynamic film, the breakdown is never entirely eliminated in the experiments described above.

The second effect is the rate of 'building-up' of the lubricant film. As the shaft starts rotating there is a rapid improvement in the lubrication over the first 10 or 20 revolutions of the shaft. After this, as a result of the running-in effect already described, a much slower improvement is observed. The early build-up of the hydrodynamic film at a speed of 335 r.p.m. is shown in Plate XXVII for the first 9 revolutions of the shaft and the progressive improvement in the lubrication is clearly visible. It should be emphasized that this is not due to a gradual speeding up of the shaft: the experiment was so arranged that the shaft reached full speed during the first revolution. Similar experiments at other speeds of rotation show that the early build-up of the hydrodynamic film is more rapid the higher the speed of rotation. This effect is probably due to hydrodynamic factors governing the establishment of equilibrium conditions. It is also possible that there is a more effective removal of the initial surface asperities at higher speeds of rotation.

Effect of Temperature on Lubricant Films

It is clear from the previous discussion that even under the most favourable conditions of a journal running in a bearing it is very difficult to obtain completely hydrodynamic conditions. For this reason the wear of the sliding surfaces will, in the last resort, be determined by the efficacy of the boundary film which remains when the hydrodynamic film is broken down. As we have seen, the lubricating properties of the boundary film are determined by the nature of the underlying metals, the speed of sliding, and other variables. One of the most important of these is the temperature developed between the rubbing surfaces. The boundary lubricating properties of hydrocarbons, alcohols, and fatty acids break down when the temperature is sufficiently high to produce a softening or melting of the lubricant film. This behaviour is observed with practically all boundary films, and provided the heating is so rapid that there is not sufficient time for appreciable oxidation of the lubricant to occur, the changes in friction and surface damage are reversible on cooling.

This behaviour is shown in Plate XXVIII for a typical commercial lubricant possessing 'good' boundary lubricating properties (Tabor, 1941). At room temperature the motion is smooth and the friction low, and this persists up to a temperature of about 70° C. (Plate XXVIII. 1, portion *A*). The track shows a correspondingly light wear (Plate XXVIII. 2*a*). At 70° C. the lubrication begins to break down (Plate XXVIII. 1, portion *B*), and the deterioration becomes more marked as the temperature is increased further (Plate XXVIII. 1, portion *C*). At 200° C. the friction has increased to a high value ($\mu \approx 0{\cdot}4$) with intermittent motion, while the tracks show a corresponding increase in surface damage (Plate XXVIII. 2*b*, 2*c*, and 2*d*). If now the surface is rapidly cooled (to prevent appreciable oxidation) the motion becomes smooth, the friction reverts to its original low value of $\mu \approx 0{\cdot}1$, and the surface damage again becomes light (Plate XXVIII. 2*e*).

It is clear that this type of change may have a very significant effect on the practical performance of a lubricant. This is particularly true with 'undoped' oils, since for most commercial lubricants of this type the breakdown of the boundary film occurs at moderate temperatures. Further, since the effect is reversible, increased seizure and surface damage may take place even though no detectable change occurs in the bulk properties of the lubricant.

A different type of change which is *not* reversible is observed if the heating is prolonged and more severe, so that appreciable oxidation of the oil can occur. At an early stage the oxidation products may produce an improvement in the lubricating properties of the oil, comparable to that produced by the addition of a small quantity of fatty acid. At a later stage, however, the oxidation leads to a marked deterioration in the lubrication. These effects are illustrated in Plate XXIX for a typical mineral oil on steel surfaces (Bowden, Leben, and Tabor, 1939). Plate XXIX. 1*a* shows the frictional behaviour of the mineral oil at room temperature and Plate XXIX. 1*b* the corresponding track. Plates XXIX. 2*a* and 2*b* show the effect of adding 1 per cent. caprylic acid to the oil. It is seen that the friction is decreased, the motion becomes smooth, and the surface damage is considerably reduced. (This effect has already been described in Chapter IX.)

If now the steel surface covered with a thin film of the straight mineral oil is heated, the oxidation of the oil can be followed by its effect on the frictional behaviour. For example, if the surface is maintained at 150° C. a marked change in the friction is observed after 30 minutes' heating. The friction becomes low, the motion smooth, and the surface damage slight. At 200° C. similar effects occur after only 15 minutes' heating. The friction and track are shown in Plate XXIX. 3*a* and 3*b* respectively, and it is seen that the behaviour is very similar to that observed when a small quantity of fatty acid is added to the mineral oil (Plate XXIX. 2*a* and 2*b*). At 300° C. the change occurs even more rapidly and low friction is observed after less than 2 minutes' heating. If the heating is continued for 20 minutes, however, another change takes place. The oil forms a thick gum, the friction rises to a high value, and the track shows greatly increased wear (Plate XXIX. 4*a* and 4*b*).

The fact that these effects are due to oxidation of the oil is shown by treatment of the mineral oil in bulk before use. If the oil is heated while air is bubbled through it, similar frictional changes are observed. There is a slow improvement in lubrication at first, accompanied by a faint discoloration of the oil. After prolonged heating the lubrication again deteriorates and the oil turns dark brown in colour. If the air is excluded during heating, these changes do not occur. Similarly if ozone is bubbled through the oil at 0° C. there is a gradual improvement in the frictional behaviour. These changes are all analogous to those observed when the oil film is heated on the steel surface. It may, however, be noted that in the latter case the changes occur much more rapidly. This suggests that the formation of the oxidation products may be catalysed by the steel surface itself.

These results show that when a mineral oil in contact with a steel surface is heated in air, compounds are formed which improve its lubricating properties; the friction is decreased and the amount of surface damage considerably reduced. At low temperatures the rate of formation of these compounds in the oil is very slow, but it is greatly accelerated by a rise in temperature. At temperatures up to 200° C. the oil film maintains its improved lubricating properties even after prolonged heating. At 300° C., however, further rapid changes take place. Most of the oil is driven off; the residue forms a thick gum and there is an increase in friction and wear.

The fact that these compounds are so readily formed on metal surfaces means that in practice a fresh oil may be a worse lubricant than an old one. The results suggest that if the temperature is about 200° C. the compounds favouring lubrication are produced quite rapidly and may decrease the friction and surface damage. It does not follow, of course, that the resultant effect of these compounds will be beneficial. As oxidation proceeds the acidic products may lead to corrosive wear of the metal surfaces. At a later stage oxidation combined with polymerization may produce an increase in the viscosity of the lubricating oil and the formation of products which are insoluble in the bulk of the oil. These products appear in the form of sludge or as gum-like deposits and in internal combustion engines they give rise to sticking of the piston rings and the valve stems. Further, as the experiments at 300° C. show, these products in themselves have poor boundary properties and may lead to increased friction and wear.

Thus although a small amount of oxidation may be beneficial, continuous oxidation may be disadvantageous. The position is rendered more difficult by the fact that the oxidation process appears to be autocatalytic. For this reason it is now common practice to add substances in small proportions to lubricating oils in order to inhibit or counteract the effect of oxidation. A review of these substances, which cover a very wide range of chemicals, has recently been given by Webber (1945).

REFERENCES

F. P. Bowden, L. Leben, and D. Tabor (1939), *Trans. Faraday Soc.* **35**, 900.
J. S. Courtney-Pratt and G. K. Tudor (1946), *Proc. Inst. Mech. Eng.* **155**, 293.
M. D. Hersey (1938), *Theory of Lubrication.* John Wiley, New York.
S. A. McKee (1927), *Mech. Eng.* **49**, 1335.
O. Reynolds (1886), *Proc. Roy. Soc.* **A 40**, 191.
D. Tabor (1941), *Engineering*, **152**, 178.
—— (1946), *Proc. Inst. Mech. Eng.* **155**, 317.
G. K. Tudor (1947), C.S.I.R. (Australia) Tribophysics Division Report A 155; (1949) *J.C.S.I.R.* (Australia), **21** (3), 202.
M. W. Webber (1945), *Petroleum*, **8**, 76.
C. G. Williams (1940), *Collected Researches on Cylinder Wear.* Inst. Auto. Engrs.

XIII

NATURE OF CONTACT BETWEEN COLLIDING SOLIDS [A]

In this chapter we shall consider briefly the nature of contact between colliding surfaces. This is of general interest in a number of fields involving resistance to deformation under impact. For example, it is of importance in the dynamic hardness of metals and in lubrication problems where vibrations in the moving parts may produce rapid changes in the normal force between the rubbing surfaces. The processes involved in the impact of solids are also of importance in other problems such as the initiation of explosives by impact.

The study of impact between colliding solid bodies received its first theoretical treatment at the hands of St.-Venant (1867), who suggested that the total period of the collision is determined by the time required for an elastic compression wave to travel through the solid and be reflected back again. Since the velocity of a compression wave in a metal is of the order of 10^5 cm./sec., this would give a value of a few microseconds for impacts between bodies which are a few centimetres in length. As we shall see later, the experimental results give a very much longer collision period and the evidence suggests that for small bodies the collision process is determined mainly by the deformations occurring at the regions of contact. If the deformations are elastic, the Hertzian equation may be applied (Hertz, 1881). If, as is generally the case, plastic deformation occurs the plastic equations of Chapter I must be used. In both cases the deforming processes at the region of contact are relatively protracted and the elastic compression waves have sufficient time to travel to and fro several times and dissipate themselves uniformly throughout the colliding bodies. It is only when the bodies are relatively long that the period of collision is determined by St.-Venant's relation.

In what follows we shall deal with collisions between solid bodies that are of the order of a few centimetres in length. Consequently we may ignore the problem of the compression wave and concentrate on the forces and deformations occurring at the actual regions of contact. We shall deal with spherical surfaces and flat surfaces in collision, and then consider the effect of liquid films interposed between them.

Spherical Surfaces

Suppose a hard sphere of radius r_1, the indenter, is dropped on to the horizontal flat surface of a softer metal, the anvil (Fig. 92). The impact may be divided into four main stages. At first the region of contact will be deformed elastically, and if the impact is sufficiently gentle the surfaces will then recover elastically and separate without residual deformation. The collision in this case is purely elastic, and the time of impact, mean pressures, and deformations are given by Hertz's equation. The second stage occurs if the impact is such that the mean pressure exceeds about $1 \cdot 1Y$ when a slight amount of plastic deformation will occur at a critical region and the collision will no longer be truly elastic (see Chap. I). Consequently Hertz's equations will no longer be strictly valid. It is interesting to note that this onset of plastic deformation occurs very readily in the *bulk* of the anvil, quite apart from the deformation of the surface asperities. For example, a ball of diameter 1 cm., mass 4 gm., dropped from a height of only 2 cm. will produce a permanent impression in very hard tool steel. As the energy of the impact increases, the deformation rapidly passes over to a condition of full plasticity (stage 3) and full-scale plastic deformation proceeds until the whole of the kinetic energy of the collision is consumed. Finally a release of elastic stresses in both the surfaces takes place, as a result of which rebound occurs (stage 4).

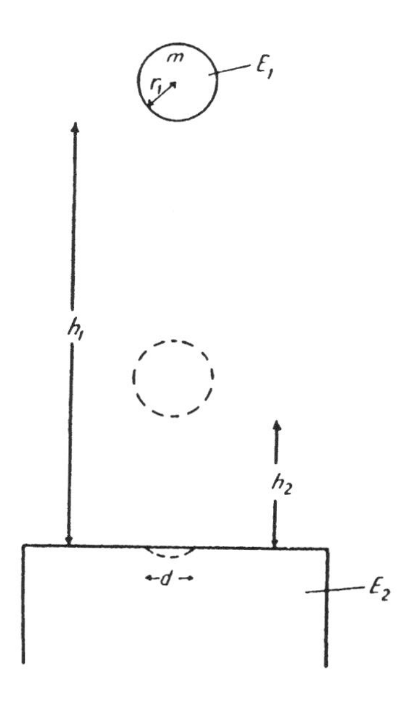

FIG. 92. A hard steel ball, radius r_1, mass m, falls from a height h_1 on to the surface of a massive anvil and rebounds to a height h_2. An indentation is left in the anvil of chordal diameter d.

A full analysis of the four phases involved in the collision process is complicated and difficult. If we restrict ourselves to a consideration of the forces involved we may simplify the analysis considerably (Tabor, 1948). Let us assume that there is a dynamic yield pressure p which is a constant and independent of the speed of impact and size of deformation. As soon as the pressure reaches a value p, plastic flow occurs and continues until the pressure falls below this value. If the indenter, after rebounding, leaves a permanent impression of volume V in the surface, the work done as plastic deformation is given by $W = pV$. If the heights of impact and rebound are respectively h_1 and h_2, the energy lost by the collision is $mg(h_1-h_2)$. Hence

$$pV = mg(h_1-h_2). \tag{1}$$

The recovered indentation is *not* a portion of a sphere of the same radius of curvature as the indenter since the occurrence of rebound implies that there has been elastic 'recovery' of the surface of the indentation. It is not difficult to calculate the extent to which the indentation has 'shallowed' as a result of this effect and to relate it to the energy of rebound mgh_2. Calculations show that if the indentation has a chordal diameter d, and E_1, E_2 are Young's moduli of the indenter and the anvil respectively and Poisson's ratio for both surfaces is 0·3,

$$mgh_2 = 0{\cdot}34p^2d^3\left(\frac{1}{E_1}+\frac{1}{E_2}\right). \tag{2}$$

We can also express the true volume V of the indentation in terms of the apparent volume V_a. This is the volume which would be obtained if the radius of curvature of the indentation were the same as that of the indenter, and is approximately equal to $\pi d^4/64r_1$. Similarly the true volume V is approximately equal to $\pi d^4/64r_2$. We may now eliminate r_2 from V by making use of the Hertzian relation which connects r_1, r_2, and d with the force $F = p\pi d^2/4$ at the end of the indentation. We have (see Chapter I, equation (1)) that

$$d = 2{\cdot}22\left[\frac{F}{2}\left(\frac{1}{r_1}-\frac{1}{r_2}\right)\left(\frac{1}{E_1}+\frac{1}{E_2}\right)\right]^{\frac{1}{3}},$$

so that

$$\frac{1}{r_2} = \frac{1}{r_1} - 5{\cdot}54\frac{F}{d^3}\left[\frac{1}{E_1}+\frac{1}{E_2}\right].$$

Consequently

$$V = V_a - 0{\cdot}21pd^3\left(\frac{1}{E_1}+\frac{1}{E_2}\right).$$

Inserting in equation (1) and combining with equation (2) we obtain

$$pV_a = mg(h_1-h_2)+\tfrac{5}{8}mgh_2.$$

Hence

$$p = \frac{mg(h_1-\frac{3}{8}h_2)}{V_a}. \tag{3}$$

This enables us to determine p in terms of the heights of impact and rebound and the apparent size of the indentation.

The validity of this analysis depends on the assumption that the internal forces involved in the actual impact are essentially the same as those involved in the analytical model just described. In particular we assume that the energy lost as elastic waves is negligible and that the temperature rise in the material around the indentation during impact is small and has a negligible effect on the strength properties of the metal. As we shall see later, both of these assumptions are reasonably valid.

Effect of variation in the yield pressure

In the above analysis we have assumed that p is constant. In practice there are two reasons why we may expect p to vary during the course of the impact. First there is a dynamic effect, analogous to a viscosity factor which makes p higher at the beginning of the indentation process where the rate of deformation of the material is a maximum. This effect is discussed later: it is, however, difficult to allow for it quantitatively. Secondly, work-hardening proceeds as the indentation is formed, so that we may expect an increase in the mean pressure resisting deformation. We may estimate the order of this effect by assuming, on analogy with the static case, that $p = kd^{n-2}$ where n is constant with a value between 2 and 2·5; see Chapter I, equation (4). If we follow the above analysis right through we obtain a relation similar to equation (3). The actual result for the mean pressure at the *end* of the indentation process is

$$p = \frac{n+2}{4}\frac{mg(h_1-\beta h_2)}{V_a}, \tag{3 a}$$

where $\beta = (2n-1)/(2n+4)$ and therefore varies from $\frac{3}{8}$ to $\frac{4}{9}$ as n varies from 2 to 2·5. This has little effect on the value of p. Similarly, the first term varies from 1 to 1·12 as n varies from 2 to 2·5. It follows that the variation of the yield pressure as a result of work-hardening produced during the process of indentation will, at most, cause an error in p, when we use equation (3), of about 10 per cent. In general the error will be appreciably less.

The validity of equations (2) *and* (3). According to equation (2), if we restrict our attention to any one material (for which we assume p and E are constant), the height of rebound h_2 should be proportional to d^3, and therefore to $(V_a)^{\frac{3}{4}}$ since, as we have seen, V_a is proportional to d^4. Some results from Edwards and Austin's paper (1923) are plotted in Fig. 93, and it is seen that h_2 is approximately proportional to $(V_a)^{\frac{3}{4}}$. Similarly, if we carry out impact experiments on a whole series of metals, the height of rebound h_2 for a fixed size of indentation (d = constant) should be proportional to $p^2\left(\frac{1}{E_1}+\frac{1}{E_2}\right)$. Results from the same authors are given in Fig. 94, where p has been calculated from equation (3). Using logarithmic ordinates and plotting $p\sqrt{\left(\frac{1}{E_1}+\frac{1}{E_2}\right)}$ against h_2 it is seen that over a very wide range of materials the points lie on a straight line of slope $\frac{1}{2}$.

Finally we may eliminate d between equations (2) and (3). The resulting relation between h_1, h_2, and p is then given by

$$p^5 = \frac{h_2^4}{(h_1-\frac{3}{8}h_2)^3}\frac{mg}{109r_1^3}\frac{1}{(1/E_1+1/E_2)^4}. \tag{4}$$

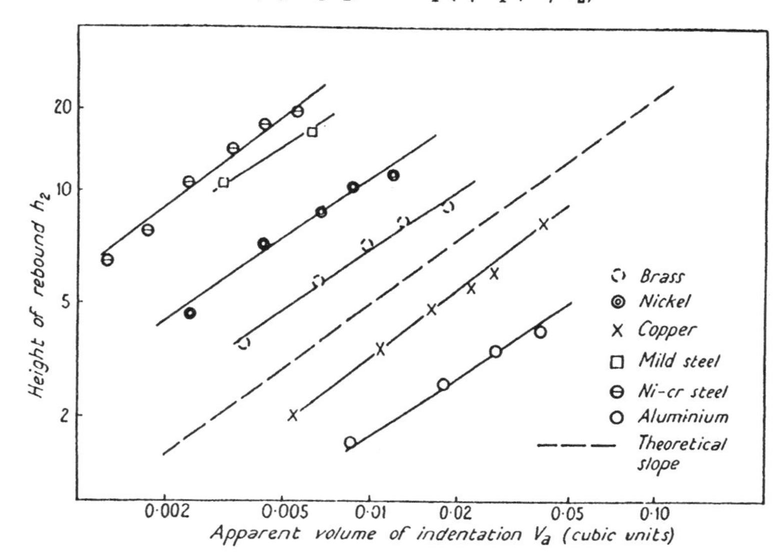

FIG. 93. Height of rebound as a function of the apparent volume of the indentation (logarithmic ordinates). For a wide range of metals there is reasonably good agreement between the observed slope and the theoretical slope.

Since the bracket involving Young's moduli does not vary greatly for most metals, we may treat this factor as a constant and plot p as a function of h_2 for a given height of fall h_1. The theoretical curve is shown in Fig. 95. If we allow for the fact that softer metals usually have a smaller Young's modulus, the curve is modified in a manner similar to that shown in the dotted curve. These theoretical curves do in fact reproduce the main characteristics observed in the practical calibration of rebound scleroscopes. It is also evident from Fig. 95 that over a wide range of experimental conditions the height of rebound, for a fixed height of fall, is almost directly proportional to the dynamic yield pressure.

The condition for elastic collisions. It is interesting to consider what happens when $h_2 = h_1$, i.e. when the collision is completely elastic. If we go back to our original equations, and calculate the maximum

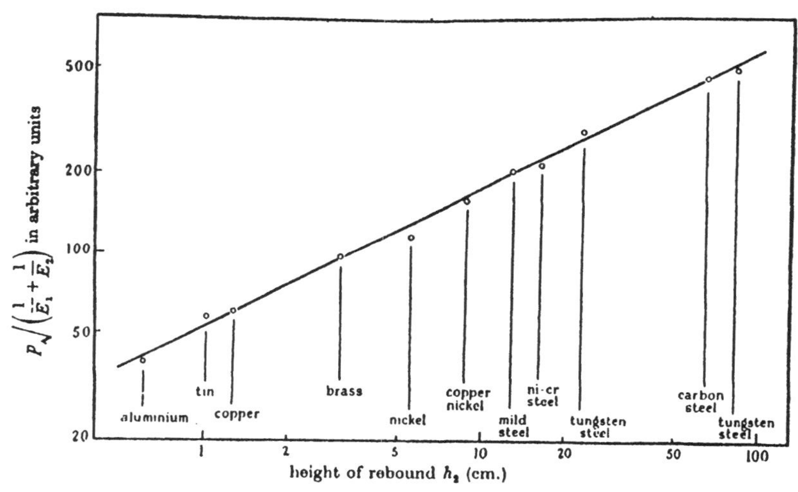

FIG. 94. The factor $p\sqrt{\left(\frac{1}{E_1}+\frac{1}{E_2}\right)}$ plotted against the height of rebound (logarithmic ordinates) for a fixed size of indentation. The theoretical relation gives a straight line of slope $\frac{1}{2}$; the observed straight line has a slope of 0·51.

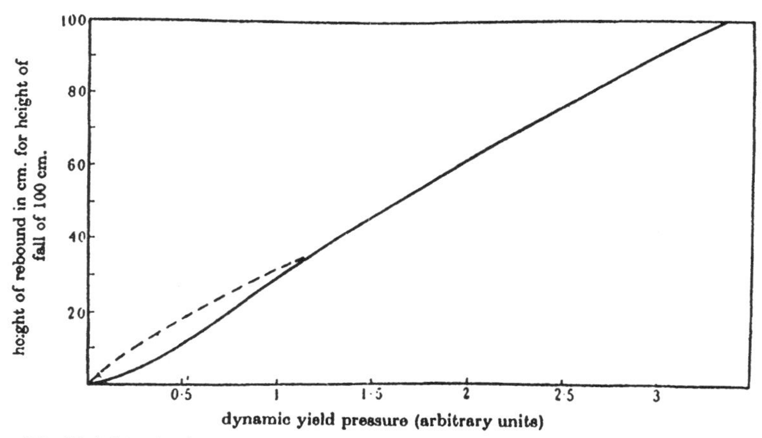

FIG. 95. Height of rebound (for a height of fall of 100 cm.) as a function of the dynamic yield pressure. The full line assumes that all materials have the same elastic constants whatever their hardness. The dotted line indicates the deviations to be expected if softer materials have small elastic constants. These curves are similar to the calibration curves of the rebound sclerometer.

pressure p_e developed between surfaces colliding elastically, we obtain

$$p_e^5 = \frac{1}{26\cdot 6}\frac{mgh_1}{r_1^3}\frac{1}{(1/E_1+1/E_2)^4}. \tag{5}$$

This is precisely the value obtained if we put $h_1 = h_2$ in equation (4).

Two conclusions follow from this result. The first is that equation (4) is valid right up to 100 per cent. rebound. Secondly, if the yield pressure of the metal is greater than p_e no plastic deformation occurs and the collision is completely elastic. If it is less than p_e, plastic deformation occurs and the value of p is the dynamic yield pressure of the material. This approach has been used by Sir Geoffrey Taylor (1946) to study the limiting conditions for plastic yielding under dynamic conditions. It is also evident that the rebound method will not discriminate between materials possessing yield stresses above p_e since they will all give a rebound of 100 per cent.† To increase the range of the method the value of p_e must be increased either by increasing h or m or by reducing r_1.

The coefficient of restitution

If v_1 is the velocity of impact and v_2 the velocity of rebound, the ratio v_2/v_1 is defined as the coefficient of restitution e. We may calculate e from equation (4) by putting $v_1^2 = 2gh_1$; $v_2^2 = 2gh_2$. If we assume that p remains essentially constant, we obtain

$$v_2 = k(v_1^2 - \tfrac{3}{8}v_2^2)^{\frac{3}{8}}. \quad (6)$$

It is clear that v_2 is not linearly dependent on v_1, so that e will not be a constant. The way in which e varies with the velocity of impact is shown in Fig. 96, where curves i, ii, iii, iv, and v respectively have been drawn for values (at an impact velocity of 450 cm./sec.) of $e = 1{\cdot}0$, $0{\cdot}8$, $0{\cdot}6$, $0{\cdot}4$, and $0{\cdot}2$. (This velocity of 450 cm./sec. corresponds to a height of fall of 100 cm.) Because p is not a constant there will be some deviation from these curves. Nevertheless the general form of these curves is fully substantiated in practice. Typical results (Tabor 1948) for cast steel and drawn brass are shown in the dotted lines and similar curves have been obtained by Raman (1918), Okubo (1922), and Andrews (1930) in experiments on the impact of spheres of similar metals.

It is clear from equation (6) and from the experimental results that, in general, the coefficient of restitution of plastic solids will not be constant. At sufficiently low velocities of impact the collisions will be purely elastic and the coefficient of restitution will be unity. This occurs even with the softest metals, as Andrews (1931) showed in his experiments with spheres of lead and tin alloys. As the velocity of impact is increased the amount of plastic deformation steadily increases and there is a corresponding decrease in the coefficient of restitution.

† In practice, of course, 100 per cent. rebound is never attainable since some energy is lost by the dissipation of elastic waves within the bodies.

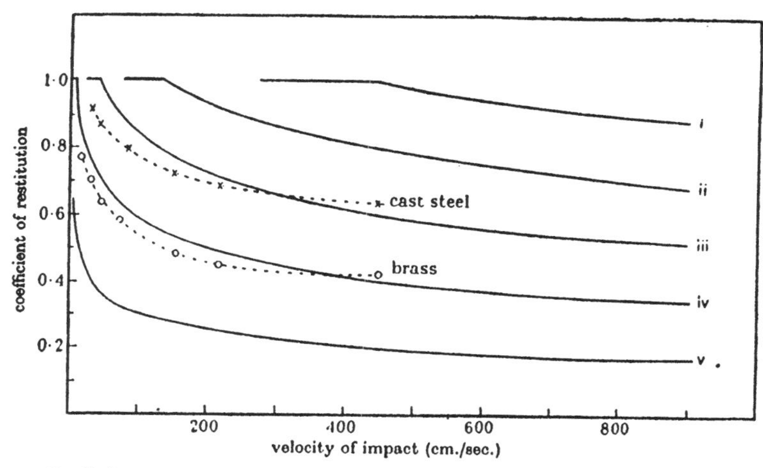

FIG. 96. Coefficient of restitution e as a function of velocity of impact v_1. Full lines represent theoretical curves.

(i) $e = 1$ when $v_1 = 450$ cm./sec.
(ii) $e = 0{\cdot}8$,, ,,
(iii) $e = 0{\cdot}6$,, ,,
(iv) $e = 0{\cdot}4$,, ,,
(v) $e = 0{\cdot}2$,, ,,

The broken lines are experimental curves. At very low velocities of impact even the softest materials give a coefficient of restitution of unity, i.e. the collision is completely elastic. At higher velocities plastic deformation occurs and the coefficient of restitution decreases.

Comparison of static and dynamic hardness

Some impact experiments were carried out on massive anvils of various metals using hard steel balls. By measuring h_1, h_2, and the chordal diameter d of the indentations formed, the dynamic yield pressure p was calculated using equation (3). Some static experiments were also carried out and determinations made of the static yield pressure p_m required to produce impressions of the same diameter as in the corresponding impact experiments. The results showed two main points. Firstly the dynamic yield pressure p was always greater than the static yield pressure p_m. This was particularly marked with soft metals, such as lead and indium. Secondly, the dynamic yield pressure is higher at greater velocities of impact. This suggests that in calculating p from equation (3), i.e. from the energy required to produce an indentation of given volume, part of the energy is used in the dynamic displacement of the metal around the indentation. This view is confirmed by a calculation of the yield pressure from the height of rebound h_2. We do this using equation (2) and give p the suffix r (p_r) to show that it is calculated from the rebound height. The results show at once that although p_r is larger than p_m it is very much closer to the static value than is p, as may be seen from a few typical values in the following table.

TABLE XXXIV

Metal	p/p_m	p_r/p_m
Steel . .	1·28	1·09
Brass . .	1·32	1·10
Al alloy . .	1·36	1·10
Lead . . .	1·58	1·11
Indium . .	5·0	1·6

Diameter of ball 0·5 cm. Height of fall about 300 cm.

The fact that the dynamic yield pressure is higher than the static and increases with the velocity of impact suggests that, in the dynamic deformation of metals, forces of a quasi-viscous nature are involved.

This is borne out by the results for the soft metals, lead and indium, where the pressures required to produce plastic deformation dynamically are very much greater than the static values. This cannot be due to the work-hardening, which may occur rapidly during the formation of the indentation since at the end of the impact, where the work-hardening would be a maximum, the effective flow pressure p_r is very much smaller than the mean dynamic flow pressure p which is involved during the course of the impact itself. It would seem that in the deformation of soft metals, where relatively large volumes of metal are displaced, appreciable forces are called into play as a result of the 'viscous' flow of the deformed material surrounding the indentation. The increase in yield pressure at very high speeds of deformation has recently been discussed by Sir Geoffrey Taylor (1946).

Finally, at the end of the impact process all plastic flow of the material has ended and there is no further bulk displacement of the metal around the indentation. All the deformation around the indenter is now of an elastic nature and any kinetic energy imparted to the material under these conditions is reversible. Consequently the pressures involved in this portion of the collision process (p_r) are only a few per cent. higher than those involved in the formation of indentations of the same size under static conditions.

The analysis given above refers to the impact between a hard sphere and a plane surface of a softer metal. The main conclusions remain valid for the collision of a soft sphere on a harder surface and for the collision of two spheres of similar metals. In all cases the collision process involves the four main phases discussed above, and in general, except for the hardest materials and the lightest impacts, the collision readily results in plastic deformation of the surfaces. The forces involved in such deformations are of the same order as those occurring in static deformations, but are invariably larger. With hard metals the dynamic yield pressure is usually a few per cent. higher than the static value, but with soft metals the difference may be very much more marked.

These results emphasize the ease with which plastic flow of colliding metal surfaces may occur. The implications of this will be discussed at greater length later.

Time of impact

The earliest method of determining the time of impact between colliding surfaces consisted in measuring by means of a ballistic galvanometer the amount of charge leaking away from a battery or charged condenser through the contacting bodies. This method is relatively simple, but suffers from the disadvantage that it merely gives the total time of impact. It does not indicate the way in which the area of contact varies during the collision itself. A more instructive method is to measure the electrical resistance or conductance between the metal surfaces during their collision. When the surfaces are separated, the conductance is zero. As they come together during the collision the conductance rises to a maximum and then falls off again as they come apart. The conductance at any instant, which may be recorded on a cathode-ray oscillograph, is a measure of the area of contact (Bowden and Tabor, 1941).

Typical results for metal spheres (radius 2 cm.) freely suspended by very fine wires are shown in Figs. 97 (*a*), (*b*) and (*c*). Fig. 97 (*a*) is a record of the conductance, using very hard steel spheres and a very low velocity of impact (10 cm./sec., equivalent to a vertical height of fall of $\frac{1}{20}$ cm.). It is seen that the conductance curve is symmetrical, indicating that the collision is essentially elastic. At higher velocities of impact, however, the results are similar to those shown in Figs. 97 (*b*) and (*c*).

Fig. 97 (*b*) is for mild steel spheres and Fig. 97 (*c*) for lead spheres impinging with a velocity of 76 cm./sec. Contact begins at the point A, and the curve AC corresponds to the range over which the elastic and initial onset of plastic deformation takes place. The curve CD corresponds to the range over which full plastic deformation occurs, while DEB represents the separation of the surfaces under the released elastic stresses. This type of conductance curve during the approach and separation of the surfaces has been shown more clearly in some recent experiments of Dr. Hirst (unpublished). The point D at which the surfaces begin to separate is not defined very precisely on a single oscillograph trace, so that the time interval from A to D can be estimated only to within a few per cent. However, this error is not important compared with the experimental variation from trace to trace.

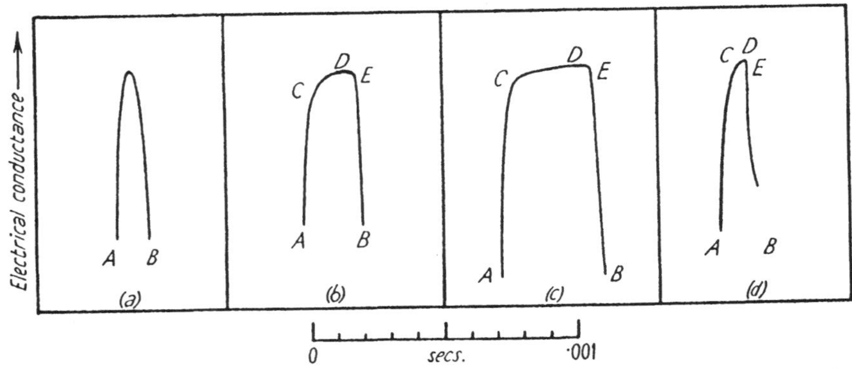

FIG. 97. Cathode-ray oscillograph records of electrical conductance between colliding metal surfaces. (*a*) Spheres of tool steel, diam. 4 cm., velocity of impact 10 cm./sec. The conductance trace is symmetrical, showing that the collision is essentially elastic. (*b*) Spheres of mild steel, diam. 4 cm.; velocity of impact 76 cm./sec. (*c*) Spheres of lead, diam. 4 cm.; velocity of impact 76 cm./sec. (*d*) Hard hemispherical surface, diam. 0·6 cm., mass 42 gm., striking a flat brass anvil with a velocity of impact 140 cm./sec. For (*b*), (*c*), and (*d*) the conductance curve is asymmetric, showing that plastic deformation has occurred. The portion AC represents the initial contact, CD the range over which plastic deformation occurs, while DEB represents the separation of the surfaces under the released elastic stresses.

Numerous collision experiments show that all impacts which involve plastic deformation of the surfaces yield the same type of asymmetric conductance curves as those shown in Figs. 97 (*b*) and (*c*). For example, some later experiments recording the impact of hemispherical indenters on flat surfaces show a similar asymmetric conductance characteristic. A typical result is reproduced in Fig. 97 (*d*) for a striker of mass 42 gm. with a hemispherical tip of radius 0·3 cm., falling with a velocity of 140 cm./sec. on to a flat brass anvil. We may note that in all the curves reproduced in Fig. 97 the time of collision is of the order of a few hundred microseconds. In this period elastic compression waves would be able to traverse the colliding bodies and be reflected some twenty times, so that we are justified in ignoring the part played by these elastic waves in the collision process.

As we saw earlier, the collision as a whole may be divided into four main parts: (i) initial elastic deformation, (ii) onset of plastic deformation, (iii) full plastic deformation, (iv) elastic rebound. An attempt to estimate the time involved in each of these processes is very complicated and the solution given by Andrews is of an admittedly approximate nature. We may, however, using a method first described by Andrews (1930), calculate in a very simple way the time involved, if we assume that the first three parts may be replaced by a single process of purely plastic deformation which takes place against a mean dynamic yield pressure p.

Consider two equal spheres, radius of curvature r, mass M. Let one be at rest and be struck by the other at a velocity v. The collision is equivalent to the impact occurring between the spheres when they collide at equal and opposite velocities $v/2$. (For complete equivalence a constant velocity $v/2$ must be superposed on both spheres.) The plane of contact in this 'equivalent' collision may thus be considered as the stationary plane during the deceleration of the spheres. Suppose $2a$ is the diameter of the circular impression formed on each sphere at any instant, the force opposing motion due to the plastic yield pressure p will be $p\pi a^2$. Each sphere is flattened by an amount x, where to a first approximation $2rx = a^2$. Since each sphere has been decelerated over the distance x under the force $p\pi a^2$, the equation of motion is

$$p\pi a^2 = -M\frac{d^2x}{dt^2}. \tag{7}$$

The solution is $x = A\sin(2p\pi r/M)^{\frac{1}{2}}t$, since when $t = 0$, $x = 0$. The velocity of the spheres becomes zero when $dx/dt = 0$, i.e. when

$$t = \frac{\pi}{2}\sqrt{\left(\frac{M}{2p\pi r}\right)}. \tag{8}$$

This is the time taken for the sphere to be brought to rest, and immediately afterwards there is a separation of the surfaces due to the relaxation of elastic stresses in both spheres. Consequently equation (8) should give the period from A to D in Fig. 97 (b), (c), and (d).†

It is seen from equation (8) that the time of collision t_{AD} should be independent of the velocity of impact. Experimental observations showed, in fact, that for all collisions in which the impact was predominantly plastic, the time of collision did not vary beyond the limits of experimental error when the velocity of impact was varied by a factor of 35. We may compare the range of experimental values for t_{AD}

† The overall elastic compression of the spheres during the collision makes t somewhat larger than that given in equation (8). For a fuller discussion see Tabor (1951, pp. 130–7).

with those calculated from equation (8), the values of p being determined from equation (3). The results are given in Table XXXV.

TABLE XXXV

Collision	Time of impact from A to D microsec.	
	Calculated	*Observed*
Steel on steel spheres . .	90	150 ± 40
Lead on lead spheres . .	400	550 ± 150
Hard hemisphere on brass anvil .	60	100

The third line refers to a similar calculation for the impact of a hard hemisphere on a brass anvil (see Fig. 97 (*d*)). The agreement in the three cases quoted is reasonable.

Temperature of impact

Some experiments were carried out to see whether any appreciable temperature rise is produced at the region of contact during collision. This temperature should be observable as a thermal e.m.f. if the colliding bodies consist of dissimilar metals. Such measurements will, of course, only show the temperature while the bodies are in contact, and they will (as we saw in Chap. II) give an integrated average value rather than a measure of the maximum temperature. The results show at once that for smooth spheres colliding with one another the temperature rise at moderate impact velocities is small. For example, for a sphere made of Wood's alloy colliding with a constantan sphere at a velocity of 100 cm./sec. the temperature rise was only a few degrees. This is probably due to the fact that the energy of deformation, which appears overwhelmingly as heat, is distributed over a relatively large volume. If, however, the constantan sphere is replaced by a constantan spike with a sharp point on which the Wood's metal sphere impinges, a considerable portion of the energy of impact is converted into frictional heat which is generated on the surface of the spike as it penetrates the Wood's metal. Consequently we may expect a considerably larger temperature rise. A typical result (reproduced in Plate XXX. 1) shows that this is the case. At a velocity of impact of about 100 cm./sec. the maximum temperature rise was of the order of 35° C., and it is seen that the temperature falls to half its value in about 10^{-3} sec.

These experiments show that the temperature generated at the point of contact of smooth metals colliding at moderate velocities will be small. If, however, sharp points are present so that the deformation is restricted to a small area, and the shape of the bodies is such that an appreciable fraction of the energy of collision is used up as frictional work, the momentary temperature rise may be high (see Chapter XVI).

Effect of lubricant film

Some impact experiments were carried out with the steel spheres when a thin film of caprylic acid had been applied to the surfaces. As before, the spheres were suspended by very fine wires and the electrical conductance between them measured during the impact. At the time these experiments were done it was thought that the presence of a good boundary lubricating film might alter the amount of metallic contact and perhaps influence the time of impact. No such effect, however, was observed. The conductance curve and the time of impact were very similar to those observed with unlubricated surfaces. It is clear that electrical contact occurs through the lubricant film. The nature and the time of the impact are not sensibly altered.

The experiment was then repeated with a thin film of medicinal paraffin oil. The viscosity of this (120 centipoise) was considerably greater than that of caprilic acid (6 centipoise). The conductance never rose above zero during the impact, showing that no *metallic contact took place*. An examination of the surfaces of the mild steel spheres showed that they had been flattened by the force of the collision. If the spheres were allowed to remain touching, or were gently rubbed together, metallic contact was made through the oil film.

These results suggest that the reason for the failure to make contact during collision is a hydrodynamic one. The prevention of metallic contact is *not* due to the adsorption of the lubricant nor to its surface properties, but simply to its viscosity. The impact time is so short that the oil is not able to escape from between the surfaces. In this time the metal surfaces are appreciably and permanently deformed and undergo plastic flow, so that the forces involved must be enormous. In the case of caprilic acid these forces are sufficient to squeeze out the film of acid completely: or if a thin film remains it is a relatively good conductor. In the case of paraffin oil, however, there is not sufficient time during the impact for the oil film to be squeezed out, and the electrical conductance remains low throughout the collision.

Some typical results recently obtained by Mr. E. Rabinowicz which show the deformation of copper surfaces under these conditions are shown in Plate XXX. 2. A steel sphere of 1 in. diameter, mass 70 gm. was dropped from a height of about 4 cm. on to a flat copper surface mounted on a heavy anvil. Impacts were observed when the surfaces were dry and when covered with an oil of viscosity about 500 centipoise. In the latter case the electrical conductance measurements showed that no metallic contact occurred through the lubricant film. The figures show, however, that although the impact is a light one, marked deformation has occurred. It is interesting to note that the deformation which occurs through the oil film is different in shape from that occurring with dry surfaces and is very suggestive of the trapping of a pool of liquid between the colliding surfaces. Further experiments show that the shape of the indentation formed depends on the viscosity of the interposed liquids. With oils less viscous than 100 centipoise metallic contact occurs, whilst with more viscous oils where no metallic contact occurs the 'dimple' is less clearly marked. Presumably as the ball approaches the surface the pressure in the oil film increases until it exceeds the yield pressure of the copper. This will always occur at the centre of the region of contact since the hydrodynamic pressure here is a maximum. The copper now yields plastically, so that there is a receding pool of oil below the centre of the ball. The region of plastic flow grows as more and more oil is squeezed out. If the oil is of suitable viscosity the rate of escape after the initial deformation may be rapid and a marked 'dimple' results. If it is too viscous the oil does not escape very rapidly and the indentation is much nearer the shape of the indenter, although there is still a suggestion of a pointed depression at the centre of the indentation.

These results again show that, even with light impacts, metals may be readily deformed through a liquid film without penetration of the liquid.

Flat Surfaces

The previous section has shown that the presence of a viscous film between colliding surfaces may profoundly affect the collision process. We may expect that these effects will be very much accentuated in the collision of flat surfaces where the approaching boundaries of the liquid film have an area that is generally very much larger. As we shall see, extremely high pressures, rates of flow, and rates of shear may be developed in the liquid film as it is expelled, and in many cases the collision may reach its end with the surfaces still separated by an appreciable film of liquid (Eirich and Tabor, 1948). Even in such cases, however, the pressures developed in the liquid may be sufficient to deform the solid surfaces.

Let us consider the case of a flat surface (the hammer) of radius R, mass M, which approaches a parallel flat surface (the anvil) with an initial velocity V_0 (Fig. 98). We suppose that the anvil is covered with a thin film of a liquid of initial thickness h_0, density ρ, viscosity η. We first assume that the solid surfaces are rigid and that the viscosity of the liquid is independent of temperature, pressure, and rate of shear.

When the hammer strikes the liquid film there is a slight energy transfer as the liquid is set in motion. This produces an instantaneous

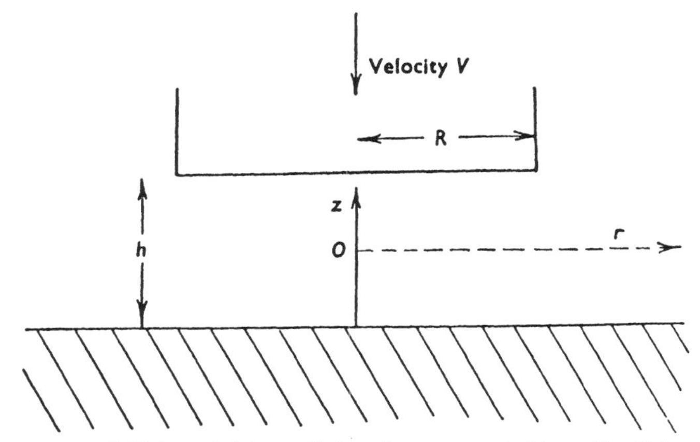

FIG. 98. Collision of flat parallel surfaces separated by a liquid layer of thickness h.

decrease in the velocity of the hammer; in general, however, this is very small and may be neglected. As the hammer continues to descend it begins to expel the liquid, and energy is dissipated. The full solution of the hydrodynamic problem involved is a very complicated one, but we may make a few assumptions which enable a simple solution to be developed.

Let us assume that once the liquid has started moving the motion is laminar and that the inertia forces are relatively unimportant. We suppose that there is no slip at the solid boundaries. Then if at any stage the hammer velocity is V, the film thickness h, the pressure in the film p, and c is the velocity of flow of the liquid in any plane z, the viscous flow of the liquid is governed by the equation

$$\frac{\partial^2 c}{\partial z^2} = \frac{1}{\eta}\frac{dp}{dr},$$

$$\therefore \quad c = \frac{1}{2\eta}\frac{dp}{dr}\left(z^2 - \frac{h^2}{4}\right). \qquad (9)$$

Thus the velocity profile is parabolic.

The volume of liquid which is squeezed out through the boundary at any instant is equal to the volume swept out by the hammer in the same instant. Simple geometry shows that the mean rate of flow $\bar{c}$ of the liquid film is therefore given by

$$\bar{c} = \frac{rV}{2h}. \tag{10}$$

Consequently by integrating equation (9)

$$p = \frac{3\eta V(R^2-r^2)}{h^3}, \tag{11}$$

so that the maximum pressure occurs at the centre of the hammer and is given by

$$p_{\text{max}} = \frac{3\eta VR^2}{h^3}. \tag{11 a}$$

We may determine the velocity of the hammer as the collision proceeds since in any short interval of time the kinetic energy lost by the hammer $MV\,dV$ is equal to the work done on the liquid film in that interval, viz. $\int 2\pi rp\,dr$. The solution of the differential equation yields

$$V = V_0 - \frac{3\pi\eta R^4}{4M}\left(\frac{1}{h^2}-\frac{1}{h_0^2}\right). \tag{12}$$

Since $1/h_0^2$ is usually small compared with $1/h^2$ we may neglect it. Consequently the velocity of the hammer is reduced to zero when

$$h = \sqrt{\left(\frac{3\pi\eta R^4}{4MV_0}\right)}. \tag{13}$$

Under most experimental conditions this thickness is of the same order as the surface roughness. If, however, an appreciable film of liquid still remains at this stage, the hammer sinks through the remaining film under its own weight. A simple integration similar to that carried out in equation (12) shows that the time $t_{1,2}$ for the hammer to sink from a height h_1 to a height h_2 is given by

$$t_{1,2} = \frac{3\pi\eta R^4}{4Mg}\left(\frac{1}{h_2^2}-\frac{1}{h_1^2}\right). \tag{14}$$

This is the same as the time taken for the surfaces to be pulled apart (by the same force Mg) from a height h_2 to a height h_1. Thus if parallel surfaces are initially separated by a film of liquid of thickness h_2 the time required to pull them completely apart ($h_1 = \infty$) is proportional to η/Mgh_2^2. This relation was first derived by Stefan (1874) and Reynolds (1886). It is used by Michell (1923) in his ball viscometer and by Heidebroek (1941) in his 'break-off' viscosity experiments.

Using equation (12) we may express the maximum pressure, rate of flow, rate of shear (defined as $s = dc/dz$) in terms of V_0 and h. The results obtained neglecting the term involving $1/h_0^2$ are given in Table XXXVI.

TABLE XXXVI

	General equation	*Maximum value*	*Film thickness at which max. value occurs*
p_{max}	$\frac{3\eta R^2}{h^3}V$	$\frac{0\cdot154}{R^4}\sqrt{\left(\frac{M^3V_0^5}{\eta}\right)}$	$\frac{5}{4}\sqrt{\left(\frac{\pi\eta R^4}{MV_0}\right)}$
$\bar{c}$	$\frac{R}{2h}V$	$\frac{0\cdot125}{R}\sqrt{\left(\frac{MV_0^3}{\eta}\right)}$	$\frac{9}{4}\sqrt{\left(\frac{\pi\eta R^4}{MV_0}\right)}$
s_{max}	$\frac{3R}{h^2}V$	$\frac{0\cdot318}{R^3}\sqrt{\left(\frac{MV_0^2}{\eta}\right)}$	$\frac{3}{2}\sqrt{\left(\frac{\pi\eta R^4}{MV_0}\right)}$

The maximum rate of shear given in the table occurs at the edge of the hammer, very close to the solid surfaces, i.e. $z = \pm\frac{1}{2}h$, $r = R$.

It is seen that the pressure, velocity of flow, and rate of shear increase with decreasing radius of hammer. Calculations show that the maximum values rise rapidly as the film thicknesses approach the dimensions of the surface roughness.

Pressure developed in the liquid film

The general characteristics of the equations for p_{max}, s_{max}, and $\bar{c}$ are similar. They all increase more and more rapidly as the liquid film is squeezed out, reach a maximum, and then rapidly fall to zero as the hammer is brought to rest. Some typical results of p_{max} for five ranges of impacts are plotted in Fig. 99, η being assumed to be 0·25 poise (the value for a light machine oil or for nitroglycerine). It is seen that the pressures for the lighter impacts reach values of the order of several thousand atmospheres. For the heaviest impact the pressure is over 10^6 atmospheres, but it is clear that long before this value is reached even the hardest steels would yield plastically. As we shall now see, even elastic deformation of the surfaces may profoundly affect the flow conditions.

Let us assume that the pressure distribution in the liquid film is still parabolic,† i.e.

$$p = p_{max}\left(1-\frac{r^2}{R^2}\right). \tag{15}$$

† Actually the distortion of the anvil will modify the flow conditions in the liquid so that the pressure distribution will no longer be parabolic. However, the result is not greatly altered if we assume a modified pressure distribution.

We suppose again that the hammer is rigid, but that as a result of p the surface of the anvil is deflected by an amount ω. Then at any point Q in the surface distant a from an element of area $dxdy$ where the

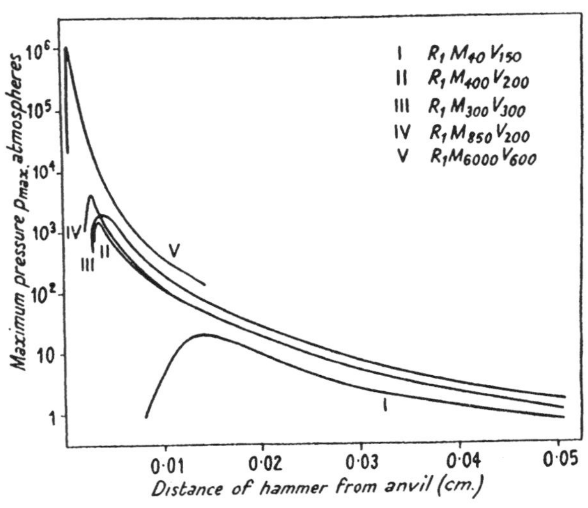

FIG. 99. Maximum pressures developed at the centre of a liquid film trapped between colliding parallel surfaces. As the liquid is expressed the pressure rises more and more rapidly, reaches a maximum, and then falls to zero as the surfaces are brought to rest. Initial thickness of film 0·05 cm., viscosity 25 centipoise, radius of hammer 1 cm., mass M and velocity V of hammer as shown in figure in gm. and cm./sec. respectively.

pressure is p, the deflexion ω perpendicular to the surface is given by

$$\omega = \frac{1-\sigma^2}{\pi E} \int\int \frac{p\,dxdy}{a} \tag{16}$$

(Prescott, 1927), where E is Young's modulus, and σ Poisson's ratio for the material of the anvil (see Fig. 100).

Substituting for p from equation (15) and following the procedure described in Prescott (1927), we find that

$$\omega = \frac{4(1-\sigma^2)}{3E} p_{\max}\left\{1-\frac{3r^2}{3R^2}+\frac{9r^4}{64R^4}\cdots\right\}. \tag{17}$$

The work W expended in deforming the anvil by this amount is the integral of $2\pi r\,dr\,p\,d\omega$. If we take the first three terms in the expansion of ω and assume that $\sigma = 0{\cdot}3$, we obtain

$$W = \frac{0{\cdot}8R^3}{E}(p_{\max})^2. \tag{18}$$

Thus if *all* the energy of collision is expended in deforming the anvil plastically the pressure cannot exceed the value

$$p_{\text{max}} = \left\{\frac{E}{0{\cdot}8R^3} \times \text{energy of impact}\right\}^{\frac{1}{2}}. \tag{19}$$

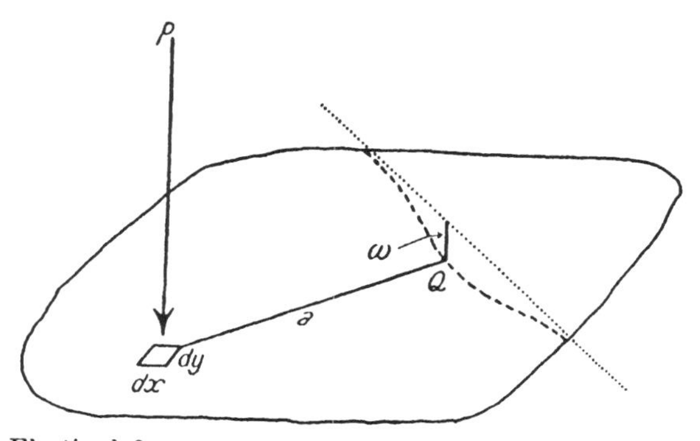

FIG. 100. Elastic deformation of the surface of the anvil by the pressure developed in the liquid film.

As a matter of interest the maximum possible values of p_{max} for the five hammers are calculated in Table XXXVII for brass anvils, assuming $E = 10^{12}$ dynes/cm.^{2} and $\sigma = 0{\cdot}3$.

TABLE XXXVII

Maximum pressures at centre of hammer

Hammer	*Energy of impact* 10^8 *ergs*	*Max. pressure* p_{max} 10^6 *dynes/cm.*^{2} *From elastic energy*	*From flow equations*
I	0·0045	750	21
II	0·08	3,200	1,360
III	0·135	4,100	2,440
IV	0·17	4,600	4,210
V	10·8	37,000	10^6

In the last column the values of p_{max} obtained from the viscous flow equations are inserted. It is seen that for the three lightest impacts the pressure that could be developed in deforming the anvil is considerably higher than that calculated for viscous flow. This means that the elastic deformation of the anvil will not appreciably affect the maximum pressure reached during impact. For hammer IV the values are comparable, which means that the elastic deformation of the anvil may have an appreciable effect on the maximum pressure reached. With the heaviest hammer, however, the maximum pressure that can be reached is thirty times less than that calculated from viscous flow. This means that if all the energy of impact went into elastic deformation of the anvil, the pressure instead of exceeding 1,200,000 atmospheres could not exceed 40,000 atmospheres. If we allow for the fact that a considerable amount of the energy is expended during the collision in causing viscous flow, it is evident that the maximum pressure that can be reached will be

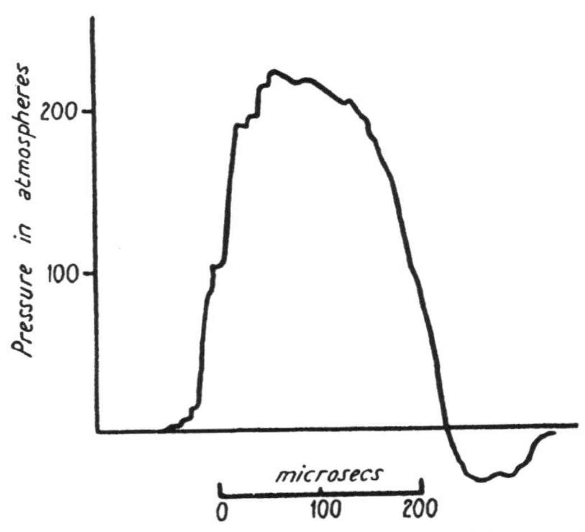

Fig. 101. Cathode-ray record of the pressure developed in a confined volume of liquid as the result of impact. Piezo-electric method. An impact energy of 5,000 gm./cm. produces a pressure rise of over 200 atmospheres.

even less. Professor T. M. Cherry (1945), who has carried out a detailed investigation for hammer V, has shown that, in fact, the pressure will not exceed about 7,000 atmospheres. It follows that if the impact is carried out on an anvil made of a softer metal, e.g. lead or copper, this pressure will be sufficient to deform the anvil plastically, *through* the lubricant film.

If, of course, there is any impediment which hinders the escape of the liquid, high pressures are reached with even lighter impacts. Direct experimental confirmation of this has been obtained by Dr. Gray (1948), who measured the pressure developed by a piezo-electric method. An experimental curve showing the rise in pressure for a light impact is shown in Fig. 101. The impact was produced by a 25-gm. hammer falling 20 cm. (impact energy 5,000 gm. cm.) on to a cylindrical plunger 0·2 sq. cm. area resting on a confined volume of liquid. It is seen that the momentary pressure rise produced in the liquid is over 200 atmospheres and the maximum is reached in about 50 microseconds.

Velocity of flow and rate of shear

Reverting to the case of an unconfined film of liquid between a hammer and anvil, we may use the relations given in Table XXXVI to calculate the velocities of flow and rates of shear in the liquid film. The main results are summarized in Table XXXVIII.

TABLE XXXVIII

Maximum pressure p_{max}, rate of flow $\bar{c}$, and rate of shear s_{max} developed in liquid film during impact between rigid surfaces

	Hammer				
Index	*Mass* *gm.*	*Velocity* *cm./sec.*	p_{max} *atmos.*	$\bar{c}$ *metres/sec.*	s_{max} *cm./sec./cm.*
I	40	150	21	30	$1\cdot2\times10^7$
II	400	200	1,360	142	$2\cdot0\times10^7$
III	300	300	2,440	226	$3\cdot5\times10^7$
IV	850	200	4,210	207	$4\cdot4\times10^7$
V	6,000	600	10^6	2,860	$280\cdot0\times10^7$

It is seen that for the four lighter impacts the velocities of flow lie between 30 and 200 metres/sec., while the maximum rates of shear (about 10^7 cm./sec./cm.) imply a relative velocity between neighbouring molecular layers of about 1 cm./sec. This is very much less than the mean thermal velocity of the liquid molecules. With the heavy impact the values of $\bar{c}$ and s_{max} are considerably larger. However, the deformation of the anvil produces a depression in the anvil which recedes beneath the approaching surface of the hammer. As a result, the rate at which the liquid is extruded is greatly reduced, and a detailed analysis by Cherry shows that $\bar{c}$ and s_{max} will never be very much larger than the values obtained for the lighter impacts.

Temperature developed in the liquid film

The energy used in overcoming the viscous forces in the liquid film is dissipated as heat. Since the velocity profile is parabolic the heating is not uniform but is greatest at those portions of the film where the rate of shear is maximal. In addition the effect is cumulative, so that it will depend on the total period during which the portion of the liquid is in shear. The maximum heating will therefore occur at the periphery of the hammer very close to the solid surfaces, where

$$z = \pm\tfrac{1}{2}h, \quad r = R.$$

Here the rate of shear is a maximum. In addition the radial velocity is zero, so that the element remains in shear for the whole of the interval of impact.

The rate of dissipation of energy H by viscous flow per unit volume of liquid per second is given by

$$\frac{dH}{dt} = \eta\left(\frac{dc}{dz}\right)^2. \tag{20}$$

We substitute for the value of dc/dz and replace dH/dt by

$$\frac{dH}{dt} = \frac{dH}{dh}\cdot\frac{dh}{dt} = -\frac{dH}{dh}V.$$

Equation (20) may then be integrated from the initial instant of impact to the end of the impact when the hammer is brought to rest and h is given by equation (13). We obtain

$$H = \frac{0{\cdot}31}{R^4}\left(\frac{M^3V_0^5}{\eta}\right)^{\frac{1}{2}}. \tag{21}$$

We see that the heat dissipation increases very rapidly as the radius of the hammer decreases. Further, if the viscosity decreases appreciably as a result of the heating, this is likely to increase the resultant value of H.

The temperature rise T may be calculated if we assume that the amount of heat carried away by the flow of the liquid (convection), and by conduction into the solid surfaces of the hammer and anvil may be neglected. If J is Joule's equivalent and γ the specific heat of the liquid,

$$T = \frac{H}{J\gamma\rho}. \tag{22}$$

Typical results for the temperature rise produced by viscous heating are given in Table XXXIX.

TABLE XXXIX

Hammer	$H.10^7$ *ergs*	$T°$ *C.*
I	4·3	1·5
II	281	98
III	502	174
IV	867	300
V	254,000	88,000

It is seen that the temperature rise for the four lighter impacts lies between 2° and 300° C. If allowance is made for convection and conduction, Professor Cherry has shown that the temperature rise is reduced by about one-half and occurs at a small distance from the solid surfaces. This means that the highest temperatures developed with these light impacts will not exceed 150° C.

With the heavy impact the true value is very greatly reduced from that given in Table XXXIX, on account of the deformation of the anvil. Cherry has shown that if allowance is made for this factor and for conduction and convection the temperature rise will not exceed about 3,000° C.

It follows from these results that the assumption of constant viscosity in the liquid film is very far from valid. For many liquids the viscosity increases with pressure and at pressures of the order of 1,000 atmospheres the viscosity may be as much as five or ten times the normal value. A more important effect, which operates in the opposite direction, is that of temperature. For many liquids the viscosity decreases by a factor of 2 for an increase in temperature of 10° to 20° C. Consequently the viscosity of the liquid will be greatly reduced near the solid boundaries where the heating is a maximum. This may have a marked effect on the resulting pressures, rates of shear, and temperature rise. We may, however, note that a decrease in viscosity near the solid boundaries will lead to an increase in the rate of shear of the liquid at these regions. As far as the temperature rise is concerned, this may partially compensate for the reduction in viscosity. In spite of these uncertainties, we may expect that the main characteristics described in the above analysis will still be descriptively correct, and that for lighter impacts the quantitative values will be valid to about a single order of magnitude.

Practical Implications

These results show that when metal surfaces collide the forces developed between the surfaces may be expressed in terms of the elastic and plastic constants of the metals. The time of collision is usually very short (of the order of 10^{-4} sec.) and the pressures developed at the region of contact may be very high indeed. Even for relatively light impacts these pressures may be sufficient to cause plastic flow of the metals. The yield pressures under conditions of impact are of the same order of magnitude as those involved in static deformation, but are always larger and tend to increase with the velocity of the collision.

These results emphasize the ease with which plastic flow of colliding metal surfaces may occur. It is clear that this is of considerable importance in the design of ball races and other types of bearings operating under impulsive loading or under conditions where vibration may occur. In ball races the impact of the ball on the race may produce a 'brinelling' of the race or a flattening of the ball with subsequent damage to the whole bearing, as the work of Jones (1946) has recently shown. As we saw in equation (5), the energy (mgh_1) required to initiate plastic deformation in a metal by impact is proportional to r^3, where r is the radius of the sphere.† Consequently the danger of damage to the bearing is considerably lessened by using races containing balls of larger radius, provided that the loading on the ball is not thereby increased unduly.

Similar considerations apply to the design of other types of bearings such as roller and journal bearings, but since in these bearings the contact is spread over a larger area the range within which the deformations remain elastic will be appreciably higher.

The results also show that although the presence of a lubricant film between the colliding surfaces may affect the impact process, high pressures and plastic deformation may readily occur. In general as the lubricant film is extruded during the impact, high rates of flow, rates of shear, and pressures are developed. If the surfaces are spherical these pressures may be sufficient to cause permanent deformation and damage to the colliding surfaces at relatively small velocities of impact even though no actual metallic contact takes place. With flat colliding surfaces the pressures will be sufficient to deform the surfaces elastically at relatively light impacts, but in general considerably heavier impacts may be necessary to produce plastic deformation of the surfaces through the liquid film. When plastic flow occurs through the film the surface damage will generally be less than would be the case in the absence of the liquid film, since a considerable portion of the collision energy is expended in squeezing out the liquid. However, under many practical conditions it may still be a serious factor.

The analysis shows in addition that, other conditions being constant, the maximum pressure developed in a liquid film trapped between colliding flat surfaces is proportional to $\sqrt{(1/\eta)}$ (see Table XXXVI), and a similar type of relation appears to hold for spherical surfaces. Thus if there is heavy impulsive loading or intense vibration the presence of a liquid film of very high viscosity may prevent the generation of pressures sufficiently high to cause plastic flow of the surfaces. For this reason it might be desirable to select as a lubricant a liquid, the viscosity of which rises to a high value when the pressures or rates of flow are great. Some practical evidence for the importance of viscosity under conditions of impact has recently been obtained by Blok (1948) in a study of the lubrication of gears. He has shown that the failure of gear

† In contrast, the static load required to initiate plastic flow is proportional to r^3 since $W^{\frac{1}{3}} = kp_m r^{\frac{2}{3}}$. (See Chap. I, equation (4).)

teeth under exaggerated impulsive loading depends far more on the viscosity of the lubricant than on its 'oiliness' or extreme pressure properties.

Finally we may consider the temperatures produced by impact. If no liquid film is present and plastic deformation of the surfaces occurs, the greater part of the energy of deformation appears as heat. The deformation process is, however, distributed over a relatively large volume so that the actual temperature rise is low. If, however, sharp points or hard sharp asperities are present so that an appreciable fraction of the energy of collision is used up as frictional work, the temperature rise during impact may be considerably higher. If the bodies are not metallic and possess a low thermal conductivity, or if particles of sand or grit are present between the surfaces, we may expect the local temperature rise to be correspondingly greater.

If the surfaces are separated by a liquid film, the heating due to the plastic deformation of the metal itself will be very greatly reduced or eliminated. However, the high rates of shear in the liquid film may produce appreciable viscous heating in the liquid. For moderate impacts this may amount to as much as 200° C., whilst for severe impacts the theory indicates that it could exceed several thousand °C. These high temperatures occur in those liquid layers which are close to the metal surfaces. This would mean, first, that the viscosity is considerably reduced, with a consequent increase in the chances of plastic deformation through the liquid film and subsequent metallic contact. Secondly, the surface layers may well lose their boundary lubricating properties as a result of the high temperatures. If therefore there is any vibration so that the system is alternating between boundary and fluid lubrication, the viscous heating may produce a very serious deterioration in the lubrication. The disadvantages of viscous heating have long been recognized in the lubrication of journal bearings where at high *speeds* of rotation the rates of shear in the lubricant may be very high and may so produce excessive heating. The present discussion suggests that even if the *speed* of rotation is low, rapid vibrations or heavy impulsive loading may produce similar effects (Tabor, 1949).

The damage of metal surfaces by pressures developed in the lubricant film may become of increasing practical importance under modern conditions. With high-speed ball races, for example, which are subjected to vibration, a flattening of the balls and damage of the race may occur while the fluid film is still complete. It may also be important for fast-moving gears, roller bearings, journal bearings, and other mechanisms, particularly if they are operating under conditions of impulsive loading or excessive vibration. Although many examples are available of this type of damage, of erosive damage produced by rapidly flowing liquids, and of the damage produced by cavitation, the subject has not perhaps received the attention it deserves. A more systematic theoretical and experimental study of the damage of metals by forces transmitted through the liquid might, with advantage, be made.

REFERENCES

J. P. Andrews (1929), *Phil. Mag.* **8**, 781; (1930) ibid. **9**, 593; (1931) *Proc. Phys. Soc. Lond.* **43**, 8.
H. Blok (1948), Summer Conf. on Mechanical Wear, Massachusetts Institute of Technology.
F. P. Bowden and D. Tabor (1941), *Engineer* (London), **172**, 380.
T. M. Cherry (1945), C.S.I.R. (Australia) Tribophysics Division Report A 116.
C. A. Edwards and C. R. Austin (1923), *J. Iron & Steel Inst.* **107**, 324.
F. W. Eirich and D. Tabor (1948), *Proc. Camb. Phil. Soc.* **44**, 566.
P. Gray (1948), Colloquium, *La Cinétique et le mécanisme des réactions d'inflammation et de combustion en phase gazeuse.* Paris.
E. Heidebroek (1941). See Tingle (1947), B.I.O.S. Report No. 1610.
H. Hertz (1881), *J. reine angew. Math.* **92**, 156.
A. B. Jones (1946), A.S.T.M. **46**, Preprint No. 45.
A. G. M. Michell (1923), *Mechanical Properties of Fluids*, chap. iii. Blackie & Son, Ltd.
J. Okubo (1922), *Sci. Reports Tôhoku Univ.* **11**, 445.
J. Prescott (1927), *Applied Elasticity*. London.
C. V. Raman (1918), *Phys. Rev.* **12**, 442.
O. Reynolds (1886), *Phil. Trans. Roy. Soc.* **177**, 157.
B. de St.-Venant (1867), *J. de Math. Liouville, Paris*, Series 2, **12**. See also A. E. H. Love (1934), *Mathematical Theory of Electricity*, article 284, Cambridge Univ. Press.
J. Stefan (1874), *Sitz. Ber. Akad. Wiss. Wien*, **69**, 713.
D. Tabor (1948), *Proc. Roy. Soc.* **A 192**, 247.
—— (1949), *Engineering*, **167**, 145.
—— (1951), *The Hardness of Metals*, Clarendon Press.
Sir G. I. Taylor (1946), *J. Inst. Civil Engrs.* **26**, 486 (James Forrest Lecture).

XIV

THE NATURE OF METALLIC WEAR [*A*]

LOCAL ADHESION AND WEAR

THE wear of solid surfaces is a complex process which under many conditions includes both chemical attack and physical damage. Any slight change in the operating conditions may change the whole nature of the wear process. For this reason practical 'tests' seldom give reproducible or conclusive results and considerable caution must also be exercised in interpreting even the most carefully controlled laboratory experiments. It is also clear that any theoretical treatment of the problem of wear must make sweeping simplifications. In this chapter we shall consider the part played by the continual formation and shearing of metallic junctions in producing wear between rubbing surfaces and discuss briefly some of the factors which, in practice, complicate the wear process.

When a metallic junction is formed between sliding surfaces the shearing may occur in four different ways. If the junction is weaker than the metals themselves, shearing will occur at the actual interface where the junction is formed. Consequently the amount of metal removed from either surface will be very small even though the friction may be relatively high. This occurs, for example, with a tin-base alloy sliding on steel, where the coefficient of friction is about $\mu = 0{\cdot}7$ and the amount of alloy worn away and smeared on to the steel is so slight that even after traversing the same track 400 times the lapping marks on the steel are still clearly visible (see Chap. VI, Pl. XVII. 4). This type of junction is usually formed in the presence of tough oxide films and sometimes in the presence of thin sulphide or chloride films. These surface layers, even if they are only of molecular dimensions, may hinder the formation of strong metallic junctions provided the films are not broken up by the deformation of the underlying metal.

If the junction is stronger than one of the metals, shearing will often occur within the bulk of the weaker metal and fragments of the softer metal will be left adhering to the harder surface. Under these conditions the amount of material removed from the softer metal may be very large even though the friction is similar to that observed in cases where little wear occurs. This is shown, for example, when a lead-base alloy slides on steel. The coefficient of friction is about $\mu = 1{\cdot}0$, and the wear of the alloy is so marked that after a few hundred traversals of the steel surface the lapping marks on the steel surface are almost entirely obscured (see Chap. VI, Pl. XVII. 5). This type of wear gradually builds up a film of softer metal on the harder surface, so that ultimately the sliding is characteristic of similar metals. Consequently the friction, surface damage, and wear are very high.

Thirdly, if the junction is stronger than both metals, shearing will generally occur in the bulk of the weaker metal, but it will also occasionally occur within the stronger metal itself. In this case there will be considerable removal of softer metal, but there will also be a small but finite removal of harder metal during sliding. This occurs, for example, when copper slides on steel. The taper sections given in Plate VI. 3, Chapter IV, show that most of the shearing occurs within the bulk of the copper, but that a number of small pits are left in the steel surface where the junctions have proved stronger than the steel itself. The ability of a soft metal to pluck away portions of a harder metal through the formation of strong intermetallic junctions is shown in Plate XXXI. 1 for a slider of hard gun steel (V.D.H. 600) after it has traversed a copper surface for a track distance of only 1,000 cm. Although the wear of the harder metal may be far less than that of the softer metal, it will still be appreciable.

Finally we may consider the behaviour of similar metals. Here the junctions are of the same material as both surfaces, but the process of deformation and welding will work-harden them and appreciably increase their shear-strength. (A striking example of the work-hardening of the irregularities on the surface layers of copper is shown in Plate II, Chapter I.) Consequently shearing will rarely occur at the interface itself but will take place within the bulk of the metals. For this reason the surface damage of both sliding bodies will be very large. Some alloys, particularly tin-base alloys, are known to 'work-soften' (Leyman, 1937) when subjected to very heavy deformations. Junctions formed under these conditions might, therefore, be weaker than the bulk of the metal itself and shearing would occur at the interface with very little metallic transfer or surface damage. This suggests that if alloys could be prepared which show a very pronounced 'work-softening' at the points of rubbing contact, they might have good wear-resisting properties.

It is clear from this discussion of the role of metallic junctions in friction and wear that we cannot expect to find any direct relation between the coefficient of friction and the amount of wear. The amount of metal lost from the surface depends critically on the region within which shearing of the junctions occurs. The frictional force, however, may be essentially the same whether shearing occurs within the bulk of the metals or at the actual interface of contact. Consequently the amount of metal plucked from the surfaces will, in general, bear little relation to the amount of frictional work performed during sliding. The greater part of the frictional work is dissipated as heat.

These observations are of general validity and under certain conditions the type of junction formed plays a direct part in determining the amount of wear. This will be discussed in the following section in relation to the wear-resisting properties of hard metallic films. In general, however, the actual wear behaviour in any practical case is profoundly affected by other factors. As we shall see, chemical reaction at the rubbing surfaces and in particular the formation of oxide films may be of major importance. This is, of course, especially marked in the absence of a lubricant. In the presence of lubricant films the wear, though greatly reduced, is appreciable and refined methods of measurement show that it may depend on the chemical nature of the lubricant.

Wear-reducing Properties of Thin Metallic Films

The wear between metal surfaces may be decreased by covering one of the surfaces with a thin film of a metal which in itself has good wear-resisting properties. For example, rhodium and chromium are hard metals which are very little worn by other metals during sliding. Thin films of these metals may therefore be used to protect surfaces from heavy wear. In addition chromium has a very strong oxide film which is not easily disrupted since it is backed by a hard metal. As a result very few intimate metallic junctions are formed through the oxide layer; most of the junctions are formed with the oxide film itself, and these junctions tend to shear in the actual surface of contact. Consequently with chromium the wear of the other surface may also be considerably reduced.

Rhodium and chromium films deposited by electroplating have been used with success in the protection of standard screw gauges, test plates, drawing-dies, cutting-tools, cylinder liners, etc. Some experiments (Moore and Tabor, 1942) show in a striking way the wear-reducing properties of these films and some of the factors which are of practical importance in their application. For example, Plate XXXI. 2 shows a steel slider that has traversed a brass surface several hundred times. (The slider had a radius of curvature of 3 mm., the load was 4 kg., and the sliding speed about 10 cm./sec.) The adhesion of the brass to the steel is very marked. If the brass is etched away the underlying steel surface shows considerable plucking and wear (Plate XXXI. 3). If a similar brass slider is coated with rhodium there is again some pick-up, though it is less marked (Plate XXXI. 4). On etching away the brass the appearance of the surface depends on the *thickness* of the rhodium film. If the film is less than 10^{-4} cm. in thickness the film is cracked and partially torn away (Plate XXXI. 5). If it is thicker than 2×10^{-4} cm. the damage is negligible (Plate XXXII. 2). With the chromium film 5×10^{-4} cm. thick the amount of pick-up is still less (Plate XXXII. 3), and when this is removed by etching, the chromium shows no sign of surface damage (Plate XXXII. 4).

The importance of film thickness was again demonstrated in some measurements of the wear of cast iron sliders (Tabor, 1942). The wear of the cast iron on a chrome molybdenum steel surface is very heavy indeed. If the cast iron slider is coated with a film of chromium less than about 10^{-4} cm. in thickness the chromium is unable to withstand the distortion of the underlying metal and is rapidly broken up (Plate XXXII. 5). If, on the other hand, it is more than about 5×10^{-4} cm. in thickness it has sufficient mechanical strength under the given experimental conditions and it provides very good wear protection (see Plate XXXII. 6).

This result may be represented in a different way. The measurements showed that over a fairly wide range of experimental conditions the wear at a fixed load is directly proportional to the distance traversed. Consequently the wear may be expressed as a loss in weight per unit length of track traversed. The results for chromium plated on to cast iron are drawn in Fig. 102, where the wear is expressed as micrograms lost per cm. of track traversed. It is seen that the rate of wear is relatively high for film thicknesses less than 5×10^{-4} cm., while for thicker films it is less than 0·005 microgm./cm. The corresponding value for unplated cast iron is about 5 microgm./cm. In all cases the coefficient of friction remained approximately constant at about $\mu = 0{\cdot}5$.

We may expect that the wear properties of chromium films will depend on the plating conditions. Experiments carried out to investigate this show that, in fact, the wear is generally low for hard bright chromium deposits and higher by a factor of about 10 for burnt or milky deposits. However, the wear of the burnt or milky deposits in these experiments was still 50 to 100 times less than that of cast iron.

Experiments carried out with chromium-plated specimens sliding on chrome molybdenum steel surfaces heated to a temperature of 250° C. gave essentially the same results. The wear of the chromium was as low as that observed at room temperature, but there was a slight increase in the amount of steel picked up.

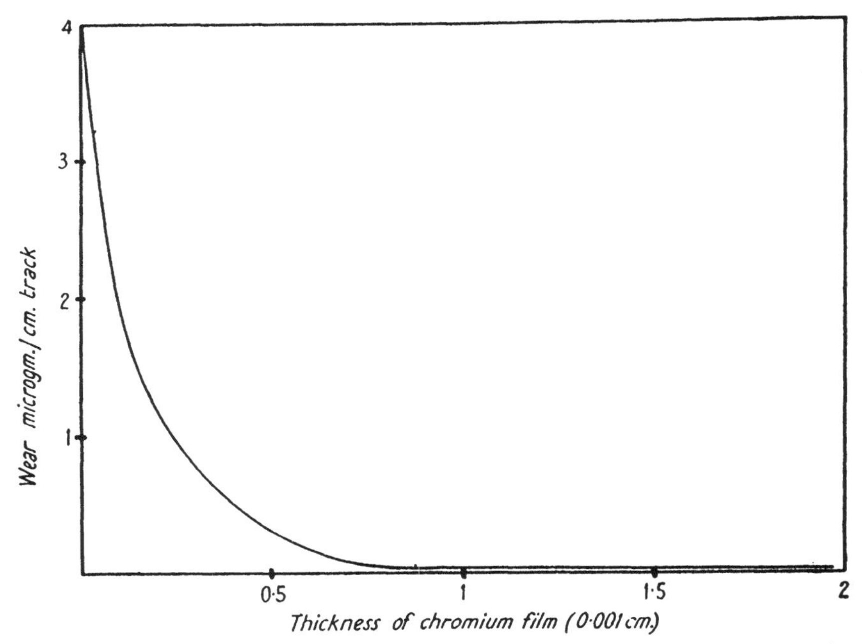

FIG. 102. Wear of chromium-plated film as a function of film thickness. The chromium is deposited on a hemispherical cast iron slider of diameter 0·6 cm. Where the film is less than 5.10^{-4} cm. thick the wear is high as a result of the breakdown of the film. If the film is thick enough to resist mechanical breakdown the wear is very low.

These experiments show that hard chromium films can be most effective in reducing the amount of wear between metal surfaces if the films have sufficient mechanical strength to withstand the distortion of the underlying metals and if the plating conditions are appropriate. In practice, however, chromium films possess one very serious disadvantage. As we saw in Chapter X, chromium is very difficult to lubricate. In fact the friction and wear of chromium surfaces are not greatly reduced by the presence of most lubricants. Consequently if in practice it is possible to lubricate the surfaces, the wear may be appreciably lower when chromium films are not used. If, on the other hand, the surfaces have to operate under high temperatures and pressures where boundary lubrication cannot hope to succeed, chromium films can produce very substantial reduction in wear. Some attempt to improve the lubrication of chromium has been made by etching the surface so that small cracks or crevices are formed. It is suggested that these cracks provide minute reservoirs which trap small quantities of the lubricant and 'feed' it on to the sliding surfaces.

Chemical Reaction and Wear [*A*]

The role of chemical as well as physical processes in the wear of sliding metals has been investigated during the last two decades by a number of workers such as Fink, Rosenberg and Jordan, Smith, Gough, Tomlinson, Williams, Donandt, Siebel, and by Dies. The work which Dies and others have carried out in Germany during the last ten years is of some general interest. Their experiments have been concerned mainly with the wear of steel surfaces sliding on a rotating hard steel disk in the absence of lubricant, at relatively high loads and speeds (Mailander and Dies, 1943). These workers make a clear distinction between what they call 'seizure-wear' and 'abrasive-wear'. Seizure-wear is said to occur when there is heavy tearing of the rubbing surfaces and marked vibrations are set up in the moving parts. With similar metals (e.g. mild steel on mild steel) it occurs even at the smallest loads and the wear is always very heavy (Fig. 103, curve I). Abrasive-wear is said to occur when the surface damage is much less severe and the surfaces can run safely until they are completely worn away. Although the distinction between these types of wear appears to be a matter of degree rather than of kind, Dies and other workers find it convenient to restrict their investigations to the field of abrasive-wear. With mild steel sliding on a hard chromium steel disk this involves working at nominal pressures below about 30 kg./cm.^{2} With this combination of metals these workers find that the wear is considerably less than for mild steel on mild steel. As the load is increased it is found that the wear at first increases, then rapidly falls, and finally rises again (Fig. 103, curve II). The critical load at which this drop in wear occurs is increased if the surfaces are cooled by a circulating system and is decreased if the speed is increased. The results suggest that the decrease in wear is a real effect which corresponds to a transition from one type of wear process to another; this transition is determined largely, though not exclusively, by the surface temperatures developed during sliding. This view is supported by the observation that when the wear decreases a corresponding change occurs in the chemical composition of the wear products. For experiments carried out in air the wear products for small loads are mainly FeO and Fe. Beyond the critical load, however, the amount of Fe becomes negligible and the major part of the wear products consists of Fe_2O_3 and FeO (Fig. 104). At the same time a significant change in the hardness of the surface layers of the steel slider occurs. Up to the critical load the surface hardness is somewhat greater than the bulk hardness: beyond the critical load it is very much

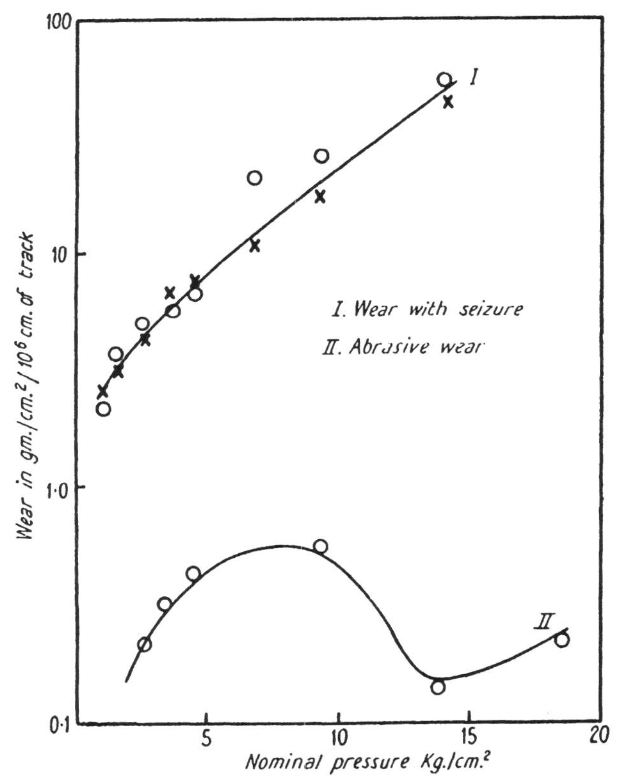

FIG. 103. Wear of unlubricated metals as a function of load (from Dies's paper). Note that the wear is on a logarithmic scale. For mild steel on mild steel (curve I) the wear is always very heavy and involves continuous seizure. For mild steel sliding on hard chromium steel (curve II) the wear is considerably less and shows a marked change at a nominal pressure of about 10 kg./sq. cm.

greater. This marked increase is attributed partly to the formation from the atmospheric nitrogen of nitrided layers which penetrate to an appreciable depth into the surface and partly to the distortion and work-hardening of the metal under the more severe conditions of operation.

Interesting confirmation of the important part played in friction and wear by films of iron oxide, formed at high speeds of sliding, has recently been described by Johnson, Godfrey, and Bisson (1948). They find that the oxide films formed on the rubbing surfaces may in themselves function as lubricants and suggest that αFe_3O_4 is more effective than Fe_2O_3.

Dies has also emphasized the importance of the surrounding atmosphere on the wear. In air, as we have already seen, chemical reactions occur, in which not only oxides but nitrides of the metal may be formed.

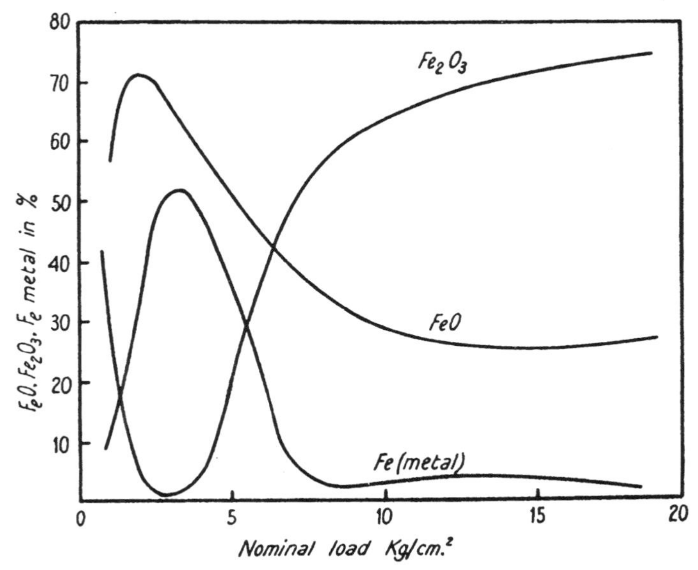

FIG. 104. The wear products of mild steel sliding on a hard chromium steel as a function of load (from Dies's paper). The composition of the wear products shows a marked change at the same load as that at which the total amount of wear changes. See Fig. 103, curve II.

Other atmospheres have their own specific effect. Water vapour appears to have the most marked effect in increasing the wear, and in air at very low pressures where the amount of oxygen and moisture is very small the wear may be twenty times smaller than that observed in the normal atmosphere (Fig. 105). These conclusions are in general agreement with similar investigations carried out in England and America. However, several investigators, whilst stressing the important part played by the atmosphere in metallic wear, have obtained results which are significantly different.

Nevertheless, there is general agreement amongst workers in the field that the wear mechanism involves a complicated interaction of chemical, physical, and mechanical processes and that small variations in the experimental conditions (e.g. ambient temperature, humidity) may have a profound effect on the wear observed. It is clear that in future wear-research much greater precautions will be needed to standardize the conditions under which the measurements are carried out.

An interesting practical conclusion from Dies's work is that with plain carbon steels rubbing on chromium steel, the wear in general decreases

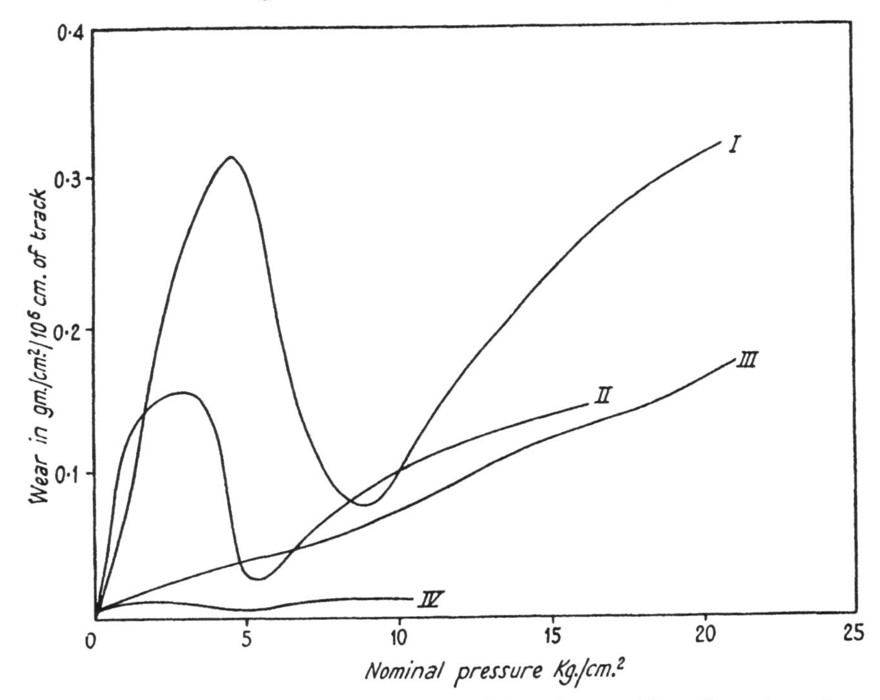

FIG. 105. The wear of mild steel sliding on a hard chromium steel in various atmospheres (from Dies's paper). I, ordinary atmosphere, humidity 4 gm./m.³; II, nitrogen; III, oxygen; IV, atmospheric air at a pressure of 3 mm. When the amount of oxygen and water vapour is very small (curve IV) the wear may be 20 times as small as that observed in the normal atmosphere.

with increasing hardness. A martensitic structure gives a smaller wear than a pearlitic structure and a laminar pearlitic steel gives a smaller wear than a granular pearlitic structure.

IMPORTANCE OF SURFACE OXIDATION

The influence of oxide films on the friction and surface damage of metals has been described in Chapter VII.

It is evident, as many workers have shown, that surface oxidation can play a major part in the wear process and this has been brought out very clearly in another interesting paper by Dies (1943). He has shown that wear is often dominated by the specific properties of the oxide films formed by frictional heating and by the way in which the oxide is attached to the underlying metal. If the oxide is very hard and embeds itself firmly into the underlying parent metal the rubbing

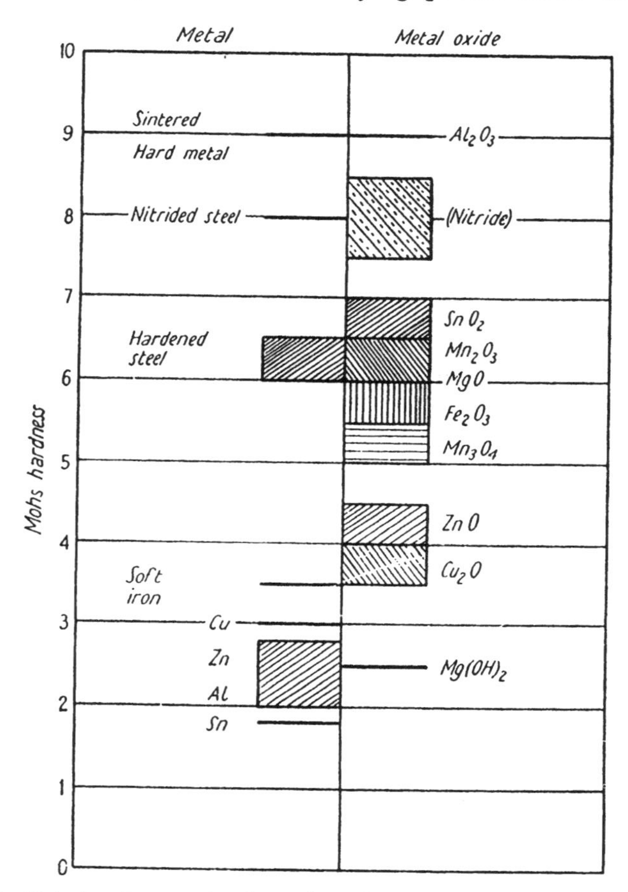

FIG. 106. Mohs hardness scale of metals and metal oxides (from Dies's paper). Under conditions favouring surface oxidation the wear of sliding metals may be determined primarily by the hardness of the corresponding metal oxide. Thus a soft metal such as tin may produce heavier wear of a hard chromium steel than a hard steel slider rubbing on the same chromium steel.

surface may become, in effect, an abrasive surface. As Finch (1935) has already shown, this readily occurs for example with aluminium and aluminium alloys, where surface layers of crystalline Al_2O_3 or alundum are readily embedded in the underlying metal. The surface of alundum does not readily weld on to the other sliding surface, but the sharp edges of the alundum crystals act as very fine, hard, cutting-tools. As a result an aluminium surface may produce very heavy wear of the hardest steel since alundum is very much harder than the oxides of iron.

On the other hand, if the oxide is soft it may play little part in abrading the other surface. Thus magnesium forms a soft hydroxide and the wear produced by magnesium on harder metals under conditions favouring the formation of the oxides is relatively small. These results may partially explain the practical observation that pistons made of magnesium alloy produce far less scoring and tearing of the cylinder liner than pistons made of aluminium alloy. If the oxide is hard but is not firmly attached to the underlying metal its behaviour will be very complicated. Nevertheless, the broad outlines of the effect of metal oxides have been well established by Dies. His diagram showing the Mohs hardness of various metals and their oxides is reproduced in Fig. 106. This figure explains his striking observation that a soft metal like tin produces heavier wear of a hard chromium steel disk (under conditions favouring oxidation) than a hard steel slider on the same chromium steel. The metal oxide may also play a decisive part in fretting corrosion. This has been confirmed by experiments due to Hampp (see Tingle, 1947) on the false brinelling of ball races during transit or during storage under vibration. The continuous stressing of the surfaces in the presence of atmospheric CO_2 and moisture leads to relatively rapid corrosion since the corrosion products are continuously broken up by the vibration. Consequently small pits are formed at the points where the balls are in contact with the race. Hampp finds that a lubricant film which can exclude CO_2, oxygen, and moisture produces a marked reduction in the amount of 'false-brinelling'. This conclusion is in general agreement with that reached by Almen (1937) in America.

Influence of Lubricant Films on Wear

As we saw in Chapter IX, a lubricant film reduces the amount of intimate metallic contact between sliding metals. As a result the amount of metallic transfer between the surfaces may be greatly reduced. This is shown by the electrographic surface analysis of steel surfaces which have been traversed once by a copper surface at a speed of a few cm./sec. (Chap. IV, Pl. IX). The amount of copper transferred on to the steel is reduced by a factor of 50 or more in the presence of a suitable lubricant. As we have seen in Chapter IX, the lubricant films themselves exhibit a marked difference in their resistance to wear. For example, fatty acid films a few molecular layers thick are much more resistant to wear than a film of cholesterol of the same thickness. The ability of the lubricant to repair the damage either by surface mobility of the film or by rapid adsorption from the bulk of the lubricant is also of great importance.

In practical wear measurements where the rubbing surfaces run at considerably higher speeds and the same track is usually traversed many times, the wear process is complicated by the temperatures developed during sliding, by the gradual change of the surfaces themselves as a result of wear and pick-up, by the incidence of quasi-hydrodynamic lubrication, and by other factors discussed in the previous sections. For these reasons it is difficult to make satisfactory wear measurements in the presence of lubricants and the results are generally difficult to interpret.

A great deal of work has been carried out by Kenyon (1946) in developing an experimental technique that will provide reproducible measurements of wear in the presence of a lubricant. He finds that reproducible results can be obtained only if one of the rubbing surfaces is so hard that its surface is virtually undamaged by the sliding process. For reasons which will be discussed below, the hard surface must not be too smooth, otherwise the wear is negligibly small. Kenyon's technique consists of rubbing a small flat tungsten-carbide slider possessing a standard degree of surface roughness on the surface of a rotating steel disk and measuring, by sensitive chemical and colorimetric methods, the amount of steel abraded during sliding. Although the flat slider is very small (about 1 mm. diameter) and its surface is accurately parallel to the rotating disk (to within about $\frac{1}{20}°$), there is evidence that hydrodynamic lubrication will occur even when the sliding speed is as low as 20 cm./sec. This is shown, for example, by the contact resistance between the rubbing surfaces. If the slider is very smooth the contact resistance readily reaches values greater than 10 ohms and the wear is negligible. This provides interesting confirmation of Beeck's work on the ease with which hydrodynamic lubrication is set up even at low speeds and is supported by Fogg's recent observations on the incidence of fluid lubrication between parallel surfaces (1945). If the slider has the appropriate degree of surface roughness the contact resistance lies between 0·01 and 1 ohm, and the observed wear is largely due to the surface irregularities which project through the hydrodynamic film. At the tips of these asperities boundary lubrication and wear occur. Consequently, as the surface roughnesses on the slider are worn away, the wear of the steel surface steadily diminishes. This effect is very marked if the slider is not sufficiently hard. With tungsten carbide sliders, however, the surface of the slider is scarcely affected by the sliding process, and the rate at which metal is removed from the steel surface (expressed as weight/cm. track) is reproducible to within about 10 per cent. Measurements indicate that over a load range of about 1–7 kg. the wear is roughly proportional to the load.

Kenyon has carried out some interesting comparisons of the wear rate in the presence of various lubricants. The standard lubricant was a highly refined white oil free from polar compounds, in which various additives were dissolved. Typical results are given in Table XL.

TABLE XL

Lubricant	*Wear* 10^{-10} *gm./cm. of track*	
	Load = 2 kg.	*Load = 5 kg.*
White oil	226	785
2% oleic acid	760	1,660
5% oleic acid	760	1,660
1% tricresyl phosphate	66	415
1·5% additive containing sulphur and barium	144	590

The results show that oleic acid produces a considerable increase in the amount of steel removed during sliding. This observation is striking in view of the fact that oleic acid in general provides good boundary lubrication and reduced friction. It is possible that this behaviour is due to the chemical attack of the surface by the fatty acid. On this view the metallic soap film is rubbed away during sliding and is continuously reformed by further chemical reaction. Kenyon points out that non-reactive detergents may produce a similar increase in wear and suggests that the additive removes, by a detergent action, the detritus on the surface which otherwise would act as a protective film. This increase in wear with a good lubricant has been observed by other workers. Halder (see Tingle, 1947), for example, finds that under comparable conditions, lubricants giving a lower coefficient of friction often give a higher abrasive wear. This is particularly marked with lubricants containing oxygen and sulphur, where the friction is low but the amount of chemically combined metal removed from the rubbing surfaces is relatively high. These conclusions may, however, only be considered as a general trend. In some cases the presence of sulphur (as in the last line of Table XL) may lead to reduced wear. The reduced wear obtained with tricresyl phosphate on steel surfaces has been generally observed by other workers. For example, Beeck, Givens, and Williams (1940) have suggested that this effect is due to a chemical attack which produces a phosphide eutectic at the tips of the rubbing asperities; the eutectic has a relatively low melting-point and is readily softened by the high temperatures developed at the points of contact. As a result the rubbing action involved in the sliding process itself is sufficient to wipe away the eutectic from the larger asperities. This leads to a rapid smoothing of the surface and the amount of steel abraded during sliding is considerably reduced.

It is clear from these considerations that in nearly all practical cases the wear will depend on a delicate balance of chemical reactivity and lubricant-film breakdown, as well as on the detailed mechanism of the interaction between the surfaces.

REFERENCES

J. O. Almen (1937), *Mech. Eng.* **59**, 415.
O. Beeck, J. W. Givens, and E. C. Williams (1940), *Proc. Roy. Soc.* **A 177**, 103.
K. Dies (1943), *Archiv für das Eisenhüttenwesen*, **10**, 399.
G. I. Finch, A. G. Quarrell, and H. Wilman (1935), *Trans. Faraday Soc.* **31**, 1051.
A. Fogg (1945), *Engineering*, **159**, 138.
R. L. Johnson, D. Godfrey, and E. E. Bisson (1948), N.A.C.A. Tech. Note No. 1578.
H. F. Kenyon (1946), Thornton Research Laboratories (Shell): private communication.
R. E. Leyman (1937), International Tin Research and Development Council, Series A, No. 53.
R. Mailander and K. Dies (1943), *Archiv. für das Eisenhüttenwesen*, **10**, 385.
A. J. W. Moore and D. Tabor (1942), C.S.I.R. (Australia) Tribophysics Division Report A 46.
D. Tabor (1942), C.S.I.R. (Australia) Tribophysics Division Report **A 55**.
Useful references to wear will be found in:
I.A.E. Report No. 1945/T/1, *Wear of Metals*. Collection of Abstracts 1930–1945.
M.I.T. Summer Conference on Mechanical Wear, 1948.
E. Tingle (1947), B.I.O.S. Report No. 1610. *Fundamental Work on Friction, Lubrication and Wear in Germany*.
C. G. Williams (1940), *Collected Researches on Cylinder Wear*. Inst. Auto. Engineers.

XV

ADHESION BETWEEN SOLID SURFACES: THE INFLUENCE OF LIQUID FILMS

EARLIER chapters have shown that the friction between metal surfaces is due to the shearing of metallic junctions formed by adhesion or welding at the points of intimate contact. We might, at first thought, expect that when the load is removed these junctions would still remain and an appreciable *normal* force would be needed to separate the surfaces. It is therefore of interest to examine the normal adhesion between solids and to determine the effect of interposed liquid films on the adhesion. Earlier work by Holm (1946), Shaw and Leavey (1930), and Jacob (1912) has shown that strong adhesion may occur between surfaces which have been cleaned and freed of adsorbed surface films, but if the surfaces are exposed to air no adhesion is observed. The work of McBain (1931) and Hardy (1936) has dealt more specifically with the strength of adhesives between solid surfaces. Similarly Budgett (1911), Stone (1930), and other workers have shown that marked adhesion between solid surfaces can occur in the presence of water or other liquid films. In this chapter we shall first discuss the behaviour of a number of surfaces for which the observed adhesion is very small unless liquid films are present. In the second part we shall show that with certain materials extremely high adhesions may be obtained even when the surfaces are dry. In such cases, liquid films produce a marked *reduction* in the adhesion. Most of the experiments described here have been carried out by Dr. McFarlane (McFarlane and Tabor, 1948, 1950).

ADHESION OF HARD SURFACES: GLASS, PLATINUM, AND SILVER

A number of simple and relatively crude experiments was carried out to measure the normal adhesion between surfaces of glass, platinum, and silver after they had been pressed together. The surfaces were cleaned by the best possible laboratory methods available. Glass was cleaned in chromic acid and then in a flame, platinum by heating in a flame, and silver by careful abrasion and polishing. In all cases the normal adhesion was negligible. For this reason a sensitive pendulum type of apparatus was used. One surface is supported vertically in a rigid frame. The other surface, usually in the form of a sphere, hangs from a fine thread and rests against the first surface on to which it can be pressed with a known force. The vertical surface may then be moved away from the sphere until, under the force of the gravity component, the sphere falls away. The deflexion of the thread from the vertical gives a measure of the adhesive force. With this apparatus adhesions as small as 10^{-6} gm. could be readily measured.

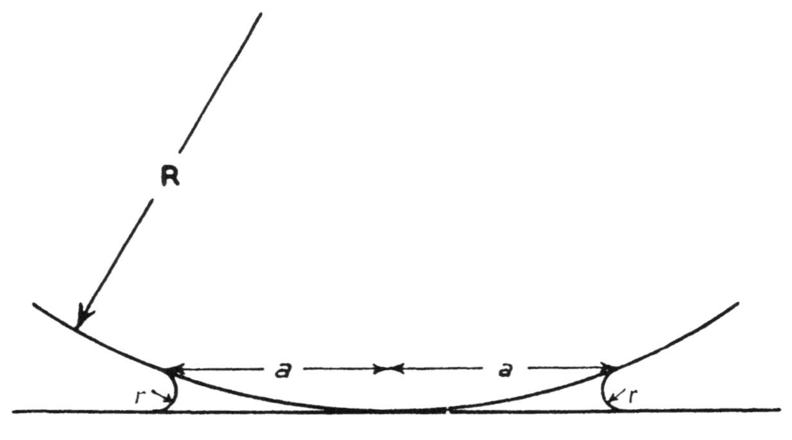

FIG. 107.

The results again showed that in clean dry air the adhesion was negligible. In a humid atmosphere, however, marked adhesion was observed, particularly with glass surfaces. The adhesion depended on the humidity, but at saturation the adhesion was the same as that observed if a small drop of water was placed between the surfaces. This suggests (see also Stone, 1930) that the observed adhesion is due to the surface tension of a thin film of water adsorbed or deposited on the glass surfaces.

A simple calculation shows that for a spherical bead on a glass plate we may express the adhesion force in terms of the surface tension. Consider a perfectly smooth sphere of radius R resting on a perfectly smooth plane covered with a thin film of water. Suppose the liquid collects to form a pool at the tip of the sphere and that the radius of curvature of the profile of the meniscus is r (Fig. 107). If the liquid completely wets the surface (i.e. the contact angle is zero), and if $r \ll R$, the pressure inside the liquid is less than atmospheric pressure by approximately T/r, where T is the surface tension of the liquid. Consequently the force over the whole of the liquid pool is equal to $\pi a^2 T/r$.

But to a first approximation $a^2 = 2R.2r$. Hence the adhesive force A is given by

$$A = 4Rr\pi T/r = 4\pi RT. \quad (1)$$

Thus the adhesion is *independent of the thickness of the liquid film* and is directly proportional to R. Experiments with glass beads of various radii of curvature fully confirm this relation. Typical results

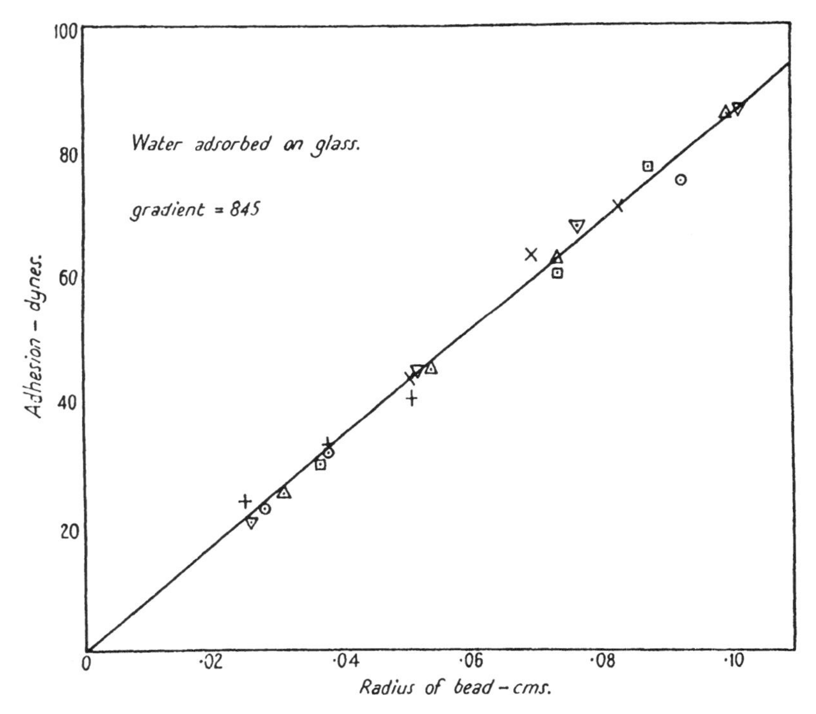

FIG. 108. Adhesion of a spherical glass bead on a flat glass surface in a humid atmosphere. The adhesive force A is directly proportional to the radius of curvature, R, of the bead. The slope of the straight line is a measure of the surface tension of the liquid between the surfaces, and the calculated value is 67 dynes/cm.

carried out in a saturated atmosphere are given in Fig. 108. The surface tension calculated from the straight line obtained is 67·3 dynes/cm., whereas the surface tension of water at the same temperature is 72·7 dynes/cm. The discrepancy may be due to some defect in the apparatus, since large-scale experiments carried out on the adhesion between a convex lens and a wet glass plate using equation (1) gave values of T of 72·5 dynes/cm. Nevertheless it seems clear that the adhesion observed in the pendulum experiments is due essentially to the surface tension forces acting in the film of water adsorbed on the surfaces. This is confirmed by experiments in which small drops of various liquids were placed between the glass surfaces. The adhesions observed gave the following values of T as calculated from equation (1):

TABLE XLI

Adhesion due to thin films of liquid on glass surfaces

Liquid	*Surface tension, dynes/cm.* Calculated from adhesion	Accepted values
Water	67·3	72·7
Glycerine	59	63·5
Decane	22·4	25
Octane	19·9	21·8

Although the values are all somewhat lower than the accepted values, it is clear that they are all close to, and vary directly with, the surface tensions of the liquids.

It is of interest to note that in the above experiments the adhesion remained at its full value even when the liquid films had thinned so far by evaporation that interference colours were no longer visible. This means that liquid films well below 1,000 A in thickness still give their 'normal' surface tension values. We may note incidentally that this provides a method of measuring the surface tension of minute quantities of liquid to within a few per cent. No adhesion was observed in atmospheres saturated with vapour of benzene or alcohol. This is probably because the adsorbed layers are too thin (see below).

Effect of surface roughness

If the glass plate was roughened by abrading with carborundum paper, the adhesion produced in an atmosphere saturated with water vapour decreased with increasing roughness. These results are summarized in Table XLII. If, however, a film of water was applied to the roughened surfaces, high adhesions were again observed.

TABLE XLII

Adhesion of glass surfaces in atmosphere at 100 per cent. humidity

Surface finish	*Mean height of surface irregularities, A*	*Adhesion as percentage of value observed with highly polished surfaces*
Highly polished	c. 150	100
500 carborundum paper	1,000	79
320 ,, ,,	4,000	51
150 ,, ,,	100,000	0

Similar results were observed with platinum surfaces, but in this case even with the smoothest polish obtainable the adhesion was never as high as that observed between glass surfaces. This is probably because the water film is appreciably thinner on platinum than on glass, and because the polished platinum surface is rougher than the fire-polished glass surface (see below).

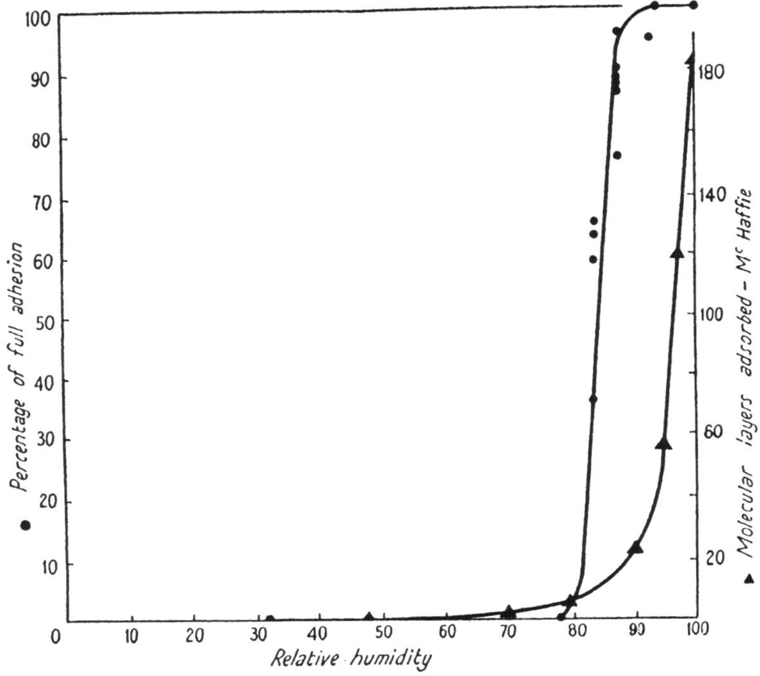

FIG. 109. Adhesion of glass surfaces as a function of the humidity of the surrounding atmosphere. The adhesion rises rapidly at humidities exceeding 80 per cent. saturation (curve ●). There is a close parallel to the results of McHaffie and Lenher (curve ▲) for the thickness of the water film adsorbed on glass surfaces.

Effect of humidity

Some experiments were carried out on the adhesion between smooth glass surfaces in atmospheres maintained at various degrees of humidity. The apparatus was surrounded by a lagged glass tank and the atmosphere inside the tank was allowed to come into equilibrium with saturated solutions of various salts. The atmosphere was thoroughly mixed before each adhesion measurement by means of a stirrer. With glass surfaces the adhesion reached its maximum value at about 88 per cent. humidity; with a glass sphere on a polished platinum surface at between 93 per cent. and 100 per cent. humidity. The adhesion results for glass are plotted in Fig. 109, where the adhesion is expressed as a percentage of the full adhesion observed at 100 per cent. humidity. On the same figure a curve has been plotted from adsorption experiments by McHaffie and Lenher (1925) showing the thickness of the layer of water adsorbed on glass surfaces at various humidities. It is seen that according to their results the thickness of the film increases rapidly at humidities greater than 90 per cent.

The above experiments show that the effect of increasing the roughness of the glass plate in a saturated atmosphere is similar to the effect obtained if the surfaces are kept smooth, but the humidity is reduced. This indicates that the adhesion depends on the height of the surface peaks and on the thickness of the adsorbed layer of water. We should, in fact, expect that the decrease in adhesion would occur when the height of the asperities is comparable with the thickness of the adsorbed film. Thus in a saturated atmosphere the adhesion of glass surfaces begins to fall off when the surface roughnesses exceed about 1,000 A. This conclusion is confirmed by the observation that the adhesion with platinum is always less than with glass, even in a saturated atmosphere. Presumably this is because the platinum is not as smooth as the glass and because the adsorbed water film on platinum, as McHaffie and Lenher found, is considerably thinner. It is clear from these experiments that water films adsorbed on the glass and platinum surfaces are responsible for the adhesion observed.

Adhesion due to surface tension and viscosity

It is interesting to consider the adhesion between flat parallel surfaces separated by a film of liquid. Suppose the surfaces are circular disks 1 in. in diameter (area 5·1 sq. cm.) and are separated by a film of oil 1,000 A thick. If the oil wets the surfaces the radius of curvature r of the meniscus at the edge of the disks is 500 A or 5×10^{-6} cm. On account of the surface tension T the pressure inside the liquid film is reduced by approximately T/r. For a typical mineral oil, T is about 30 dynes/cm., so that the pressure is 6×10^{6} dynes/cm.^{2} If the film between the surfaces is continuous the adhesive force due to surface tension is

$$5{\cdot}1\times 6\times 10^{6}\text{ dynes} = 31\text{ kg.}$$

The marked adhesion between flat surfaces when they are wetted by an interposed liquid was shown by Bastow and Bowden (1931). They observed that with an incomplete film of liquid the surface-tension forces are sufficient to cause a buckling of thick glass plates. These results suggest that the adhesion between Johannson flats and similar slip-gauges may be largely due to the surface tension of an intermediate oil film. It is interesting to note that, in experiments with flat surfaces in 1911, Budgett found that for highly polished steel surfaces (area 4·5 sq. cm.) wetted by a film of paraffin oil the adhesion was about 20 kg. This is of the same order as that calculated above.

Budgett attributed the adhesion to the cohesive forces in the liquid film. He points out, however, that if the thickness of the film is increased the pull required to separate the surfaces diminishes rapidly. It would therefore appear that the adhesion observed is not simply due to the tensile strength of the liquid, though this will clearly set an upper limit to the adhesion (see below). On the other hand, several workers have attributed the adhesion between slip-gauges to the molecular fields of the solid surfaces, acting through the liquid film (see, for example, Rolt, 1929). It should, however, be noted that for dry surfaces no adhesion is observed.

If the gap between the surfaces is completely surrounded by the liquid, the surface-tension forces should be zero. This effect is, in fact, observed. If the surfaces are completely surrounded by liquid they may be separated by the smallest normal force, provided the separation is carried out very slowly. If the rate of separation is rapid the viscosity of the liquid may become a very important factor. As the surfaces are pulled apart the liquid must flow into the space between them, and if the viscosity of the liquid is appreciable the force required to separate them in a short time-interval may be very large indeed. For example, the simple analysis given in Chapter XIII shows that if the surfaces are circular disks of radius R and the oil has a viscosity η, the time t to separate the surfaces from a separation h_1 to a separation h_2 by a force F dynes is

$$t = \frac{3\pi\eta R^4}{4F}\left(\frac{1}{h_1^2}-\frac{1}{h_2^2}\right).$$

Suppose the disks are 1 in. in diameter ($R = 1{\cdot}27$ cm.), the oil is a typical light mineral oil of $\eta = 150$ centipoise, and the original oil film thickness is 1,000 A (i.e. $h_1 = 10^{-5}$ cm.). Then the force F required to pull the surfaces apart ($h_2 = \infty$) in a time of 10 seconds against the viscous forces if the film is not ruptured is 9,500 kg., i.e. nearly 10 tons. In practice, of course, the film itself will rupture at a lower load, since its tensile strength cannot reach such a high value. If dissolved air or other gases are present the oil film will rupture more readily. Nevertheless the calculation indicates that viscous forces may produce very large apparent adhesions between parallel surfaces if the adhesive strength is measured by short-period applications of the load. The importance of viscosity in the action of adhesives has also been recently stressed by Bikerman (1947).

It is clear from these calculations and from the previous experimental results that in many cases the surface tension and viscosity of interposed liquid films may play an important part in the observed adhesion between solid surfaces. This will be particularly marked when the surfaces are very smooth and separated by a thin continuous film of liquid extending over a relatively large area.

Adhesion of Soft Metals [*A*]

The fact that in the complete absence of liquid films the normal adhesion between clean platinum, silver, and glass surfaces is negligible may appear to argue against the formation of junctions between the surfaces. There are, however, two points to be considered. The first is that even when the surfaces are most carefully cleaned in the atmosphere they will be covered with adsorbed films and these may not be readily broken through when the surfaces are placed in normal contact. When sliding occurs, however, these surface films are largely broken down by the sliding process itself. A more important point is that the solids used in the previous experiments are relatively hard. When the load is removed the elastic stresses within the bulk of the bodies are released (see Chap. I) and the resulting displacement may break any junctions that may have been formed. This view is confirmed by experiments with very soft metals such as lead and indium, where the released elastic stresses are very much less. If a clean steel ball is pressed on to a clean surface of lead, tin, or indium, very strong adhesion is observed.

For these experiments a very simple apparatus was used. It consists of an equal-armed balance, at one end of which the metal surfaces are mounted. The upper surface consists of a steel ball and the lower surface is a flat strip of the softer metal. The ball is pressed on to the lower surface with a known load for a specified time and the load is then carefully removed. Lead shot is then run into the pan at the opposite end of the arm until the surfaces are pulled apart. The balls used in these experiments had a radius of curvature of the order of a few millimetres, and the adhesions were many orders of magnitude higher than those described in the earlier part of the chapter. Consequently the surface tension of any liquid films present on the surface will have a negligible effect on the adhesion. The surface of the steel ball was prepared by polishing. The soft metal was cast in a suitable block and a fresh surface prepared immediately before the adhesion experiment by cutting a thin layer off the surface with a clean planing-tool made of steel or diamond.

Indium. With clean indium surfaces it is found that *the adhesion is of the same order as the original normal load applied*, but for any one load

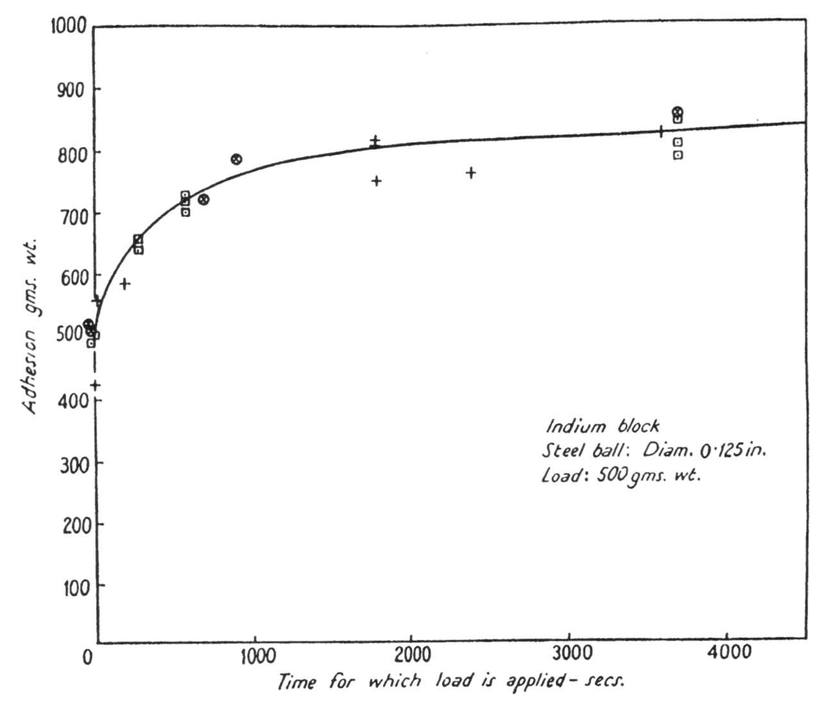

FIG. 110. Adhesion of a clean steel ball on a clean indium surface. With increased time of loading the adhesion increases, but in general the adhesive force is of the same order of magnitude as the original load, i.e. the 'coefficient of adhesion' is of the order of unity.

the adhesion increases with the time of application of the load (Fig. 110). This, however, appears to be due to a creep effect which allows the area of contact to increase steadily with time. This is shown by the fact that if the adhesion is plotted against the area of the indentation formed, i.e. against the area of contact, the result is a straight line (Fig. 111). It is of course true that the creep effect operates in the opposite direction too. That is to say, the force required to separate the surfaces is smaller if the force is applied for a long period. This effect, however, was not important under the experimental conditions used.

When the surfaces are separated a thin layer of indium is found adhering to the ball. This shows that the junctions formed between the indium and the steel surface are at least as strong as the indium itself, even though an appreciable film of oxide must be present on the steel. This view is confirmed by the observation that if the indium-coated sphere is used for further experiments on an indium surface the same

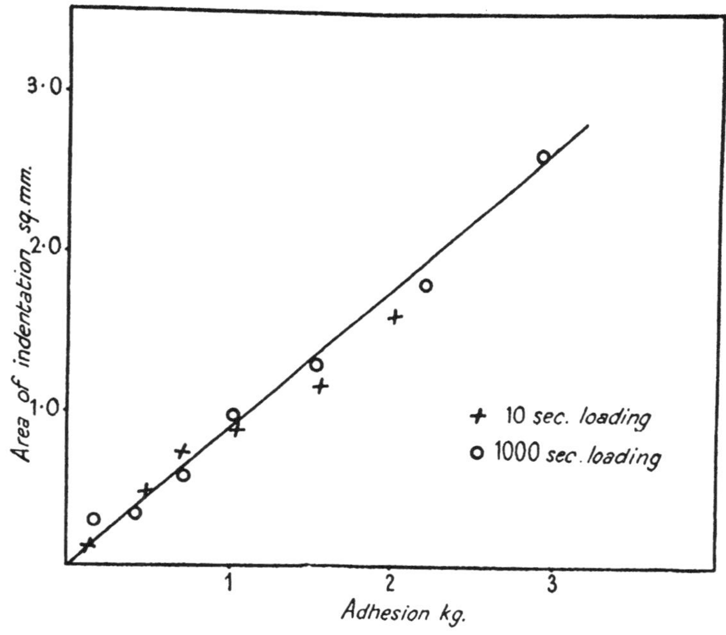

FIG. 111. Adhesion of a clean steel ball on a clean indium surface. The adhesive force is directly proportional to the size of the indentation formed, whether the time of loading is 10 sec. or 1,000 sec. This suggests that the increase in adhesion with time (Fig. 110) is essentially a creep effect.

adhesion is found as for clean steel. It is apparent from these results that intimate metallic junctions are formed at those regions of contact where the indium flows plastically. The temperature rise due to the deformation process itself is negligible (see Chapter XIII) so that this must be considered as a 'pressure-welding' or a 'cold-welding' effect. It is interesting to note that, quite recently, this effect has been used industrially for the joining of aluminium alloy strips (e.g. Tylecote (1948); Anon. (1948)).

Lead and tin. Experiments were also carried out on the adhesion between a steel ball and surfaces of lead and of tin. The behaviour was similar to that of indium. Some typical results are given in Table XLIII for a load of 2 kg. applied for a period of 1,000 sec. It is seen that for indium the adhesion is over 2 kg. If by analogy with the coefficient of friction we define the coefficient of adhesion ν as the ratio of the adhesive force to the normal load, the value for indium is about $\nu = 1{\cdot}2$. For steel on lead the value is about $\nu = 0{\cdot}7$, whilst for steel on tin it is about $\nu = 0{\cdot}4$. The adhesion for lead and tin is therefore marked but not as large as with indium.

TABLE XLIII

Adhesion of steel ball to indium, lead, and tin. Clean surfaces. Load 2 kg. Time of loading 1,000 sec.

Surface	*Adhesion, kg.*	*Coefficient of adhesion*
Indium .	2·4	1·2
Lead .	1·4	0·7
Tin .	0·8	0·4

Effect of surface oxidation

Indium is a metal which oxidizes relatively slowly. It is not surprising, therefore, to find that the adhesion of an indium surface exposed to the air falls off very slowly with time of exposure. With lead and tin, however, the adhesion falls off much more rapidly. Comparative results for indium and lead are shown in Fig. 112. It is apparent that oxide films formed on metal surfaces may produce a marked reduction in the adhesion. This is presumably because they reduce the amount of intimate metallic contact between the surfaces. This conclusion is in general agreement with frictional observations, where it is found that oxide films can produce an appreciable reduction in friction. It should, however, be borne in mind that in frictional experiments the sliding process itself may often facilitate the breakdown of the contaminant film.

Some experiments on the adhesion of metals in a high vacuum which confirm this view have been carried out by Mr. J. E. Young. If clean nickel surfaces, for example, are placed in contact *in vacuo* at room temperature, both the friction and the adhesion are high (see Chap. VII). If the surfaces are cleaned more thoroughly by prolonged outgassing the adhesion is very much greater still. Experiment shows that, if a small curved contact of nickel under a load of a few grammes is allowed to touch another nickel surface in a high vacuum, and sliding is initiated, both the coefficient of adhesion and the coefficient of friction are too great to be measured. It is necessary to prize the surfaces apart with a knife. Subsidiary experiments showed that when the initial load between the surfaces was 15 gm. the normal force required to separate the surfaces was > 400 gm. It would appear that in the absence of oxide a fitting together of the metal lattice and 'crystal growth' between the two metal surfaces can occur on quite a large scale.

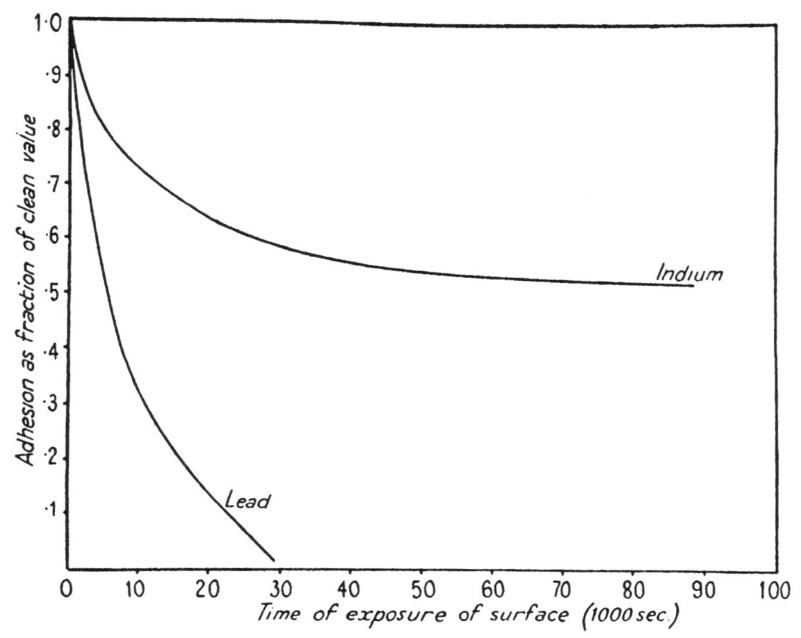

FIG. 112. Adhesion of a clean steel ball on clean indium and clean lead surfaces. The adhesion is measured after various periods of exposure of the surfaces to the air. The decrease in adhesion with time of exposure corresponds to the growth of the oxide film on the metal surface. The effect is very much more marked on lead than on indium.

Adhesion in the presence of lubricant films

We should expect that lubricant films would reduce the adhesion. Measurements were made using a non-polar lubricant (medicinal paraffin oil) and a 'good' boundary lubricant (1 per cent. lauric acid in cetane).

With indium surfaces the adhesion in the presence of paraffin oil was reduced to about one-half. When the lauric acid solution was applied a monolayer of the fatty acid was immediately adsorbed on the metal surface so that the remainder collected as small droplets which would not wet the surface. They were carefully removed. The surface on which the adhesion experiments were carried out therefore consisted of an indium surface covered by an oriented monolayer of fatty acid. It was found that the adhesion was negligibly small. If, however, heavy loads were used so that the indentation formed in the indium was sufficiently deep, some adhesion was observed. This occurred when the area of the curved surface within the indentation was about 2 per cent. greater than the original plane surface. This is shown in Fig. 113, where the coefficient of adhesion ν has been plotted against Δ, where Δ is the amount by which the curved area of the indentation exceeds the original plane

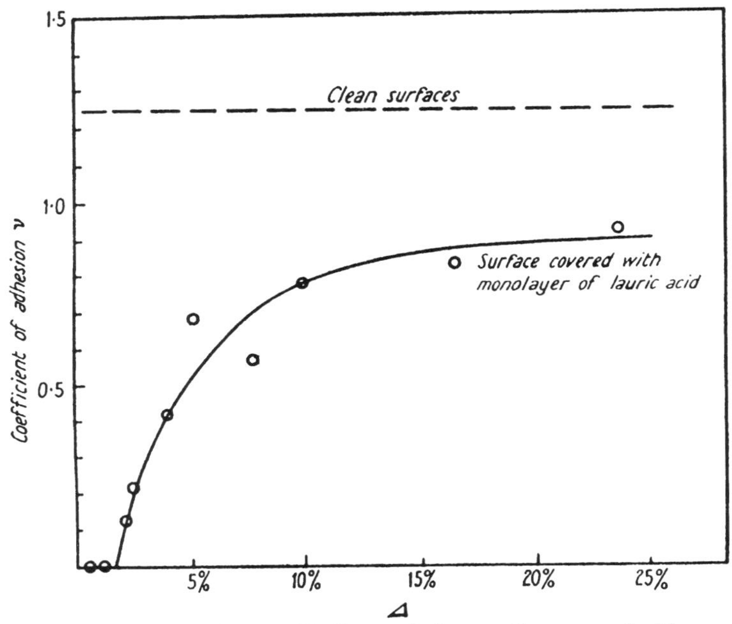

FIG. 113. Adhesion of a clean steel ball on an indium surface covered with a monolayer of lauric acid. The 'coefficient of adhesion' ν is plotted against Δ, where Δ is the percentage by which the curved area of the indentation exceeds the original plane surface area. If the monolayer is stretched by less than about 2 per cent. the adhesion is negligible. If its area is increased by more than this, there is appreciable breakdown of the monolayer and marked adhesion. The broken line gives the adhesion for clean surfaces in air.

surface. It is seen that for clean surfaces ν is constant at about $\nu = 1{\cdot}2$. For the surfaces covered with a monolayer of fatty acid, ν is zero for values of Δ below about 2 per cent., that is, for relatively small indentations. For larger indentations, however, that is, for larger values of Δ, the coefficient of adhesion ν steadily increases. This indicates that the fatty acid is still able to prevent appreciable metallic adhesion when the area of metal covered by the acid monolayer is extended or 'stretched' by a few per cent. If the area is stretched by more than about 2 per cent., however, appreciable adhesion through the monolayer occurs.

This view is supported by the observation that if a monolayer of the fatty acid is deposited on the *steel* ball the adhesion is negligible whatever the size of the indentation formed in the indium surface.

Similar results were obtained with lead and tin surfaces. The main results for small indentations are summarized in Table XLIV, where the adhesion is expressed as a fraction of that observed for clean surfaces.

TABLE XLIV

Effect of lubricant films on adhesion

Metal	*Adhesion/Adhesion clean*		
	Clean	*Paraffin oil*	*1% lauric acid in paraffin*
Indium . .	1	0·6	0
Lead . .	1	0·15	0
Tin . . .	1	..	0

It is seen that with all these metals the adhesion is appreciably reduced by paraffin oil and is reduced to a value which is too small to be measured by a monolayer of the fatty acid. It does not follow, however, that no metallic contact occurs through the fatty acid monolayer. The results mean rather that if any metallic junctions are formed through the fatty acid film, they are broken as the load is removed, possibly by the release of elastic stresses around the indentation. This explanation is not inconsistent with the large adhesions observed with clean surfaces, since in this case adhesion occurs over the greater part of the region of contact and the released stresses are not sufficient to break the welded junctions.

ADHESION AND FRICTION [*A*]

A few experiments have been carried out on the delicate apparatus described in Chapter IV. This apparatus is capable of measuring both the friction and normal adhesion between surfaces. The upper surface was a clean steel ball $\frac{1}{8}$ in. diameter, and the lower surface a freshly scraped flat bar of indium. The loads used were less than 50 gm., so that the deformation of the indium was slight and the ploughing term in the frictional measurements negligibly small. If a load of 15 gm. was applied for a few seconds, the normal adhesion was of the same order. If now the indium surface was set in motion at a very slow speed of sliding, the frictional force on the steel ball increased rapidly to a steady upper value of about 70 gm., i.e. a coefficient of friction of $\mu \approx 5$. If at this stage the tangential restoring force was removed and the *normal* adhesion was measured, it was found to have increased from about 15 gm. to about 100 gm., i.e. the coefficient of adhesion increased from about $\nu = 1$ to about $\nu = 7$. Thus the high frictional force is associated with a large adhesion between the surfaces. This effect is shown in Fig. 114 which also shows the way in which the friction and adhesion increase as the lower surface is set in motion. It is seen that before

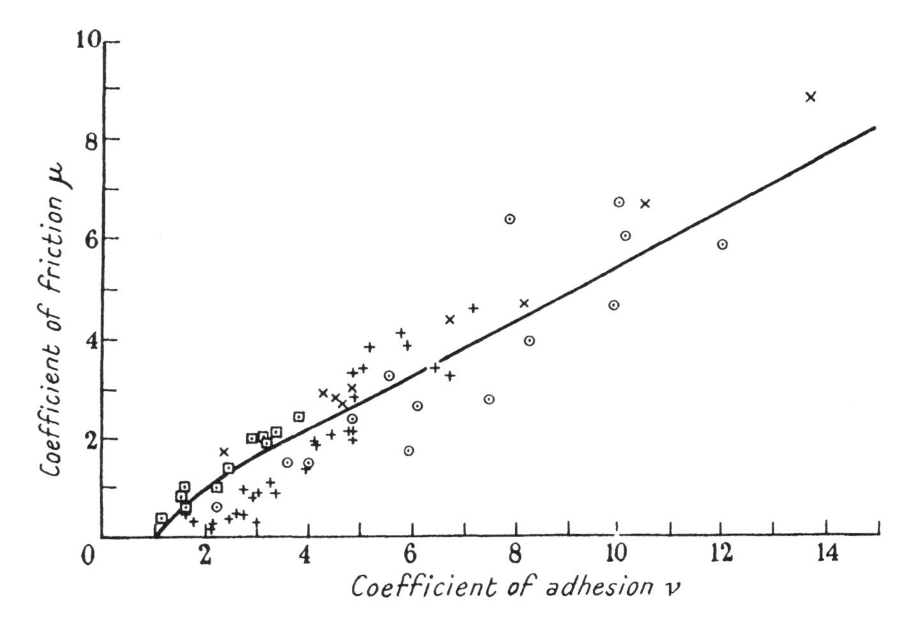

FIG. 114. Coefficient of friction (μ) and coefficient of adhesion (ν) for steel on indium as relative motion takes place between the surfaces. The points plotted are for experiments carried out at various loads: ⊙ Load ~ 6 gm.; + Load ~ 15 gm.; ⊡ Load 100 gm.; × various other loads. The heavy line is the theoretical curve $\mu^2 = 0{\cdot}3\nu^2 - 0{\cdot}3$ derived from von Mises's criterion for plastic flow under combined normal and tangential stresses. When the surfaces are first placed in contact, the smallest tangential force can initiate relative motion (microslip). As microslip proceeds the tangential force rapidly increases and ultimately an upper steady value is reached; here macroscopic sliding occurs. At this stage μ is of the same order of magnitude as ν.

sliding commences, the adhesive force is approximately equal to the original load. To initiate relative motion between the slider and the indium surface, a very small tangential force is required since the junctions are already plastic under the applied load. As relative motion proceeds there is a steady rise both in the tangential and in the adhesive force. It is evident that increased plastic flow of the indium has occurred with a corresponding increase in the area over which welded junctions are formed. This enlarged area is responsible for both the large adhesion and the large tangential force observed. It should be noted that the relative displacement of the surfaces during this stage of the sliding process is so small that unless refined methods of measurement are used no sliding is observed. Relative movement *is* occurring, but only on a microscopic scale. An upper steady state is reached, however, where the tangential force increases more rapidly than the growth of the junctions and sliding on a macroscopic scale occurs. The coefficient of friction is large and the coefficient of adhesion is of the same order of magnitude. It is apparent from these experiments that if the sliding produces a marked increase in welding, the frictional force when macroscopic sliding occurs may bear little relation to the original applied load.

The detailed behaviour of the junctions up to the point at which macroscopic sliding occurs may be expressed in terms of von Mises's criterion for plastic flow under combined normal and tangential stresses. If, for example, equation (14), Chapter V, is used, and reasonable values adopted for the arbitrary coefficients, there is fairly good agreement between the theoretical relation and the experimental results plotted in Fig. 114. These results support the more detailed theory of metallic friction discussed in Chapter V. They show in a very direct way the reality of the cold-welding process at the points of contact and the role of the metallic junctions so formed in the adhesion and friction of metals. More recent work also shows that lubricant films, by diminishing the amount of intimate metallic contact, reduce both the friction and the adhesion between the surfaces. A fuller discussion of these factors is given by McFarlane and Tabor (1950).

REFERENCES

ANON. (1948), *Engineering*, **165**, 535.
J. J. BIKERMAN (1947), *J. Colloid Science*, **2**, 163.
F. P. BOWDEN and S. H. BASTOW (1931), *Proc. Roy. Soc.* **A 134**, 404.
H. M. BUDGETT (1911), ibid. **A 86**, 25.
SIR W. HARDY (1936), *Collected Works*, Camb. Univ. Press.
R. HOLM (1946), *Electrical Contacts*, Almquist and Wiksells, Stockholm.
C. JACOB (1912), *Ann. Phys. Lpz.* **38**, 126.
J. W. MCBAIN (1931), D.S.I.R. Third Report of Adhesives Research Committee.
J. S. MCFARLANE and D. TABOR (1948), *VIIth International Congress for Applied Mechanics*, London, Vol. IV, 31; (1950), *Proc. Roy. Soc.* **A 202**, 224; 244. See also J. S. McFarlane (1949), Ph.D. Dissertation, Cambridge.
I. R. MCHAFFIE and S. LENHER (1925), *J. Chem. Soc.* **127**, 1559.
F. H. ROLT (1929), *Gauges and Fine Measurements*, **1**. MacMillan.
P. E. SHAW and E. W. LEAVEY (1930), *Phil. Mag.* **10**, 809.
W. STONE (1930), ibid. **9**, 610.
R. F. TYLECOTE (1948), Sheet and Strip Metal Users' Technical Association, Winter Conference.

XVI

CHEMICAL REACTION PRODUCED BY FRICTION AND IMPACT [*A*]

INFLUENCE OF PRESSURE, OF SHEAR, AND OF SURFACE TEMPERATURE

IT is well known that chemical reaction may be brought about by friction and by impact. Early work was concerned with the decomposition of solids by high pressure and by grinding in a mortar with a pestle, where the solid is subjected to both pressure and shear. Carey-Lea (1891) investigated the decomposition of solids such as $AgCl \rightarrow Ag$, $HgO \rightarrow Hg$, $KMnO_4 \rightarrow MnO_2$, and was able to show that decomposition by pressure was facilitated by a shearing motion. During the shearing, frictional heat is developed, but he considered that this heat played little or no part in the decomposition. Parker (1914) suggested that, when solids are ground together, reactions of the type

$$HgCl_2 + 2KI \rightarrow HgI_2 + 2KCl$$

take place owing to the local or surface melting of the solid which results from the stress. In these experiments dry salts must be used since the presence of minute traces of water vapour affects the results considerably.

In a series of papers P. W. Bridgman (1935–47) has described experiments in which he subjected many compounds to hydrostatic pressures up to 50,000 kg./cm.^{2} (or 50,000 atm.) combined with a shearing stress up to the plastic flow pressure of the material. Under these extreme conditions many compounds decomposed explosively, e.g. iodoform, silver nitrate, lead dioxide. Reactions between copper and sulphur, and silicon and magnesium oxide, also took place with explosive violence.

Two phenomena which are of importance in connexion with frictional decomposition are (i) the effect of pressure on the melting-point of solids, and (ii) the production of local high-temperature flashes at surface boundaries when two solids are rubbed together. In discussing the effect of pressure on the melting-point of solids, Johnston and Adams (1913) have pointed out that the application of a uniform pressure to a solid–liquid phase has a comparatively slight effect on the melting-point and, in fact, usually raises it by *c.* 10–30° per 1,000 atm. On the other hand, in a system where there is a non-uniform pressure on the solid–liquid phase, i.e. where there is an excess pressure on the solid phase, a lowering of the melting-point is always obtained (see also H. Jeffreys, 1935). Non-uniform compression may be visualized, for example, in the grinding of solids in a mortar with a pestle. This grinding results in a melting of the surface layers at the crystal boundaries where the reaction occurs. The liquid formed by melting flows into interstitial spaces and in this way becomes subjected to a smaller pressure than the adjacent solid particles. The reaction products are then removed during the grinding, and fresh surfaces of the reactants are continually exposed.

We have seen in Chapter II that local high temperatures are readily produced on the surface of rubbing solids, and there is strong experimental evidence that this can be a major factor responsible for the chemical decomposition. The reduction of polishing powders such as red lead and lead dioxide and the decomposition of calcium carbonate when these powders are used to polish metal surfaces may be ascribed to hot spots produced during the polishing (see Chap. III). When metals such as iron, copper, and nickel are polished, rolled, or rubbed together, oxidation of the metal surface occurs and this oxidation is accelerated by the localized surface heating. It is probable that the mechanical distortion, deformation, and breakdown of the protective oxide film also accelerates the surface attack. Chemical reactions which occur during the machining of metals have recently been described by Shaw (1948).

It has been suggested that the chemical decomposition produced by friction and impact is 'tribochemical' in origin. The meaning of this is not altogether clear, but it has been suggested that under the suddenly applied pressure of the impact, or of the rubbing, the crystals are subjected to a normal stress forcing them into more intimate contact and to tangential stresses tending to shear the crystals apart. As a result of the normal stresses, the molecular fields of the surface molecules are thrust together and linkages formed. These are almost immediately ruptured under the tangential stresses with the result that the surface molecules are left in a highly activated state.

Mechanical rupture or decomposition of the molecule may perhaps occur during the rapid flow of high polymers such as polyisobutylene when dissolved in low-boiling solvents (Frenkel, 1944; Zvetkov and Frisman, 1945; Rehner, 1945; Morris and Schnurmann, 1947; Hargrave, 1947). Provided the molecular weight of the polymer is sufficiently high, breakdown takes place under the action of shearing forces to give polymers of lower molecular weight. Hess and co-workers (1942) suggest that the depolymerization of substances of high molecular weight such as cellulose and polystyrene during grinding is due to the stresses and strains set up in the macromolecules. Similar effects have also been reported by Schmid (1940) and Szent-Gyorgyi (1933), who observed depolymerization in solution due to the action of supersonic vibrations. Chemical decomposition may also be produced by supersonic vibrations. Bobolev and Chariton (1937) have shown that nitrogen chloride could be decomposed by this means. It is also possible, when the conditions are appropriate, for the generation and subsequent discharge of frictional electricity to initiate the reaction.

Other factors such as the exposure of fresh surfaces by sliding can be important. If the sliding speed is low or the movement is small, the local temperature rise will be inappreciable. Nevertheless there may be increased chemical attack. An example of this is provided by 'fretting corrosion', where a small movement due to vibration between two metal surfaces in contact may lead to a greatly increased chemical attack or surface oxidation. When a load is applied to a metal surface the underlying metal will be plastically deformed and the consequent stretching, deformation, and damage of the protective oxide film will render the metal more accessible to oxygen and enable the attack to proceed more rapidly. If sliding occurs the continual formation and shearing of the small metallic junctions will facilitate the reaction by continuously exposing a fresh surface.

Effect of Friction on Photographic Plates

Mr. A. C. Moore (unpublished) has studied the effect of friction on photographic plates and has shown that a latent image is formed in the track of a slider which is rubbed over the emulsion. He has some evidence that this is due to local hot spots formed by friction between the grains of silver halide in the emulsion. He has also observed an interesting desensitization produced by friction. This is apparently due to the production of internal faults within the silver halide crystals. These act as electron traps and cause the subsequent latent image to be formed inside the crystal itself where it is not accessible to an ordinary chemical developer.

Decomposition of Explosives

Initiation by friction

A study of the initiation of explosives throws an interesting light on the mechanism of chemical decomposition produced by friction and by impact. Recent experimental studies have shown that, in general, the initiation is not tribochemical in origin but is due to local 'hot spots' of finite size in or on the surface of the explosive. These hot spots bring about the initiation of the explosives by the normal process of thermal ignition. These hot spots may be formed in a number of ways. With the frictional initiation of liquid explosive such as nitroglycerine, for example, they are formed at the points of rubbing contact between the solid surfaces which confine the explosive. This may be shown by a simultaneous measurement of the local surface temperature flashes (using the methods described in Chap. II) and the incidence of explosion. If the confining surfaces are of two different metals and the thermo-electric method of measuring surface temperature is used, it is found (Bowden, Stone, and Tudor, 1947) that initiation does not occur until the surface temperature is 480° C. or more. It is also found that the initiation of the explosive is dependent upon the thermal conductivity and upon the melting-point of the surfaces. Unless the melting-point of the confining solids exceeds 480° C. no initiation can occur.

Using the series of alloys and metals listed in Table V, Chapter II, it was found that no explosions could be obtained with metals melting below 450° C. With an alloy melting at 480° C. and with all metals and alloys melting above this, however, initiation occurred. Experiment showed that the hardness of the rubbing solids (which determines the pressure applied to the explosive) and the rate of shear were relatively unimportant compared with the thermal conductivity, and the melting-point of the confining solids. It is the possibility of forming hot spots of a temperature *c.* 450° C. or greater which determines the initiation.

Initiation by impact

A second and very important source of a hot spot is the presence in the explosive of small gas bubbles or gas spaces (Bowden, Mulcahy, Vines, and Yoffe, 1947). These gas spaces, which can be very minute, are present between the crystals of a solid and they readily occur or are easily introduced into liquids. If the explosive is subjected to impact, or indeed to any sudden change in pressure, the gas bubble is heated by adiabatic compression and will fire the explosive. Very small bubbles of radius about 2×10^{-3} cm. at N.T.P. can be effective. The presence of these tiny bubbles can confer a very high sensitivity on the majority of liquid, gelatinous, and plastic explosives, so that ignition can occur with the gentlest of blows. It has been shown, for example, that nitro-glycerine which contains a small bubble can be exploded by the impact of 40 gm. falling half a centimetre, i.e. an impact energy of 20 gm. cm.

There is evidence that initiation can occur when the compression ratio on the gas space is sufficient to develop a temperature of about 450–500° C. Under moderate conditions of impact the temperature of the bubbles is usually much greater than this. The effectiveness of the gas is dependent upon its physical properties (particularly upon the ratio of the specific heats γ) and also upon its chemical nature. Dr. Yoffe's (1949) experiments have shown that included gas can also play an important part in the initiation of most solid secondary explosives.

Small gas spaces are frequently present in explosives and it is not always easy to remove them. If this is done with nitroglycerine, for example, it is found that very high impact energies are necessary (of the order of 10^5 to 10^6 gm. cm.) to cause initiation, and there is evidence that in this case the initiation is due to the viscous heating of the rapidly flowing explosive as it escapes between the impacting surfaces (see Chap. XIII).

Friction between particles

Another method of generating hot spots is by friction on a particle of grit or, with solid explosives, by friction between the grains themselves. A clear indication that the initiation is thermal in origin may be obtained by adding to the explosive a few grit particles of known melting-point and subjecting it to friction and to impact. Some typical results (Bowden and Gurton, 1948) with the solid secondary explosive pentaerythritol tetranitrate (P.E.T.N.) are given in Table XLV.

Over the range of hardness given in this work the hardness of a 'grit' particle was relatively unimportant (see, however, Ubbelohde, 1948). Its effectiveness in initiating an explosion both under friction and under impact depended on its melting-point. All the substances melting above 430° C. were effective, while all those melting below 400° C. had no effect.

It is clear that although the maximum hot-spot temperature is fixed by the melting-point of the particle, the *ease* with which the hot spot is formed will be very dependent upon the hardness. With a hard sharp particle the stresses will be concentrated at one or two points, so that it will require a much smaller energy under conditions both of impact and friction to produce a localized temperature rise of the necessary magnitude. If the particle is soft it will be plastically deformed or crushed so that this local concentration of the energy is not possible. For this reason we should expect that hard particles would be much more effective than soft ones provided that the melting-point of the particles is above the critical value, and experiment shows that this is so. Equally, we should expect that the thermal conductivity of the particles would be of great importance. The general relation between the thermal conductivity and the formation of hot spots has already been established in the earlier work (Chap. II). It is much more difficult to get visible hot spots on rubbing surfaces and also to initiate the explosion if the surfaces are good thermal conductors.

TABLE XLV

Initiation of P.E.T.N. in the presence of foreign particles

'Grit' added	*Hardness (Mohs scale)*	*Melting-point (°C.)*	*Explosion efficiency % (friction)*	*Explosion efficiency % (impact)*
No grit	..	..	0	2
Ammonium nitrate	2–3	169·6	0	2·5
Potassium hydrogen sulphate	3	210	0	2·5
Silver nitrate	2–3	212	0	2
Sodium dichromate	2–3	320	0	0
Sodium acetate	1·5	324	0	0
Potassium nitrate	2–3	334	0	0
Potassium dichromate	2–3	398	0	0
Silver bromide	2–3	434	50	6
Lead chloride	2–3	501	60	27
Silver iodide	2–3	550	100	..
Borax	3–4	560	100	30
Bismuthinite	2–2·5	685	100	42
Glass	7	800	100	100
Chalcocite	3–3·5	1,100	100	50
Galena	2·5–2·7	1,114	100	60
Calcite	3	1,339	100	43

The hardness of P.E.T.N. is *c.* 1·8 (Mohs scale) and its melting-point 141° C.

Intergranular friction between the grains of the explosive itself can also be important, and an interesting difference may be noted between the majority of secondary explosives and the primary explosives. Most of the secondary explosives such as P.E.T.N. melt at a comparatively low temperature *before* violent explosive decomposition sets in. With these explosives, intergranular friction merely causes a surface melting at the points of rubbing contact. It is necessary to form the hot spots in some other way such as the adiabatic compression of entrapped gas or the friction of a grit particle on the confining surface. Under very extreme conditions of impact it might be brought about by viscous heating of the rapidly flowing explosive (see Chap. XIII). With the primary explosives, however, such as lead azide and mercury fulminate, which explode below their melting-point the hot spots formed by intergranular friction can cause initiation.

High-speed camera studies (Bowden, Mulcahy, Vines, and Yoffe, 1947; Bowden and Gurton, 1948) of the birth and growth of the explosion show that it begins as a comparatively slow burning before it accelerates to a detonation, and confirm the view that initiation by friction and by impact is essentially thermal in origin, and is not due to a direct tribochemical excitation. The mechanical energy is concentrated at some local point and is degraded into heat to form a hot spot of small but finite size. (The size of these hot spots varies, but they may be of the order of 10^{-3}–10^{-5} cm. in diameter.) Initiation of the chemical reaction then occurs at this hot spot.

REFERENCES

U. Bobolev and J. V. Chariton (1937), *Acta Physicochim. U.R.S.S.* **7**, 416.
F. P. Bowden and O. A. Gurton (1948), *Nature*, **161**, 348; ibid. **162**, 654; (1949) *Proc. Roy. Soc.* A **198**, 350.
—— M. F. R. Mulcahy, R. G. Vines, and A. Yoffe (1947), ibid. A **188**, 291, 311.
—— M. A. Stone, and G. K. Tudor (1947), ibid. A **188**, 329.
P. W. Bridgman (1935), *Physical Rev.* **48**, 825; (1938), *Amer. J. Sci.* **36**, 81; (1947), *J. Chem. Phys.* **15**, 311.
M. Carey-Lea (1891), *Phil. Mag.* **34**, 46; (1893), ibid. **36**, 351; (1894), ibid. **37**, 31, 470.
J. Frenkel (1944), *Acta Physicochim. U.R.S.S.* **19**, 51.
K. K. Hargrave (1947), *Trans. Faraday Soc.* 'The Labile Molecule', 404.
K. Hess, *et alia* (1942), *Kolloid Z.* **98**, 148, 290.
H. Jeffreys (1935), *Phil. Mag.* **19**, 840.
J. Johnston and L. H. Adams (1913), *Amer. J. Sci.* **35**, 205.
W. J. Morris and R. Schnurmann (1947), *Nature*, **160**, 674.
L. H. Parker (1918), *J. Chem. Soc.* **113**, 396.
J. Rehner (1945), *J. Chem. Phys.* **13**, 450.
D. Schmid (1940), *Z. phys. Chem.* A **186**, 113.
M. C. Shaw (1948), *J. Applied Mechanics*, **15**, 37.
A. Szent-Gyorgyi (1933), *Nature*, **131**, 278.
A. R. Ubbelohde (1948), *Phil. Trans.* **241**, 280.
A. Yoffe (1949), *Proc. Roy. Soc.* A **198**, 373.
V. N. Zvetkov and E. Frisman (1945), *Acta Physicochim. U.R.S.S.* **20**, 61.

APPENDIX

SOME TYPICAL VALUES OF FRICTION†

I. METAL SURFACES—UNLUBRICATED

I. *Friction of metals (spectroscopically pure) outgassed in vacuum. When clean there is gross seizure*

Metals	Coefficient of friction after admitting		
	H_2 or N_2	Air or O_2	Water vapour
Aluminium on aluminium	..	1·9	1·1
Copper on copper	4	1·6	1·6
Gold on gold	4	2·8	2·5
Iron on iron	..	1·2	1·2
Molybdenum on molybdenum	..	0·8	0·8
Nickel on nickel	5	3	1·6
Platinum on platinum	..	3	3
Silver on silver	..	1·5	1·5

II. *Friction of clean soft metals*

Under more rigorous conditions the friction between clean metals in vacuum may be very much higher still. Values of $\mu \approx 100$ or more may be observed and with many metals, e.g. Ni, Pt, Fe, complete seizure occurs.

III. *Static friction of unlubricated metals prepared 'grease free' in air*

(*a*) *Pure metals on themselves.* With pure metals on themselves the friction is usually of the order of $\mu = 1$ to 1·5. It is generally lower with hard metals; with chromium, for example, $\mu = 0·4$.

With some metals, such as copper, the friction is much lower ($\mu \approx 0·5$) if the surface oxide film is not appreciably ruptured. This is readily observed at very light loads. At heavier loads, when the oxide films are broken through, the friction is of the order $\mu = 1·5$.

(*b*) *Different metals on one another.* Here the friction is somewhat less than that observed with similar metals. If there is appreciable pick-up of one metal by the other the frictional behaviour soon becomes characteristic of similar metals.

† The values given in this Appendix are based on measurements made under particular experimental conditions. As we have seen, slight changes in these conditions lead to large variations in the coefficient of friction.

(*c*) *Friction of soft metals.* With grease-free soft metals in air, very high coefficients of friction may be observed as a result of the sliding process itself. For example with steel on indium at a load of 2 gm. $\mu \approx 20$.

IV. *Friction of metallic films (unlubricated)*

With thin metallic films Amontons's law is not obeyed and the friction may be very low.

Friction of thin films of indium, lead, and copper deposited on various metallic substrates (lower surface). Upper surface hemispherical steel slider, diameter 0·6 cm. Thickness of films 10^{-3} to 10^{-4} cm.

	Coefficient of static friction μ_s			
Load gm.	*Indium film on steel*	*Indium film on silver*	*Lead film on copper*	*Copper film on steel*
4,000	0·08	0·1	0·18	0·3
8,000	0·04	0·07	0·12	0·2

V. *Static friction of alloys on steel (unlubricated)*

Alloy	μ_s	*Alloy*	μ_s
Copper–lead (dendritic)	0·22	Aluminium-bronze	0·45
Copper–lead (non-dendritic)	0·22	Brass	0·35
White metal (tin base)	0·8	Constantan	0·4
White metal (lead base)	0·55	Steel	0·8
Wood's alloy	0·7	Cast iron	0·4
Phosphor-bronze	0·35		

VI. *Kinetic friction of unlubricated solids*

In general the kinetic friction is lower than the static friction but the results may depend on the experimental conditions.

II. Metal Surfaces—Lubricated

VII. *Lubrication of steel surfaces. Static friction at room temperature*

Lubricant	*L*	*m.p.* °*C.*	μ_s	*Transition temp.*	*Lubricant*	*L*	*m.p.* °*C.*	μ_s	*Transition temp.*
(*a*) *Paraffins*					(*c*) *Fatty acids*				
Nonane .	C_9	−54	*0·26	..	Acetic .	C_2	16	*0·5	..
Decane .	C_{10}	−30	*0·23	..	Propionic .	C_3	−22	*0·4	..
Hexadecane	C_{16}	17	*0·16	17	Valeric .	C_5	−35	*0·17	..
Docosane .	C_{22}	44	0·11	44	Caproic .	C_6	−2	0·12	80
Triacontane	C_{30}	66	0·11	66	Pelargonic .	C_9	12	0·11	90
					Capric .	C_{10}	31	0·11	95
(*b*) *Alcohols*					Lauric .	C_{12}	44	0·11	120
					Myristic .	C_{14}	58	0·11	125
Butyl. .	C_4	−89	*0·3	..	Palmitic .	C_{16}	64	0·11	130
Octyl. .	C_8	−16	*0·23	..	Stearic .	C_{18}	69	0·10	140
Decanol .	C_{10}	7	*0·16	7					
Cetyl . .	C_{16}	49	0·10	49					

L = length of the molecular chain.

With paraffins and alcohols on steel, the transition temperature corresponds to the melting-point. Only those which are solid at room temperature give smooth sliding: those which are liquid give a jerky intermittent motion (marked with an asterisk *).

The shorter chain fatty acids cause marked corrosion of the steel surfaces and give intermittent motion (marked with an asterisk *). For the longer chain acids the transition temperature is about 70° C. or more above the melting-point.

The values of μ_s in this table are sensibly independent of load, but the transition temperature is affected by the experimental conditions.

VIII. *Lubrication of steel surfaces. Static friction*

Lubricant	*Static friction* μ_s 20° C.	100° C.	*Lubricant*	*Static friction* μ_s 20° C.	100° C.
Clean steel. .	0·58	..	*Mineral oil*		
Vegetable oils			Light machine .	0·16	0·19
			Thick gear . .	0·125	0·15
Castor . .	0·095	0·105	Solvent refined .	0·15	0·2
Rape . .	0·105	0·105	Heavy motor .	0·195	0·205
Olive . .	0·105	0·105	B.P. Paraffin .	0·18	0·22
Coconut . .	0·08	0·08	Extreme pressure	0·09–0·1	0·09–0·1
			Graphited oil .	0·13	0·15
Animal oils			Oleic acid . .	0·08	0·08
Sperm . .	0·10	0·10	Trichlorethylene .	0·33	..
Pale whale. .	0·095	0·095	Alcohol . .	0·43	..
Neatsfoot . .	0·095	0·095	Benzene . .	0·48	..
Lard . . .	0·085	0·085	Glycerine . .	0·2	0·25

From Fogg and Hunswicks, *J. Inst. Pet. Techn.* **26** (1940).

IX. *Lubrication of metals on steel. Static friction*

Bearing surface	Rape oil μ_s	Castor oil μ_s	Mineral oil μ_s	Long-chain fatty acids μ_s
Hard steel	0·14	0·12	0·16	0·09
Cast iron	0·11	0·15	0·21	..
Gun metal	0·15	0·16	0·21	..
Bronze	0·12	0·12	0·16	..
Pure lead	..	..	0·5	0·22
Lead-base white metal	..	..	0·1	0·08
Pure tin	..	..	0·6	0·21
Tin-base white metal	..	..	0·11	0·07
Sintered bronze	..	..	0·13	..
Brass	..	0·11	0·19	0·13

Results taken mainly from Deeley.

X. *Lubricating properties and chemical reactivity*

Lubrication of various metals by 1 per cent. lauric acid (m.p. 44° C.) in paraffin oil.

Metal	Coefficient of friction μ_s		Transition temperature °C.	% acid† reacting with metal
	Clean	Lubricated (20°)		
Zinc	c. 0·6	0·04	94	10·0
Cadmium	c. 0·5	0·05	103	9·3
Copper	c. 1·0	0·08	97	4·6
Magnesium	c. 0·6	0·08	80	Trace
Iron	1·0	0·15–0·20	c. 40–50	Trace
Platinum	1·2	*0·25	20	0·0
Nickel	0·7	*0·28	20	0·0
Aluminium	1·35	*0·30	20	0·0
Chromium	0·41	*0·34‡	20	Trace
Glass	c. 1·0	*0·3–0·4	20	0·0
Silver	1·4	*0·55	20	0·0

Intermittent motion is marked with an asterisk *.
† Estimated amount of acid involved in the reaction assuming formation of a normal salt.
‡ Chromium surface freshly scraped.

XI. *Kinetic friction of lubricated surfaces*

In general the kinetic friction is less than the static friction, e.g. for paraffin oil on steel $\mu_s \approx 0{\cdot}2$ while $\mu_k \approx 0{\cdot}15$. At higher sliding speeds there may be a further decrease in μ_k. With fatty acids on steel (at slow speeds of sliding) $\mu_k = \mu_s \approx 0{\cdot}1$, but at higher speeds μ_k falls to $\mu_k = 0{\cdot}07$ or less. The effect of speed is complicated by the onset of hydrodynamic lubrication which occurs at higher speeds.

III. Ice and Snow

XII. *Friction of ice and snow as function of temperature*

On ice and snow there is an appreciable deviation from Amontons's law. There is evidence that the low kinetic friction is due in part to a thin layer of water formed by frictional heating. At low temperatures the friction rises considerably, and good thermal conductors such as metals give a higher friction than poor conductors such as ebonite.

(*a*) *Static friction ice on ice*

Temp. °C.	0	−12	−71	−82	−110
μ_s . .	0·05–0·15	0·3	0·5	0·5	0·5

(*b*) *Kinetic friction on snow; speed 0·1 m./sec.*

Temp. °C.	0	−3	−10	−40
Waxed hickory	0·04	0·09	0·18	0·4

On wet snow the friction is higher, for waxed hickory $\mu_k = 0{\cdot}14$. If the slider or ski is running on untracked snow, an additional term (which may be considerable) must be added for the work of ploughing through and compressing the snow.

(*c*) *Kinetic friction on ice* (*speed 4 m./sec.*)

Other surface	*Coefficient of kinetic friction* μ_k					
	0° C.	*−10° C.*	*−20° C.*	*−40° C.*	*−60° C.*	*−80° C.*
Ice	0·02	0·035	0·050	0·075	0·085	0·09
Ebonite	0·02	0·050	0·065	0·085	0·10	0·11
Brass	0·02	0·075	0·085	0·115	0·14	0·15

Unwaxed hickory at −3° C., $\mu_k = 0{\cdot}08$. Waxed hickory $\mu_k = 0{\cdot}03$.

IV. Non-metals—Clean and Lubricated

XIII. *Static friction of non-metals—clean and lubricated*

Material	μ_s
Glass on glass, clean	0·9–1·0
,, ,, paraffin oil	0·5–0·6
,, ,, liquid fatty acids	0·3–0·6
,, ,, solid hydrocarbons, alcohols, or fatty acids	0·1
Metal on glass, clean	0·5–0·7
,, ,, lubricated	0·2–0·3
Diamond on diamond, clean	0·1
,, ,, lubricated	0·05–0·1
Diamond on metal, clean	0·1–0·15
,, ,, lubricated	0·1
Sapphire on sapphire, clean and lubricated	0·2
Sapphire on steel, clean and lubricated	0·15
Hard carbon on carbon, clean	0·16
,, ,, lubricated	0·12–0·14
Hard carbon on steel, clean	0·14
,, ,, lubricated	0·11–0·14
Graphite on graphite, clean and lubricated	0·1
Graphite on graphite, outgassed	0·5–0·8
Graphite on steel, clean and lubricated	0·1
Steel on graphite, clean	0·1
Mica on mica, freshly cleaved	1·0
,, ,, contaminated	0·2–0·4
Crystals of $NaNO_3$, KNO_3, NH_4Cl on self, clean	0·5
Crystals of $NaNO_3$, KNO_3, NH_4Cl on self, lubricated long-chain polar compounds	0·12
Tungsten carbide on itself, clean	0·2–0·25
,, ,, lubricated	0·12
Tungsten carbide on steel, clean	0·40–0·6
,, ,, lubricated	0·1–0·2

Material	μ_s
Plastics	
Perspex on self	0·8
,, ,, steel	0·4–0·5
Polystyrene on self	0·5
,, ,, steel	0·3–0·35
Polythene on self	0·2
,, ,, steel	0·2
Teflon on self	0·04
,, steel	0·04
The friction is essentially the same for clean and lubricated surfaces.	
Fibres†	
Wool fibre on horn, cleaned	$\mu_2 =$ 0·8–1; $\mu_1 =$ 0·4–0·6
,, ,, greasy	$\mu_2 =$ 0·5–0·8; $\mu_1 =$ 0·3–0·4
Commercial condition:	
Nylon on nylon	0·15–0·25
Silk on silk	0·2–0·3
Cotton on cotton (thread)	0·3
Cotton on cotton (from cotton wool)	0·6
Solids on rubber, clean	1–4
Wood on wood, clean	0·25–0·5
,, ,, wet	0·2
Wood on metals, clean	0·2–0·6
,, ,, wet	0·2
Brick on wood, clean	0·6
Leather on wood, clean	0·3–0·4
Leather on metal, clean	0·6
,, ,, wet	0·4
,, ,, greasy	0·2
Brake material on cast iron, clean	0·4
,, ,, wet	0·2
,, ,, mineral oil	0·1

† For fibres, μ_2 against scales, μ_1 with scales.

ADDENDA

CHAPTER I. AREA OF CONTACT BETWEEN SOLIDS

Pages 6–8. Optical interference

Optical interference methods have been widely used in the last few years to study surface contours and have been particularly effective in demonstrating the existence of growth spirals on crystal faces. Dr. J. S. Courtney-Pratt (1950) has applied these methods to a study of the thickness of fatty acid layers adsorbed on mica cleavage faces by retraction from the melt. The results show that such layers are unimolecular, their thickness being very close to the length of the fatty acid molecule.

Recently, Miss Anita Bailey has used multiple-beam interferometry to study the interaction between sheets of mica which are molecularly smooth over an appreciable area. The mica sheets are bent into cylindrical form and two such cylinders are pressed together with their axes at right angles. The region of contact may be examined interferometrically and measurements may be made of the adhesive force and of the tangential force required to produce sliding. This provides perhaps the only means of studying the interaction between molecularly smooth surfaces. Monolayers may be deposited on the mica surfaces and the interaction between these may also be studied. The results are briefly discussed in the addenda to Chap. VII below.

Pp. 8–9. Electron microscope

In transmission electron microscopy it is generally necessary to prepare transparent replicas of the surface. Apart from the fact that replica techniques are often a tricky business, it is never certain that they follow the original contour exactly, nor is the interpretation of the transmission micrograph obtained from the replica always unequivocal. These difficulties do not arise when the electron microscope is used as a reflection instrument. The general principle is to direct the electron beam at glancing incidence on to the surface under examination and to focus the scattered electrons in the usual way (von Borries, 1940; Kushnir *et al.*, 1951; Cosslett, 1952). The resultant electron micrograph reveals the surface contours somewhat in the same way as a pedestrian sees the surface irregularities of a road illuminated by oncoming car headlights. Long shadows are cast by surface irregularities but they are foreshortened because of the small angle of viewing. It is, however, not difficult to calculate the true heights of the surface features revealed. The great advantage of this technique is that replicas are not necessary: the surface itself is examined directly by the beam. (If the surface is a non-metal a thin film of silver must be evaporated on to it to prevent charging up.) Recently Dr. Menter has shown that extremely simple modifications to the standard Metropolitan-Vickers E.M.3 electron microscope enable it to take reflection micrographs of great clarity, high resolution, and considerable depth of focus (Menter, 1952).

The resolution obtained at present by reflection is not as high as that obtained by transmission, particularly if the new carbon replicas are used (D. E. Bradley, unpublished), but by the reflection method any changes in the surface can be followed as they occur. Recently Mr. Chapman and Dr. Menter have used this method to study the surface structure of fibres and other non-metallic substances, while Dr. Menter and Mr. Seal have examined the structure of diamond surfaces.

Other methods of surface examination

Other optical methods for surface study not mentioned in Chap. I have recently been described in some detail. Phase contrast microscopy (Cuckow, 1949) is of great value in examing surface features *less* than a few hundred angstroms in height. The main defect of this method is that although it very readily reveals contrasts for small differences in height, it does not measure the heights quantitatively; these can only be estimated. The reflecting microscope using mirrors instead of lenses eliminates chromatic aberration and can give high magnifications with a long working distance to the surface under examination. Polarized light in metallographic studies gives marked contrasts for different orientations of surface crystallites, particularly with hexagonal or tetragonal metals (Mott and Haines, 1951; Conn and Bradshaw, 1952).

Recently Tolansky (1952) has described a simple and effective method of examining surface contours, using the 'light profile' technique. This consists in projecting the image of a straight opaque line at some angle to the surface under examination. When viewed appropriately the image of the line follows the surface irregularities and gives a contour line similar to the oblique or taper sections described on p. 9. Under suitable conditions the resolution is better than 2,000 Å both in depth and extension. An account of this method and a brief review of other microscopical techniques is given by Tolansky in a symposium organized by the Institute of Metals on 'Properties of Metallic Surfaces' (1953). Another survey dealing critically with phase-contrast microscopy, reflecting microscopy, and multiple-beam interferometry is contained in papers contributed to a conference on 'The Examination of Metals by Optical Methods' convened by the British Iron and Steel Research Association in May 1949.

In the last few years G. P. Thompson's school has shown that electron diffraction can be used to provide information not only about the structure of solid surfaces but also about the shape and orientation of surface asperities. These developments are surveyed in a paper presented by Sir George Thompson at a Conference on 'The Structure and Properties of Solid Surfaces' organized by the National Research Council of America (1952).

Area of contact between solids

A fuller account of this part of Chap. I and its implications in hardness measurements are described in a recent monograph (Tabor, 1951).

REFERENCES

B. von Borries (1940), *Zeit. Physik*, **116**, 370.
F. W. Cuckow (1949), *J. Iron Steel Inst.* **161**, 1.
G. K. T. Conn and F. J. Bradshaw (1952), *Polarised Light in Metallography*. Butterworths.
V. E. Cosslett (1952), *Nature*, **170**, 861.
J. S. Courtney-Pratt (1950), ibid. **165**, 346.
U. M. Kushnir, L. M. Biberman, and N. P. Levkin (1951), *Izvest. Akad. Nauk S.S.S.R. (Fiz.)* **15**, 306.
J. W. Menter (1952), *J. Inst. Metals*, **81**, 163; (1953) *J. Photogr. Science*, **1**, 12.
B. W. Mott and H. R. Haines (1951), *Research*, **4**, 24; ibid. 63.
D. Tabor (1951), *The Hardness of Metals*. Clarendon Press.
S. Tolansky (1952), *Nature*, **169**, 445.

CONFERENCE PROCEEDINGS

'The Examination of Metals by Optical Methods' (1949). *B.I.S.R.A.*
'Properties of Metallic Surfaces' (1953). *J. Inst. Metals.*
'Structure and Properties of Solid Surfaces' (1952). *N.R.C. (U.S.A.)*.

CHAPTER II. SURFACE TEMPERATURE OF RUBBING SOLIDS

Pp. 49–52. Measurement of transient hot spots

A fuller account of the work using infra-red sensitive photocells has now been prepared (Bowden and Thomas, 1954). The results confirm the earlier conclusions. In general the surface temperature cannot exceed the bulk melting-point of the more fusible of the sliding bodies. Occasionally, however, higher temperature flashes may occur. These are observed with metals like aluminium and magnesium which readily oxidize in air, and they are apparently due to the heat of chemical oxidation of the metals when freshly exposed to oxygen.

REFERENCE

F. P. Bowden and P. H. Thomas (1954), *Proc. Roy. Soc.* **A 223**, 29.

CHAPTER III. EFFECT OF FRICTIONAL HEATING ON SURFACE FLOW

Pp. 58–65. **Polishing and surface flow**

Some recent experiments by Dr. A. J. W. Moore and W. J. McG. Tegart (1951) help to explain the structure of the Beilby layer. In these experiments they rubbed sliders of steel or diamond over clean copper surfaces. Taper sections of the tracks formed showed that during repeated sliding over the same track oxide was embedded to a considerable depth beneath the surface. The surfaces were subsequently annealed at high temperatures and it was found that the included oxide particles prevented grain growth in the surface layers. The experiments support the view that during the polishing process minute fragments of oxide or even abrasive material are embedded in the surface; these inclusions inhibit crystal growth so that the surface layer is left with an extremely fine crystalline structure approximating to an amorphous material.

A brief account of this work is given in the full-day *Discussion on Friction* organized by the Royal Society in April 1951 and published in 1952.

Pp. 65–71. **Friction on snow and ice**

A further study has been made of the friction of real ski and of smaller models sliding on snow and ice at various temperatures (Bowden, 1953). On cold snow the static friction is high. When the sliding speed is appreciable the friction falls to a low value, and experiments support the view, put forward earlier, that this low friction is due to a localized

TABLE I

Static friction of plastics on snow at various temperatures

Snow temperature	*Perspex*	*Terylene*	*Nylon*	*P.T.F.E.*
−32° C.	0·4	..	..	0·1
−10° C.	0·34	0·38	0·3	0·08
{0° C. / Air temp. +5° C.}	0·3	0·35	0·3	0·02
Slush	0·5	0·5	0·4	0·05

surface melting produced by frictional heating. Measurements have been made on a variety of surfaces including metals, synthetic polymers, and waxes. The contact angle which water makes with the surface is important and there is evidence that this can decrease during sliding. In general, the solids with a high contact angle give a lower friction. The behaviour is also influenced by the relative hardness of ice and of the ski surface at the temperature of sliding. Polytetrafluoroethylene (P.T.F.E. or Fluon) gives a very low friction on snow and ice under all conditions.

Some typical values are given in Tables I and II.

TABLE II

Static friction of waxes and other solids on snow at various temperatures

Snow temperature	*Aluminium*	*Paraffin wax*	*Swiss wax*	*Ski wax*	*Ski lacquer*
− 32° C.	..	0·4	0·2	0·2	0·4
− 10° C.	0·38	0·35	0·2	0·2	0·4
{0° C. Air temp. + 5° C.}	0·35	0·06	0·03	0·04	0·1
Slush	0·4	0·06	0·05	0·1	0·2

REFERENCES

F. P. BOWDEN (1953), *Proc. Roy. Soc.* **A 217**, 462.
A. J. W. MOORE and W. J. McG. TEGART (1951), *Aust. J. Sci. Res.* **A 42**, 181.
—— —— (1952), *Proc. Roy. Soc.* **A 212**, 458.

CONFERENCE PROCEEDINGS

'Discussion on Friction' (1952). *Proc. Roy. Soc.* **A 212**, 439.

CHAPTER IV. FRICTION AND SURFACE DAMAGE OF SLIDING METALS

Pp. 82–83. CHEMICAL AND RADIOACTIVE DETECTION OF METAL TRANSFER

A fuller investigation of the metal transfer occurring between clean and lubricated surfaces has been carried out (Rabinowicz and Tabor, 1951) using the autoradiographic technique originally described by Mr. Gregory. The results for unlubricated surfaces show that under normal loading a small but finite transfer occurs. Once sliding commences the transfer is very greatly increased, probably because of the break through of the surface oxide film which occurs just when relative movement is initiated (see also Cocks, 1952). The amount of transfer is roughly proportional to the load, the evidence suggesting that an increase in load produces a proportional increase in the number of junctions formed rather than in their size. The transfer is greatly dependent on the nature of the metals. Thus for similar metals the pick-up is 50–100 times greater than for dissimilar metals. On the other hand there is little change in the friction; for similar metals $\mu \simeq 0{\cdot}6\text{–}1{\cdot}5$, for dissimilar metals $\mu \simeq 0{\cdot}4\text{–}0{\cdot}7$. This shows that the friction itself is not a very critical measure of the amount of transfer or metallic damage. The results for lubricated surfaces are discussed in Chap. X below.

Pp. 83–87. FRICTION AND SURFACE DAMAGE AT LIGHT LOADS

The earlier work of Whitehead (1950) has been extended by Dr. R. Wilson (1952) who has also measured the electrical resistance between the sliding surfaces. See Chap. VII below.

REFERENCES

M. COCKS (1952), *Nature*, **170**, 203.
E. RABINOWICZ and D. TABOR (1951), *Proc. Roy. Soc.* A **208**, 455.
R. J. WHITEHEAD (1950), ibid. A **201**, 109.
R. WILSON (1952), ibid. A **212**, 450; (1953) 'Properties of Metallic Surfaces', *Inst. Metals*, p. 356.

CHAPTER V. MECHANISM OF METALLIC FRICTION

Recently I. Ming Feng (1952) has suggested that when metal surfaces are placed in contact the interface undergoes minute deformations which have a ratchet-like structure. There is negligible adhesion between these surfaces but the ratchets themselves interlock and resist sliding. On this view sliding takes place by shearing a short distance away from the interface itself, the high temperatures developed during slip being then sufficient to facilitate welding at the interface. There are many objections to this mechanism. It cannot explain the marked welding observed at speeds of sliding so slow that friction heating cannot account for a temperature rise of more than a few degrees. Nor can it explain the frictional behaviour of non-crystalline materials.

Mechanism of Rolling Friction

A study has recently been made of the mechanism of rolling friction (Eldredge, 1952). The investigation has not been concerned with the characteristics of real ball-bearings but with the detailed processes involved when a ball rolls between two flat surfaces. The behaviour may be divided into two parts. When rolling first commences, even at small loads, there is marked plastic flow, and the resistance to rolling is due primarily to the plastic displacement of metal ahead of the ball. With successive traversals of the same track there is an increase in track width and a diminution in rolling resistance. An equilibrium state is ultimately reached where no further increase in track width occurs so that the deformations are essentially elastic. The rolling resistance is now constant and is unaffected by boundary lubricants.

A detailed study of the rolling resistance in the elastic range has not yet been completed but it seems that there are two main factors involved. The first, which has received little attention in the literature, is that due to elastic hysteresis losses (Tabor, 1952). As the ball rolls forward it depresses the front portion of the ellipse of contact and releases the back portion. Under perfectly elastic conditions the energy recovered from the back would be exactly equal to the energy expended in the front. Because of hysteresis losses, however, this is not so and there is an overall loss of energy. In effect, the hysteresis loss throws the centre of pressure ahead of the centre of the ellipse of contact and so introduces a retarding torque. There are few data on hysteresis losses at the large strains involved in these rolling experiments, but a preliminary investigation gives losses of the order of several per cent. for metals such as brass and copper.

In cylindrical rolling hysteresis is probably the main source of energy loss, the microslip suggested by Reynolds (1875) appearing to be negligibly small. With a ball rolling in a groove, however, a second factor is involved (Heathcote, 1921). Since different points on the ball in the region of contact are at different distances from the axis of rotation they will measure out different peripheral lengths for each revolution of the ball. Unless this can be taken up by differential stretching of the elements of the contact ellipse this must lead to micro-skidding between the ball and the groove. A simple theory has been developed which expresses the rolling couple produced in terms of the load, the coefficient of skidding friction between the ball and the groove and the geometry of the ball and track. Assuming a reasonable (constant) value for the coefficient of friction the calculated values are in fair agreement with the observed results. (Eldredge and Tabor unpublished).

Further work is still needed, but with most rolling mechanisms we may expect the total rolling resistance to be the sum of both the skid factor and the hysteresis factor. It is important to realize that high hysteresis losses can in some cases completely dominate the behaviour. Thus a 'low-friction' material like P.T.F.E. gives a higher rolling resistance than a typical 'high-friction' material like copper.

REFERENCES

K. R. Eldredge (1952), Ph.D. Dissertation, Cambridge.
H. L. Heathcote (1921), *Proc. Inst. Aut. Engrs.* **15**, 1569.
I. Ming-Feng (1952), *J. App. Phys.* **23**, 1011.
O. Reynolds (1875), *Phil. Trans.* **166**, 1.
D. Tabor (1952), *Phil. Mag.* **43**, 1055.

CHAPTER VI. ACTION OF BEARING ALLOYS

Little further work has been published in recent years on the fundamental mechanism of bearing alloys. A semi-empirical approach has recently been described by Lunn (1952) who measures the electrical resistance between lubricated sliding surfaces undergoing reciprocating motion. The average resistance is taken as a measure of the amount of breakdown of the lubricant film. Lunn claims that this approach provides a useful figure of merit for various metal-oil combinations. The method, however, does not really shed much light on the detailed mechanism involved during sliding.

A different approach which makes use of fundamental work on the frictional properties of relatively new substances has led to the development of new bearing materials. Thus polytetrafluoroethylene (P.T.F.E., Fluon or Teflon) has exceptionally good frictional properties (see Chap. VIII) but it is mechanically weak and has a poor thermal conductivity. If, however, it is incorporated in a porous material such as sintered copper we have a material which has the mechanical properties and thermal conductivity of copper, but since the surface is covered with a thin film of P.T.F.E. it has the frictional properties of the plastic and the coefficient of friction is as low as $\mu = 0{\cdot}05$ (Bowden, 1950*a*). Moreover the act of sliding or of wear may continue to feed the plastic on to the surface in much the same way as occurs with the copper-lead type of bearing alloy. Practical tests show that bearings made in this way can be used without any lubrication, and function effectively at high temperatures where conventional bearings fail after a few minutes (Love, 1952).

Another material which may be used is molybdenum disulphide. If metallic molybdenum is used as one of the bearing surfaces, it may be attacked chemically to form molybdenum disulphide *in situ* (Bowden, 1950*b*). The disulphide is then firmly attached and if the molybdenum is porous it may penetrate to a considerable depth below the surface. Even when running red hot the coefficient of friction is only $\mu = 0{\cdot}07$, and since the molybdenum disulphide is an integral part of the surface it is replenished during repeated sliding (Bowden, 1952). Molybdenum disulphide when incorporated in plastics and other non-metallic solids also gives a low friction (Bowden, 1950 *c*).

REFERENCES

F. P. Bowden (1950 *a*), *Research*, **3**, 147.
—— (1950 *b*), ibid. **3**, 383.
—— (1950 *c*), ibid. **3**, 384.
—— (1952), *Proc. Roy. Soc.* **A 212**, 440.
P. P. Love (1952), ibid. **A 212**, 484.
B. Lunn (1952), *Trans. Dan. Acad. Techn. Sci.* No. 2.

CHAPTER VII. FRICTION OF CLEAN SURFACES: EFFECT OF CONTAMINANT FILMS

Pp. 146–9. Effect of adsorbed gases on metallic friction

The work of Bowden and Young (1951) has been extended by Mr. Rowe who has devised an apparatus which measures not only the friction of outgassed metals but also the normal adhesion between the surfaces at any stage of the sliding process. The results show that the very large frictional forces observed with clean metals are always associated with large adhesions and there is, in fact, a fairly linear relation between the tangential force and adhesive force at any stage of the sliding process. This confirms the view that the attempt to slide the surfaces itself produces an appreciable growth in the area over which strong metallic junctions are formed. The behaviour is not unlike that described for clean indium surfaces in the atmosphere (see pp. 212–14). Contaminant vapours which reduce the friction produce a very much greater reduction in the adhesion. Thus a contaminated copper surface giving a coefficient of friction of $\mu = 2{\cdot}5$ may show a negligibly small normal adhesion (Rowe, 1953).

These observations are generally supported by the work of Gwathmey *et al.* (1952) in their investigations of the friction and adhesion of clean single crystals of copper. The oxide is removed by heating the crystals in hydrogen at 500° C. The friction is found to be greater between two (110) faces in contact than between (111) faces. Similar cohesion experiments show that the junctions formed between the crystals have the same strength as the parent metal.

Pp. 149–51. Influence of oxide films on friction

The work of Dr. Whitehead has been extended by Dr. Wilson (1952) who has made simultaneous measurements of the frictional force and the electrical resistance between the sliding surfaces. The results in general support Whitehead's conclusion but it is found that in some cases a break through of the oxide film, as shown by resistance measurements, is not necessarily accompanied by an appreciable increase in friction. With most metals the natural oxide film is sufficient to prevent metallic contact at very small loads. The degree of protection depends on a number of factors such as the surface roughness and the thickness of the oxide film, but the most important factor appears to be the relative hardness of the oxide and the metal substrate. This is shown by some typical values in Table I (see Bowden and Tabor, 1952).

TABLE I

Breakdown of oxide film on metals during sliding as shown by electrical resistance measurements

Metal	Vickers Hardness Kg./mm.^{2}		Load at which appreciable metallic contact occurs, g.
	Metal	Oxide	
Gold	20	..	0
Silver	26	..	0·003
Tin	5	1650	0·02
Aluminium	15	1800	0·2
Zinc	35	200	0·5
Copper	40	130	1·0
Iron	120	150	10
Chromium	800	..	> 1,000

This view is also born out by some experiments by Dr. A. J. W. Moore and W. J. McG. Tegart (1952) on the friction of a copper beryllium alloy treated in various ways to give a material varying in Vickers hardness from 100 to 400 Kg./mm.^{2} The harder alloys always give a lower friction. It would seem that the softer alloys undergo more deformation and this facilitates the disruption of the oxide film. The early stage of oxide disruption as relative motion between the surfaces is initiated has recently been described by Cocks (1952). He finds that when copper surfaces are first placed in normal contact the oxide film is intact. Breakdown begins after a displacement of the order of 10^{-4} to 10^{-3} cm.

Pp. 158–9. FRICTION OF GRAPHITE

Further work on the friction of outgassed graphite is discussed in the addendum to Chap. VIII below.

REFERENCES

F. P. BOWDEN and D. TABOR (1953), 'Properties of Metallic Surfaces', *J. Inst. Metals*, 197.

—— and J. E. YOUNG (1951), *Proc. Roy. Soc.* A **208**, 311.

M. COCKS (1952), *Nature*, **170**, 203.

A. T. GWATHMEY, *et al.* (1952), *Proc. Roy. Soc.* A **212**, 464.

A. J. W. MOORE, and W. J. McG. TEGART (1952), ibid. A **212**, 452.

G. W. ROWE (1953), Ph.D. Dissertation, Cambridge.

R. W. WILSON (1952), *Proc. Roy. Soc.* A **212**, 450.

CHAPTER VIII. FRICTION OF NON-METALS

A good deal of work has been carried out in the last few years on the friction of non-metals. An account of this is given by Bowden in the Redwood Lecture for 1953 to the Institute of Petroleum. The results show that an adhesion mechanism similar to that described for metals is, in general, applicable. Thus Moore and Tabor (1952) showed that very marked normal adhesion could be obtained between indium and a wide range of non-metallic materials (see Table II).

TABLE II

Adhesion of indium to various materials

Material	*Coefficient of adhesion*
Diamond	0·9–1
Glass	1
Tungsten carbide	1
Metals: Fe, Cd, Zn, Co, Ag, Pb, Cu, Au	1
Thick oxides of copper or silver	1
Rock salt	0·7
Plastics: polystyrene, Perspex	0·5–0·7
Plastics: P.V.C., polythene	0·02
Plastics: P.T.F.E.	0

Pp. 161–2. Crystalline solids

A detailed study of the friction and strength properties of crystalline materials such as rock-salt, lead sulphide, and ice has been made by Dr. R. F. King (1952). When a hard hemispherical steel slider or the pointed corner of a rock-salt crystal is slid over the flat cleavage face of another rock-salt crystal, a well-defined groove is formed, the width of which gives an approximate measure of the area of intimate contact between the sliding surfaces. From the frictional measurements it is thus possible to calculate the specific shear strength of the material in the contact region, and it is found that this is nearly ten times as great as the bulk shear strength of a rock-salt single crystal. The reason for this discrepancy becomes clear when it is realized that the material in the region of contact is under a very high hydrostatic pressure. Independent compression experiments show that under these conditions rock-salt ceases to be brittle and can undergo marked plastic deformation, the stresses that the crystal can withstand before it yields plastically being about ten times higher than the stresses required to produce fracture in an uncompressed specimen. Further, if two rock-salt specimens are pressed together between rigid anvils so that there is appreciable plastic flow, marked adhesion occurs between the crystals, and the interface is almost as strong as a single crystal. These compression experiments can be carried out between glass anvils and it is then seen that the criss-cross cracks which traverse the crystal during the early stages of the compression are healed as the pressure is increased. In this way the flaws which normally act as regions of high-stress concentrations are closed up by the high hydrostatic pressures and the rock-salt becomes relatively ductile. These experiments explain the observation that the track formed during the sliding of rock-salt surfaces shows considerable plastic deformation, and the results fit in with the view that the friction is primarily due, as with metals, to strong interfacial adhesions. Outgassing in a vacuum does not produce gross seizure as with metals, and this appears to be a fairly general observation with relatively non-ductile materials (King and Tabor, 1954).

Pp. 162–4. **Sapphire, diamond, carbon, and graphite**

Further work has been carried out on the frictional behaviour of thoroughly cleaned surfaces of carbon, graphite, diamond, and sapphire (Bowden and Young 1951; Bowden, Young, and Rowe, 1952; Rowe 1953). The results with carbon and graphite show that the friction rises by a factor of about three (from $\mu = \cdot 2$ to $\mu = \cdot 6$) when the surface films normally present are removed by outgassing. A small pressure of oxygen or water vapour (about 10^{-3} mm.) restores the friction to its normal value. The change is reversible on pumping out without heating, suggesting that a major factor is the physical adsorption of the surface films. These results are compatible with the work of Savage (1948), who found that a pressure of about 6 mm. of oxygen or water vapour prevents the onset of very rapid *wear* of carbon brushes. The higher pressures necessary in his experiments are probably due to the greater loads and speeds of sliding resulting in high surface temperatures and a shorter time available for effective transport of the vapour to the actual rubbing interface.

With diamond the friction is increased from about 0·05 in the atmosphere to about $\mu = 0{\cdot}4$ after outgassing. The coefficient of friction falls off with increasing load and the behaviour is consistent with the view that the contact is primarily elastic. A calculation of the area of real contact, using Hertz's equations for elastic deformation, indicates that the effective shear strength of the contact region is comparable with that of bulk diamond. Similar results have been obtained with sapphire surfaces.

It would seem from these observations that the friction between clean surfaces of carbon, graphite, diamond, and sapphire is again due, as with metals, to strong bonds formed by adhesion at the interface. The fact that, in contrast to metals, gross seizure does not occur seems to be due to the very small ductility (if any) of these materials. This view is supported by a study of the friction of metals sliding on outgassed graphite, diamond, and sapphire. Here the metal is able to flow plastically as a result of the sliding process itself (McFarlane and Tabor, 1950), and this leads to an increase in the real area of contact. As a result very high values of the coefficient of friction are observed. Thus for copper on graphite the friction reaches a value of $\mu = 1{\cdot}6$, and for diamond on platinum $\mu = 3{\cdot}6$.

It is interesting to note that when copper slides on clean graphite there is a very marked transfer of copper to the graphite and the copper fragments are attached very strongly indeed. Some transfer is even observed in the atmosphere, showing that there must be very strong adhesion between graphite and copper (Mr. D. M. Kenyon, unpublished).

P. 164. Molybdenum disulphide

Experiments by Mr. G. W. Rowe show that molybdenum disulphide gives a very low coefficient of friction at temperatures up to 800° C. or more. In the presence of oxygen, however, oxidation rapidly occurs and the low-friction properties are vitiated. The incorporation of molybdenum disulphide in bearing materials is discussed in the addendum to Chap. VI. Molybdenum is not the only sulphide which shows very low frictional properties. Similar effects are observed with uranium and tungsten sulphides. Both of these materials (like MoS_2) are believed to have a lamellar structure.

P. 164. Mica

In the addendum to Chap. I, it has been pointed out that interferometric techniques have recently been applied by Miss Bailey to a study of the interaction between atomically smooth mica sheets. Preliminary results show that for clean mica surfaces the friction is extremely high ($\mu = 8$), and from the area of intimate contact it may be estimated that under the conditions of her experiment the shear strength of the contact region is *c.* 10 Kg./mm.^{2} Similar measurements of the adhesion between the surfaces show that as the load is removed the contact area does not fall to zero. A finite normal force is necessary to pull the surfaces apart and the tensile strength of the contact region at this stage is also very high. This value depends on the experimental conditions but it may be of the same order of magnitude as the shear strength of the interface during sliding.

The effect of adding a monolayer of a soap is very striking—a large reduction in the area of intimate contact being immediately observed. There is a correspondingly large reduction in the friction and adhesion. In some cases there is negligible mica-mica interaction and the friction measurements thus provide a measure of the shear strength of the monolayer itself. The value obtained is of the order of 250 gm./mm.^{2} for monolayers of calcium stearate when the average pressure over the region of contact is of the order of a few gm./mm.^{2} For a preliminary account of the work see Bowden and Tabor (1952) and for earlier work on mica see Macaulay (1927).

Pp. 164–8. **Plastics**

A study has been made (Shooter, 1952; Shooter and Tabor, 1952) of the frictional properties of a group of linear polymers of the type summarized in Table XXII, p. 165. At light loads Amontons's law breaks down and the friction tends to increase with decreasing load. This may be because the deformation at these loads is essentially elastic (see for example Lincoln, 1952) but the issue is still unsettled. At loads above a few gm. the coefficient of friction is constant and almost independent of the size and shape of the surfaces. Using a hard curved slider which gives a well-defined track the area of intimate contact during sliding can be estimated and from the frictional force the shear force per unit area of contact can be calculated. For a wide range of plastics this is roughly equal to the specific shear strength of the plastic itself. This suggests that strong adhesion occurs over the region of contact and that during sliding shearing occurs within the bulk of the plastic a short distance from the interface, rather than in the interface itself. This is supported by the observation that marked transfer of plastic occurs. When metals slide on plastics, autoradiographic experiments show that there is also appreciable transfer of metal to the plastic even when the metal has a much greater strength than the plastic (Shooter and Rabinowicz, 1952). The frictional behaviour of these plastics is thus similar to that of metals, the coefficient of friction being approximately equal to the ratio of the bulk sheer strength s to the bulk yield pressure p.

This view is supported by some later experiments of Dr. R. F. King, who has measured the friction and the bulk strength properties of plastics over a temperature range from $-100°$ C. to $80°$ C. He finds that the friction varies with temperature in the same way as the ratio s/p (King and Tabor, 1953). This behaviour is similar to that observed with metals (Simon, McMahon, and Bowen, 1951).

The frictional behaviour of P.T.F.E. (Fluon or Teflon) differs from the other plastics in that the adhesion during sliding is small and the shear force per unit area of contact resisting sliding is much less than the shear strength of the plastic itself. However, here too, the friction varies in the same way as the ratio s/p over a wide temperature range. Although the adhesion to P.T.F.E. is small it is finite, and with a sufficiently soft metal such as freshly cut sodium there is marked adhesion and transfer of metal to the plastic. The low specific adhesion and the corresponding low coefficient of friction usually observed are not due to surface films. Even if P.T.F.E. is heated to its softening-point in a vacuum, the plastic on cooling gives the same low value of μ (King, 1952). The behaviour is due to the structure and chemical nature of the polymer. The polymer chains are screened by the large fluorine atoms so that their individual interaction is very small. On this view the bulk strength of P.T.F.E. is primarily due to an interlocking of the polymer chains (Hanford and Joyce, 1946). The exceptionally low coefficient of friction of P.T.F.E. ($\mu = 0{\cdot}05$) up to 300° C. has led to its use as a new bearing material. See addendum to Chap. VI, above.

P. 168. Tungsten carbide

An interesting study of the wear or 'cratering' of carbide cutting tools (see p. 241) which confirms the importance of high temperatures has recently been described by Trent (1952). His experiments were carried out under practical machining conditions where the speeds of sliding, and therefore the surface temperatures, are very high. He finds that with ferrous materials the tool wear is due essentially to the dissolving of the tungsten carbide in the metal chip. This readily occurs if the temperature at the rubbing interface exceeds 1,300° C. Under similar conditions titanium carbide does not dissolve in ferrous materials or dissolves at a very much slower rate. Consequently the wear of carbide tools, in the machining of ferrous materials at high speeds, is very greatly reduced if the tools contain titanium carbide.

Pp. 168–9. Glass

Recent experiments on glass show that in rolling (Eldredge, 1952) and in sliding (King, 1952) appreciable flow of the glass occurs. Here again, as with rock-salt, it would seem that the high localized pressures inhibit brittle fracture and the shear strength of the junctions is very much greater than the bulk shear strength of uncompressed glass.

P. 169. Rubber

Experiments on rubber (Schallamach, 1952*a*) show that the friction increases as the load is reduced, and it has been suggested that this is due to the fact that the area of intimate contact is determined by the elastic deformation of the rubber. Schallamach has also described an interesting series of investigations on the way in which rubber is torn and worn under various sliding conditions (Schallamach, 1952 *b*, 1953).

The friction and wear of rubber are of great importance in all forms of road transport, where it is necessary to combine a reasonably high coefficient of friction between the tyre and the road with a low rate of wear. It has been pointed out that the friction and wear properties of a rubber tyre may be crucially affected by high temperatures generated at the surface of the tyre by frictional heating and possibly by the adiabatic compression of air bubbles trapped in the tyre itself (Bowden, 1951). Some of these problems were discussed at an International Conference on Abrasion and Wear convened by the Rubber Stichting at Delft, Holland, in November 1951.

Pp. 169–73. **Fibres**

A wide interest in the friction of synthetic and natural fibres has been shown in the last few years. The fields of investigation fall into two main groups. The first deals with the development of techniques for the determination of the coefficient of friction. One method which has been widely used consists in measuring the initial tension T_1 and the drawing tension T_2 in a fibre which is pulled over a cylindrical surface. If the angle subtended by the fibre at the centre of curvature is θ, and if μ is constant, the relation $T_2/T_1 = e^{\mu\theta}$ is obeyed. In this method it is found that the friction is reduced if the surface of the cylinder or the fibre is roughened (Reicher and Bradbury, 1952). Another method, devised by Gralèn (1952) consists in twisting two fibres together and finding the force required to pull one fibre out of the other. A relation of the same type as that quoted above has been derived. One advantage of these methods is that they both give a measure of the friction over a considerable length of contact and so give a good average of a great number of individual points of contact. This has, of course, a corresponding disadvantage in that the basic mechanism of friction is more easily investigated if contact occurs over a single well-defined region. A second disadvantage is that the derived equations assume that μ is constant, whereas in fact μ depends on the load and this varies progressively along the length of the line of contact. The value of μ obtained from the relevant equation is therefore a special average value. These methods are, however, very useful in investigating the effects of various surface treatments on the friction of natural and synthetic fibres.

The second field of investigation has been the dependence of friction on load. In their study of the friction of plastics, Shooter and Tabor (1952) found that below a certain load the coefficient of friction μ increases with decreasing load W. This effect has also been found with fibres. Working with nylon fibres Lincoln (1952) finds that the coefficient of friction falls off exactly as $W^{\frac{1}{3}}$ and therefore considers that the behaviour is due entirely to the fact that the contact area is determined purely by elastic deformation. Most other workers do not, however, find so simple a relationship between μ and W. Thus Gralèn finds that a relation of the type $\mu = a+b/W^n$, where n is between 0·6 and 1 fits his results, the best fit being obtained for a value $n = 0{\cdot}7$.
A similar relation is found by Howell (1951), and Shooter's results also fit a relation of this type. Mrs. Makinson (1952) recently suggested that the increasing friction at low loads may be due to surface films which, by surface tension forces, produce an additional normal force between the surfaces (see Chap. XV, pp. 299–305). Shooter has, however, shown that neither surface tension forces nor electrostatic charges can be responsible for the rise in friction observed in his experiments. It is clear that further work on this is still necessary. Apart from this interesting problem there is, as Mrs. Makinson has pointed out, 'much work to be done in the textile field in explaining the behaviour of fibres in terms of their observed frictional properties, even apart from the question of the basic nature of this friction'.

Frictional electricity

During the last few years there has been an increasing interest in the electrification produced when solids (or liquids) are placed in contact or slide over one another. This is particularly important in the textile industry where high electrostatic charges on the fibres can produce large, undesirable forces between them. A recent conference convened by the Institute of Physics has emphasized the experimental difficulties involved in investigating the mechanism of frictional electrification. A major difficulty arises from the fact that it is almost impossible to prepare solid surfaces (even of the same materials) which are identical in all respects. This point has been stressed by Harper in his work on the electrification produced when similar metals are pressed together. A further complication is the apparent difference between the electrification which occurs in contact without sliding, and that produced during sliding itself. Thus body A pressed on body B may leave a positive charge on B, but if rubbed on B it may leave a negative charge. Henry has suggested that this is very probably due to the asymmetry of the local hot spots, the hot spots on A coming into repeated contact with the cold stationary surface B. A short early survey is given by Ward (1937).

REFERENCES

F. P. BOWDEN (1951), *Engineering*, **172**, 818.
—— and D. TABOR (1952), Contribution entitled 'Adhesion of Solids', N.R.C. Conference on 'The Structure and Properties of Solid Surfaces'.
—— and J. E. YOUNG (1951), *Proc. Roy. Soc.* **A 208**, 444.
—— —— and G. W. ROWE (1952), ibid. **A 212**, 485.
K. R. ELDREDGE (1952), Ph.D. Dissertation, Cambridge.
N. GRALÈN (1952), *Proc. Phys. Soc.* **A 212**, 491.
W. E. HANFORD and R. M. JOYCE (1946), *J. Amer. Chem. Soc.* **68**, 2082.
H. G. HOWELL (1951), *J. Textile Inst.* **42**, 12.
R. F. KING (1952), Ph.D. Dissertation, Cambridge.
R. F. KING and D. TABOR (1953), *Proc. Phys. Soc.* **B 66**, 728.
—— —— (1954), *Proc. Roy. Soc.* **A 223**.
B. LINCOLN (1952), *Brit. J. App. Phys.* **3**, 260.
J. M. MACAULAY (1927), *J. Roy. Techn. Coll. Glasgow*, No. 4, 5.
J. S. MCFARLANE and D. TABOR (1950), *Proc. Roy. Soc.* **A 202**, 244.
K. R. MAKINSON (1952), ibid. **A 212**, 495.
A. C. MOORE and D. TABOR (1952), *Brit. J. App. Phys.* **3**, 299.
A. REICHER and E. BRADBURY (1952), *J. Textile Inst.* **43**, T350.
G. W. ROWE (1953), Ph.D. Dissertation, Cambridge.
R. H. SAVAGE (1948), *J. App. Phys.* **19**, 654.
A. SCHALLAMACH (1952 *a*), *Proc. Phys. Soc.* **B 65**, 657.
—— (1952 *b*), *J. Polymer Sci.* **9**, 385.
—— (1953), *Proc. Phys. Soc.* **B 66**, 386; ibid. 817.
K. V. SHOOTER (1952), *Proc. Roy. Soc.* **A 212**, 488.
—— and E. RABINOWICZ (1952), *Proc. Phys. Soc.* **B 65**, 671.
—— and D. TABOR (1952), ibid. **B 65**, 661.
I. SIMON, H. O. MCMAHON, and R. J. BOWEN (1951), *J. App. Phys.* **22**, 177.
E. M. TRENT (1952), *Proc. Roy. Soc.* **A 212**, 467.
W. H. WARD, (1937), *Report on Progress in Physics IV*, 247.

CONFERENCE PROCEEDINGS

'International Conference on Abrasion and Wear', Rubber Stichting (1951). See *Engineering* (1951), **172**, 694, 724, 758, 790, 818.
'Static Electrification.' Conference convened by The Institute of Physics, March 1953. Published in *Brit. J. App. Phys.* Suppl. No. 2.

CHAPTER IX. BOUNDARY FRICTION OF LUBRICATED METALS

Pp. 181–3. Effect of temperature

The autoradiographic study of metal transfer for clean and lubricated surfaces has now been published (Rabinowicz and Tabor, 1951). The results show, as mentioned in earlier work, that the metallic transfer is fragmentary rather than continuous. An additional observation is that a lubricant reduces the size of the pick-up fragments rather than their number. With a poor boundary lubricant such as a liquid mineral oil (or a soap above its melting-point) the friction is reduced by a factor of 2 or 3, and the pick-up by a factor of several hundred. With fatty acids which react with the surface, or with a soap directly applied, the friction at room temperature is reduced by a factor of about 20 (from $\mu = 1$ to $\mu = 0{\cdot}05$) whilst the pick-up may be diminished by a factor of 20,000 or more. Typical results are given in Table III and it is evident that metallic transfer is immensely more sensitive to changes in surface conditions than is the coefficient of friction.

TABLE III

Friction and transfer for cadmium surfaces at room temperature. Load 2 Kg. Sliding speed 0·01 cm./sec.

Lubricant	*Coefficient of friction*	*Pick up 10^{-9} gm./cm. of track*
None	0·8	50,000
Cetane $C_{15}H_{31}CH_3$	0·6	500
Cetyl alcohol $C_{15}H_{31}CH_2OH$	0·4	100
Palmitic acid ($C_{15}H_{31}COOH$):		
Poor reaction	0·09	50
Heavy reaction	0·07	1
Copper palmitate $(C_{15}H_{31}COO)_2Cu$	0·05	1

The lowest friction and pick-up are observed when the lubricant film is solid. As the temperature is raised there is a marked increase in friction and pick-up at a temperature close to the melting-point of the film as mentioned on p. 182. A new and striking observation is that at still higher temperatures a second deterioration in lubricating properties occurs; the friction and pick-up are the same as for unlubricated surfaces although lubricant is visibly present on the surface. At this stage the lubricant molecule is desorbed or is rendered mobile and the change is again reversible on cooling.

These changes in the state of the lubricant film as the temperature is raised have been confirmed by the electron diffraction studies, previously mentioned. Below the melting-point of the boundary film the molecules are in a strongly oriented, close-packed array. At the melting-point the lateral attraction between the chains is overcome by thermal motion and the film loses its strong orientation. On cooling, the oriented pattern reappears showing that the effect is due to a physical and not a chemical change in the state of the film. At a higher temperature desorption occurs and, since the experiment is carried out in a high vacuum, the film does not reappear on cooling. The behaviour of lubricants in the solid, liquid, and desorbed or mobile states is summarized in Table IV.

TABLE IV

Lubrication of cadmium surfaces by palmitic acid

Temperature	State of lubricant film	Coefficient of friction	Pick-up 10^{-9} gm./cm. of track
20° C.	Solid	0·05–0·1	2–3
130° C.	Liquid	·3	500–1,000
160° C.	Desorbed or Mobile	·6	20,000–30,000
—	Unlubricated	·6–·8	30,000

Pp. 189–91. **Minimum film thickness for effective lubrication**

Experiments have recently been carried out by Hirst, Kerridge, and Lancaster (1952) on the effect of load and surface finish on boundary lubrication. Their results show that at small loads a single monolayer is sufficient to give effective lubrication, that is a low coefficient of friction ($\mu = 0 \cdot 1$ or less) and very slight metallic transfer. As the load is increased a stage is reached at which a large increase in the friction and in the amount of pick-up is observed. If rough surfaces are used this breakdown of the lubricant film occurs at lower loads. These experiments again suggest that a primary influence on the breakdown of a boundary film is the extent of the deformation of the underlying metal surfaces during sliding. These conclusions are similar to those already described on the breakdown of surface oxide films.

REFERENCES

W. HIRST, M. KERRIDGE, and J. K. LANCASTER (1952), *Proc. Roy. Soc.* A **212**, 516.
E. RABINOWICZ and D. TABOR (1951), ibid. A **208**, 455.

CONFERENCE PROCEEDINGS

'Physics of Lubrication' (1951). *Brit. J. App. Phys.* Suppl. No. 1.

CHAPTER X. MECHANISM OF BOUNDARY LUBRICATION

Pp. 207–11. STRUCTURE OF THE LUBRICATING LAYER: ELECTRON DIFFRACTION EXPERIMENTS

A fuller account of the electron diffraction study of lubricant films has now been published (Menter and Tabor, 1950; Sanders and Tabor, 1950). The paper by Menter and Tabor deals primarily with the structure and orientation of thin films of fatty acids and metallic soaps. As has already been pointed out (pp. 209–11), there is direct evidence for the formation of metallic soaps when the fatty acid is deposited on a reactive metal (see also Spink, 1950). These films disorient at a temperature close to the softening or melting-point of the *soap*. If the surface is cooled the oriented pattern reappears, showing that the effect is essentially due to a change in the physical state of the film. As mentioned above, if the surface is heated to about 50° C. above the disorientation temperature, the oriented pattern does not reappear on cooling. It would seem that at these elevated temperatures the surface film is desorbed.

The paper by Sanders and Tabor deals primarily with the structure of alcohols and esters deposited on metal surfaces. The results show that the first molecular layer is generally oriented with the hydrocarbon chains normal to the surface. With alcohols the monolayer loses its orientation at a temperature close to the bulk melting-point of the alcohol and the disorientation temperature does not depend appreciably on the nature of the metallic substrate. Similar results are observed with esters on non-reactive metals. On reactive metals, however, some of the esters give good orientation up to temperatures well above their bulk melting-point, and the evidence points to the formation of metallic soaps. These observations are thus complementary to the radio-active tracer experiments described on p. 215. There is evidence that the reaction with the surface is due to a trace of fatty acid impurity; it is also possible, when water is present, that hydrolysis of the ester could occur.

Courtel has (1952) recently examined the structure of clean metal surfaces prepared by grinding them *in situ* in the body of the electron diffraction camera itself. He finds that on such surfaces the vapours of fatty materials are very rapidly adsorbed to form well-oriented films.

Pp. 211–14. Mechanism of Soap Formation; Influence of Water

Further experiments by Lancaster and Rouse (1951) on heats of adsorption confirm the earlier conclusion that the reaction rate between a metal oxide and a fatty acid depends on the amount of water present. Under sufficiently dry conditions chemical reaction does not occur.

Pp. 214–15. Investigation of Surface Adsorption by Radio-active Methods

The work on the radio-active tracer experiments has now been published (Bowden and Moore, 1951) and supports the general picture of chemical attack and of the build-up of polymolecular layers.

Pp. 216–19. ADSORPTION OF FATTY ACIDS, ALCOHOLS, AND ESTERS ON METALS

Part of the work of Daniel quoted in Chap. X has now been published in detail (Daniel, 1951).

Thickness of adsorbed films of water and other vapours

The adsorption of water and other vapours on solid surfaces has recently been studied by Bowden and Throssell (1951) using two independent methods. The first was by a direct weighing of the surface in a microbalance. This gives the total amount of liquid or vapour taken up by adsorption. The second was by the reflection of polarized light, the ellipticity produced being a measure of the thickness of a uniform surface film of adsorbate.

With vapours of hydrocarbons both methods agree in showing that the adsorption is light and that even near saturation the adsorbed layer is not more than one or two molecules thick. With water, however, the weighing method shows that the adsorption on a platinum surface carefully cleaned by ordinary methods is very heavy indeed. This result agrees with earlier workers (see for example Henniker, 1949; McBain, 1950) and suggests that the adsorbed films are very thick (20 molecular layers) when the relative vapour pressure is only 0·8. If, however, the platinum surface is cleaned by heating to red heat in the vacuum of the microbalance itself, heavy adsorption is no longer observed at room temperature, and near saturation the water film is approximately two molecular layers thick. The polarized light method, even with conventionally cleaned platinum surfaces again shows that the adsorbed surface film is only one or two molecules thick near saturation. This suggests that the heavy adsorption observed by the weighing method (and reported by other workers) is due to traces of hygroscopic impurities on the surfaces. The relatively large patches of water which will form around these contaminants will merely scatter light in the optical experiment and will contribute nothing to the rotation of the plane of polarization. This is effected only by the thin uniform monomolecular (or possibly bimolecular) film of water adsorbed over the major part of the surface.

These results show that the physical adsorption of water vapour on very clean surfaces shows no anomaly. It is similar to that of other vapours and corresponds, at pressures near saturation, to one or two molecular layers. A trace of a suitable contaminant, however (and it is not easy to prepare surfaces free from such contaminants), will lead to the presence of relatively large quantities of water on the surface even at pressures well below saturation. This water may affect a number of the physical and chemical properties of surfaces and may therefore be important in friction and lubrication. For example it probably has a direct bearing on the lubrication of metals by fatty acids (see pp. 211–14).

Studies of the contact between flat surfaces (Bastow and Bowden, 1931), of the viscosity of very thin films (Bastow and Bowden, 1935), and of the adsorption of water vapour mentioned above provide no evidence for the long-range surface effects extending for thousands of angstroms which have been reported by many workers.

Pp. 219–26. MECHANISM OF BOUNDARY LUBRICATION

The autoradiographic method provides, for the first time, an approximate method of estimating the actual area of the metallic junctions formed through the lubricant film (Rabinowicz and Tabor, 1951). The calculation involves a number of rather drastic assumptions but the results are probably of the right order of magnitude. The calculations indicate that for a metallic soap the metallic junctions contribute only about 2 per cent. to the observed friction, for less effective solid films 5 to 10 per cent., for films in the liquid state between 15 and 25 per cent., and for desorbed films between 80 and 100 per cent.

It follows that the relation

$$F = A\{\alpha s_m + (1-\alpha)s_l\}$$

is still valid but that for good boundary lubricants α is very small. The metallic junctions contribute little to the frictional force but they are, of course, responsible for the wear. This means that good lubricants may give amounts of pick-up varying by a factor of say 20 without showing any appreciable difference in the coefficient of friction. Poorer lubricants provide less protection to the surfaces and α is very much larger. Under these conditions the friction is more indicative of the amount of wear but even here it is only a crude measure of the amount of metallic interaction. It follows that pick-up measurements provide a far better measure of the protective properties of lubricant films than measurements of the coefficient of friction, particularly in the comparison of good boundary lubricants.

The fact that, with good lubricants, the resistance to sliding is due mainly to the shearing of the lubricant film itself, raises another interesting problem. If we compare the behaviour of tin surfaces (V.P.H. 7) with hard steel surfaces (V.P.H. 700) lubricated with a typical soap it is evident that the area supporting the load for the tin surfaces will be approximately 100 times the area for the steel surfaces. Thus the area of lubricant film to be sheared will be 100 times greater for tin than for steel. The observed coefficients of friction, however, do not differ by more than a factor of 2. This difficulty may be explained by the observations of Bridgman (1946) and Boyd and Robertson (1945), who showed that the shear strength of lubricant films is proportional to the pressure to which they are subjected. Support for this view is also provided by the work of Clark, Woods, and White (1951).

Properties of a boundary lubricant

This summary supersedes some of the discussion at the end of Chap. X.

The fact that a lubricant reduces the size rather than the number of pick-up fragments confirms the view that the main function of a lubricant film is (*a*) to reduce the amount of metallic interaction at those regions where the surfaces would otherwise form large metallic junctions and (*b*) to inhibit the growth of junctions when sliding occurs (see Chap. XV). The film must also be easily sheared and this restricts conventional boundary lubricants to long chain hydrocarbons and to similar organic compounds. The pick-up is never entirely eliminated but extremely low values may be obtained when the lubricant is in the solid state. So long as the lubricant is solid, its adsorption to the surface is relatively unimportant since a solid hydrocarbon which is weakly adsorbed is more effective than a fatty acid or a soap above its melting-point. Consequently, the best protection is provided by lubricant films which have strong lateral attraction between the molecular chains, high melting-points, and suitable shear properties and it is largely for these reasons that metallic soaps are such effective boundary lubricants.

As the temperature is raised the lateral attraction between the molecules is overcome by thermal motion and an appreciable rise in friction and pick-up occurs at the melting-point of the film. The film loses its orientation but is still sufficiently attached to the surface to reduce metallic interaction by a factor of several hundreds. It is interesting to note that at this stage the pick-up and the friction are almost the same for paraffins, fatty acids, or soaps.

The partial protection provided by the molten film persists with increasing temperature until the lubricant becomes desorbed or mobile. At this stage the surfaces can displace the lubricant film wherever the asperities come in contact, and the friction and surface damage are almost the same as those observed with unlubricated surfaces. The temperature at which desorption occurs will depend on the strength of attachment of the molecule to the surface. We may expect, following the work of Bigelow, Glass, and Zisman (1947) that substances with high heats of adsorption will persist as effective films to higher temperatures. It follows that apart from suitable shear properties, a high melting-point and a strong attachment to the surface at elevated temperatures are both desirable properties.

REFERENCES

S. H. Bastow and F. P. Bowden (1931), *Proc. Roy. Soc.* A **134**, 404.
—— —— (1935), ibid. A **151**, 220.
W. C. Bigelow, E. Glass, and W. A. Zisman (1947), *J. Colloid Sci.* **2**, 563.
F. P. Bowden and A. C. Moore (1951), *Trans. Faraday Soc.* **47**, 900.
F. P. Bowden and W. R. Throssell (1951), *Proc. Roy. Soc.* A **209**, 297.
J. Boyd and B. P. Robertson (1945), *Trans. Amer. Soc. Mech. Engrs.* **67**, 51.
P. W. Bridgman (1946), *Rev. Mod. Phys.* **18**, 1.
O. H. Clark, W. W. Woods, and J. R. White (1951), *J. App. Phys.* **22**, 474.
R. Courtel (1952), *Proc. Roy. Soc.* A **212**, 459.
S. G. Daniel (1951), *Trans. Faraday Soc.* **47**, 1345.
J. C. Henniker (1949), *Rev. Mod. Phys.* **21**, 322.
J. K. Lancaster and R. L. Rouse (1951), *Research*, **4**, 44.
J. W. McBain (1950), *Colloid Science*, Boston, Heath & Co.
J. W. Menter and D. Tabor (1950), *Proc. Roy. Soc.* A **204**, 514.
E. Rabinowicz and D. Tabor (1951), ibid. A **208**, 455.
J. V. Sanders and D. Tabor (1950), ibid. A **204**, 525.
A. J. Spink (1950), *Nature*, **165**, 613.

CHAPTER XI. ACTION OF EXTREME PRESSURE LUBRICANTS

Pp. 239–40. Reactivity of Extreme Pressure Additives

Breakdown temperature of E.P. lubricants

The failure of boundary lubricants under extreme conditions is primarily due to the high temperatures developed at the rubbing interface rather than to the magnitude of the pressures to which the lubricant is subjected. For this reason the temperatures at which various E.P. lubricant films break down are of particular interest. Recently Bowden and Young (1951) have examined the frictional behaviour of thoroughly outgassed iron surfaces which have been attacked with H_2S or Cl_2 to form the sulphide or chloride film *in situ*. They find that the chloride provides good lubrication up to about 300° C. ($\mu = 0{\cdot}4$) but by 400° C. the friction has risen to $\mu = 1{\cdot}6$. With the sulphide the friction is always greater ($\mu = 0{\cdot}8$) but protection persists to a higher temperature and breakdown is only marked at temperatures above about 850° C. (e.g. at 930° $\mu = 2$).

Ferric chloride is a comparatively soft wax-like material having a low shear strength, so we may expect it to give a low friction. Ferric chloride melts at about 300° C and would be driven off the surfaces at somewhat higher temperatures. On the other hand ferrous sulphide is very much harder and gives a higher friction but it is more stable and remains on the surface to a much higher temperature.

These more recent observations support the view that the breakdown of the lubricant film is essentially due to a rise in temperature. These high temperatures will cause the film to melt, to decompose, or to be driven off the surface. We may, therefore, apply the temperature-rise calculations of Chap. II to the behaviour of an E.P. lubricant film. Both the simple treatment (equation (5)) and the more exact calculation (equation (11)) show that the temperature T of the rubbing hot spot can be written

$$(T-T_0)/\mu = f(W, v, k_1, k_2, \epsilon, ...),$$

where T_0 is the ambient temperature, μ the coefficient of friction, and f some function of the load W, the speed v, the conductivities k_1 k_2, the emissivities ϵ, etc. If the breakdown temperature of the surface compound is T_c it is clear that the greater T_c-T_0 the more severe the conditions the film can withstand. Similarly, a low coefficient of friction (which will be determined by the plastic properties of the E.P. film) means that the surfaces can run under more extreme conditions before the critical temperature is reached. For these reasons Blok (1939) has proposed that the parameter $(T_c-T_0)/\mu$ can be used as a measure of the effectiveness of an E.P. additive; in this case, as Williams (1952) has suggested, T_c now represents the melting-point of the surface compound formed and μ its yield stress or some other appropriate plastic property. More work must be carried out on the functioning of various E.P. additives before its usefulness is fully confirmed.

Pp. 240–5. Extreme Pressure Lubricants in the Cutting and Drawing of Metals

Mechanism of metal cutting

It is not the purpose of this book to discuss in detail the mechanism of metal cutting, but one or two recent developments are of some interest. The work of Ernst and Merchant (1940) which stimulated so many of the recent investigations dealt with the shearing stress in the chip and the friction between the chip and the tool as separate forces which could be compounded to obtain the equilibrium conditions at the tool nose. Recently Lee and Shaffer (1951) have attempted a more sophisticated treatment which considers the combined effect of the shear stress and the friction in producing plastic flow in the material at the nose of the tool. The results obtained are promising, but further work is needed to confirm and extend the analysis. Another development is the work of Backer, Marshall, and Shaw (1952). They find that for very thin cuts the metal appears to be extremely strong, and for cuts of the order of 10^{-4} cm. in depth the strength approaches the theoretical strength of the metal. This is in agreement with the observation of Galt and Herring (1952) that thin whiskers of tin are exceptionally strong: but here again more work is desirable.

Shaw and Smith (1952) have also described an analysis of the factors involved in tool-wear. They suggest that at low speeds, pressure welding between the tool and the work predominates. As a result, the metal at the interface is stronger than the bulk of the metal and rupture occurs within the work rather than at the interface. This leads to the formation of a built-up edge of work-hardened material. When the built-up edge breaks away it gouges or ploughs out material from the clearance face of the tool. At higher speeds of sliding, when the frictional heating is large, 'temperature-welds' are formed; the interface is not work-hardened and the welds are usually sheared in the interface. No built-up edge occurs and the wear in this case is largely due to the plucking of very small particles from the tool surface. Apart from the wear produced by these two welding processes, there may also be an abrasive or lapping action by hard constituents in the matrix of the work. This effect will not depend appreciably on speed except in so far as higher speeds produce a general softening of the tool material because of the temperature rise. By making reasonable assumptions about the relative importance of these factors Shaw and Smith are able to explain the various types of tool-life curves observed in practice as a function of cutting speed.

Finally we may quote Trent's work (described in the addendum to Chap. VII above) which showed that in the high-speed cutting of ferrous materials, temperatures of the order of 1,300° C. could be developed at the tip of the tool (Trent, 1952). These high temperatures must have a very important effect on the strength properties of the work and hence on the detailed action of the cutting process as Shaw and Smith have pointed out. They also play a fundamental part, as Trent showed, in the detailed wear mechanism of sintered carbide tools.

REFERENCES

W. R. BACKER, E. R. MARSHALL, and M. C. SHAW (1952), *Trans. Amer. Soc. Mech. Engrs.* **74**, 61. See also M. C. SHAW (1952), *J. Franklin Inst.* **254**, 109.
H. BLOK (1939), *J. Soc. Aut. Eng.* **44**, 193. See also (1952), *De Ingenieur*, **39**, O52.
F. P. BOWDEN and J. E. YOUNG (1951), *Proc. Roy. Soc.* **A 208**, 311.
H. ERNST and M. E. MERCHANT (1940), *Chip Formation, Friction and Surface Finish.* Cincinnati Milling Machine Co.
J. K. GALT and C. HERRING (1952), *Phys. Rev.* **85**, 1060.
E. H. LEE and B. W. SHAFFER (1951), *J. App. Mech.* **18**, 405.
M. C. SHAW and P. A. SMITH (1952), *Machinist*, **95**, no. 49, 1868.
E. M. TRENT (1952), *Proc. Roy. Soc.* **A 212**, 467.
C. G. WILLIAMS (1952), ibid. **A 212**, 512.

CHAPTER XII. BREAKDOWN OF LUBRICANT FILMS

Resistance measurements have been used by other workers to investigate the breakdown of lubricant films. Lunn has used this approach to evaluate the performance of lubricants and bearing alloys (see addendum to Chap. VI above).

REFERENCE

B. LUNN (1952), *Trans. Dan. Acad. Techn. Sci.* No. 2.

CHAPTER XIII. NATURE OF CONTACT BETWEEN COLLIDING SURFACES

A recent investigation of interest has been described by Crook (1952). Using a piezo-electric crystal he measured the growth and decay of the force between colliding surfaces. His results confirm most of the relevant equations described in Chap. XIII. In particular his experiments show that for impacts which are well in the plastic range the yield pressure opposing deformation is almost constant over the whole period of the impact. The dynamic yield pressure is greater than the static but Crook considers that this is not due to forces of a quasi-viscous nature. It may be noted, however, that his force displacement histograms show that the bodies are still approaching one another after the maximum force between them has been passed. This is characteristic of a resistance which is partly velocity-dependent and would thus support the idea of a viscous force. The matter, however, is by no means clear-cut and further work is necessary before the issue can be settled.

The two high-speed cameras developed by Dr. Courtney-Pratt (see addendum to Chap. XVI) should have useful applications in the study of high-speed deformation and fracture; in engineering processes such as the deformation of rotating turbine blades and in the study of cavitation.

REFERENCE

A. W. CROOK (1952), *Proc. Roy. Soc.* **A 212**, 482.

CHAPTER XIV. THE NATURE OF METALLIC WEAR

The mechanism of wear

Burwell and Strang (1952) have described experiments on the wear between a cylindrical metal pin and a hardened steel disk flooded with an inert lubricant to exclude air and grit. Under steady state conditions they find that the volume of material worn away is proportional to the load and to the length of path traversed as Kenyon has already observed (p. 297). This suggests that of the welded junctions formed at the interface and sheared during sliding, a constant fraction is detached to form the wear particles. On this view an increase in load produces a proportional increase in the number of welds, each of which remains approximately of constant size. This is supported by an examination of the wear particles. The mechanism proposed by Burwell and Strang seems preferable to the atomic wear model suggested by Holm (1946) which also yields a wear equation of the same type.

At higher loads, when the average pressure exceeds about one-third of the hardness of the pin, a large increase in wear is observed. Burwell and Strang suggest that this is primarily due to the fact that the true area of contact has become such a large fraction of the apparent contact area available that a loose wear particle, once formed, is not able to get away without producing further particles in a self-accelerating process.

More recently Archard (1952) has pointed out that a wear-rate proportional to the load does not necessarily imply that the junctions are all of the same size. If the junctions are geometrically similar and d is the effective diameter of any one of them, the mass removed when the junction is sheared is proportional to d^3. But the shearing occurs in a distance d so that the mass removed per cm. of traverse is proportional to d^2 and in the plastic range this is proportional to the load supported by the junction. If the same applies to all the junctions, whatever their size, the wear-rate will be proportional to the total load. Nevertheless the autoradiographic study, and Burwell's examination of the wear detritus indicate that the shorn junctions do not vary in size over a very wide range.

Pp. 290–3. CHEMICAL REACTION AND WEAR

An interesting study of the wear of lubricated surfaces operating in a partial vacuum or controlled atmosphere has recently been described by Davies (1951). Most of the experiments have been carried out between a steel ball (or a copper ball) rubbing on a mild steel cylinder, the wear being measured in terms of the scar on the ball or the width of the track formed on the cylinder. With a neutral paraffin oil as lubricant the wear is small until the air pressure falls to about 100 mm. when heavy wear or seizure occurs. If oxygen or nitrogen is admitted the wear is greatly reduced but, with copper, oxygen is far more effective than nitrogen presumably because the nitride is unstable. Oxidized paraffin oil or the addition of an oil soluble peroxide greatly increases the running load necessary to produce seizure and small wear is obtained even in the absence of air. Similar results are observed when triphenyl phosphine or tricresyl phosphate are added to the oil. It is again evident that chemical reaction plays a very important part in the wear mechanism and that surface oxides and nitrides are particularly effective with ferrous materials in reducing the amount of wear and the susceptibility to seizure.

REFERENCES

J. F. Archard (1952), *Research*, **5**; (1953), *J. App. Phys.* **24**, 981.
J. T. Burwell and C. D. Strang (1952), *Proc. Roy. Soc.* A **212**, 470.
C. B. Davies (1951), *Annals. New York Acad. Sci.* **53**, 919.

CONFERENCES

'Mechanical Wear' (1950). Amer. Soc. Metals.
'Fundamental Aspects of Lubrication' (1951). *New York Academy of Science Annals*, vol. **53**.

CHAPTER XV. ADHESION BETWEEN SOLID SURFACES: THE INFLUENCE OF LIQUID FILMS

Pp. 306–9. Adhesion of Soft Metals

The adhesion of steel to indium as a function of time of loading has been studied in greater detail by Moore and Tabor (1952). The results show, as indicated on p. 307, that the increase may be accounted for quantitatively in terms of the creep properties of indium. In the same paper the authors show that strong adhesion also occurs between indium and many non-metals, including glass, diamond, metal-oxides, and certain plastics (see addendum to Chap. VII above).

Adhesion of indium in impact

Crook and Hirst (1950) have examined the adhesion between indium surfaces under impact, measuring piezo-electrically the force of impact and the force of adhesion during separation. They find that the adhesion developed in the impact experiment is less than one-third that observed, for an equal area of contact, in a static test. The separated surfaces also show much less tearing in the impact experiment.

This indicates that the strength of a pressure weld increases with the time of contact, as was suggested by Sampson, Morgan, Reed, and Muskat in 1943. This effect may, in part, account for the general observation that metallic friction decreases with increasing velocity of sliding.

Pp. 312–14. ADHESION AND FRICTION

The papers by McFarlane and Tabor (1950*a*, *b*) have now been published. Parker and Hatch (1950) have also described an investigation of the friction and adhesion of indium and lead surfaces on glass. Their experimental results are in very close accord with those described in Chap. XV.

In their second paper McFarlane and Tabor (1950 *b*) have examined in greater detail the relation between the friction and adhesion of steel sliding on indium. They find that if μ is the coefficient of friction and ν the coefficient of adhesion at any stage of the sliding process before macroslip occurs, the behaviour for a whole series of experiments at different loads may be represented by a single equation of the form

$$0{\cdot}3\nu^2-\mu^2 = 0{\cdot}3.$$

The result is in very satisfactory agreement with the discussion on the effects of combined stresses given on pp. 101–4.

They also find that in the presence of a good boundary lubricant such as a metal soap the adhesion is negligible; the friction is small and shows little increase as relative motion takes place. With a poor boundary lubricant such as paraffin oil, however, a different behaviour is observed. The tangential force and the adhesional force increase in the same way as for unlubricated surfaces, that is, the experimental points lie on the same μ–ν curve. But macroslip occurs at a much lower value of μ and ν. It is evident, therefore, that in contrast to a good boundary lubricant, a poor lubricant does not markedly reduce the initial area of metallic contact; its main effect is to restrict the growth of the junctions as sliding takes place.

Recently Mr. G. W. Rowe has examined the friction and adhesion of very clean surfaces (e.g. Cu, Fe, Ni, etc.). As the initial microdisplacements take place, there is a growth of the area of intimate contact, both the normal and tangential stresses combining to produce plastic flow of the junctions at the interface. The behaviour is indeed descriptively similar to that observed with indium.

Mr. E. Eisner has measured, by interferometric methods, the microdisplacements that take place as the tangential force is increased up to the stage of macroslip. He finds that measurable displacements take place even for the smallest tangential forces and that a large part of these displacements is always irreversible. With oxide-free surfaces (e.g. gold) these displacements are accompanied by a fall in electrical resistance which is also irreversible. It is evident that combined stresses are producing plastic flow and junction growth at the regions of contact.

Since this behaviour of junction-growth and very high frictions is now well established for indium, and for very clean metals, we may consider why most common metals under ordinary atmospheric conditions give a coefficient of friction of the order of $\mu = 1$ and appear to fit in with the simple theory of friction. This was discussed briefly on p. 104, but in view of the later work the suggestions then made can now be supported and clarified. It would seem that there are two possible factors involved (McFarlane and Tabor, 1950 *b*). The first is the effect of released elastic stresses. With harder metals these may strip apart the rear edge of the junctions although junction growth may still be occurring at the leading edge. The second and more important factor is the effect of oxide or other surface contaminant films. These may act in a manner similar to that observed with paraffin oil on indium surfaces. Some junction growth may occur but at an early stage of the process the growing junction is unable to penetrate the contaminant film and the enlarged junction makes scarcely any additional contribution to the shear strength of the interface. Consequently only a small increase in the tangential force is sufficient to produce macroslip. A possible third factor is the time of contact which may play some part in determining the strength of the junctions formed (Crook and Hirst, 1950). As an overall conclusion it would seem that the simple theory of friction which gives a coefficient of friction of the order of unity works fairly effectively with ordinary oxide covered metals, primarily because little junction-growth occurs before macroslip takes place.

Action of adhesives

There is a voluminous literature on the action of industrial adhesives and on various methods of testing glued joints. A good survey of the work to date is in the book edited by de Bruyne and Houwink, *Adhesion and Adhesives* (1952). There is also now available a detailed account of a three-day Conference on Adhesion organized by the Society of Chemical Industry in April 1952. In a recent discussion of the basic mechanism of adhesion, Tabor (1952) has pointed out that almost all organic materials have sufficient van der Waals attraction for solid surfaces to give theoretical adhesive strengths greater than those usually required in practice. He has also shown that, if it is valid to apply thermodynamic and surface-chemical concepts to the strength of solid adhesives, the bond at the interface will always be stronger than the material a short distance away. Consequently, adhesives will not, in general, fail at the interface but within the bulk of the adhesive itself so that the joint will appear to have a strength equal to the bulk strength of the adhesive.

In practice the behaviour of the joint is complicated by two main factors. The first is the presence of contaminant films which may prevent proper spreading of the adhesive when applied as a liquid. In addition the contaminant may persist as a thin film (a few molecules thick) of relatively weak material between the adhesive and the adherend. The joint naturally fails first at these regions and its strength is greatly diminished. The second factor is the existence of stress concentrations within the adhesive; these may be produced by air-bubbles, foreign inclusions, or surface cracks, and in particular by differential expansion or contraction when the adhesive solidifies. Recently Mylonas (1951) has shown that even in the absence of all these factors, stress concentrations may arise at the meniscus of the adhesive film for geometric reasons. The stress concentration is small if the contact angle is small, that is, if the adhesive wets the solid surface. If, however, the adhesive makes a large contact angle at the interface, stress concentration factors of the order of three may be reached at the meniscus edge. Failure will first occur here and the overall strength of the joint will be greatly reduced. It is indeed probable that the contact angle is important in practical adhesives primarily because of its influence on stress concentration and that the adhesive properties depend to a secondary extent on the more subtle physico-chemical factors involved. These conclusions show that for good adhesion contaminant films should be absent and stress concentrations should be reduced to a minimum. It is evident that there is a very close parallelism between these criteria and those relating to the adhesion between metals and other solids when they are pressed together (see Bowden and Tabor, 1952).

REFERENCES

F. P. Bowden and D. Tabor (1952), Contribution entitled 'Adhesion of Solids', N.R.C. Conference on 'The Structure and Properties of Solid Surfaces'.

A. W. Crook and W. Hirst (1950), *Research*, **3**, 432.

N. A. de Bruyne and R. Houwink (1951), *Adhesion and Adhesives*, Amsterdam, Elsevier.

J. S. McFarlane and D. Tabor (1950 *a*), *Proc. Roy. Soc.* **A 202**, 224.

—— —— (1950 *b*) ibid. **A 202**, 244.

R. HOLM (1946), *Electric Contacts*, Almquist and Wiksells, Uppsala.
A. C. MOORE and D. TABOR (1952), *Brit. J. App. Phys.*
C. MYLONAS (1951), Ph.D. Dissertation, London.
R. C. PARKER and D. HATCH (1950), *Proc. Phys. Soc. Lond.* **B 63**, 185.
J. B. SAMPSON, F. MORGAN, D. W. REED, and M. MUSKAT (1943), *J. App. Phys.* **14**, 689.
D. TABOR (1952), Article entitled 'Basic Principles of Adhesion', *Ann. Report Progress App. Chem.* 1951–2.

CONFERENCE PROCEEDINGS

'Adhesion' (1952). Soc. of Chemical Industry, London.

CHAPTER XVI. CHEMICAL REACTION PRODUCED BY FRICTION AND IMPACT

Further work has been carried out on the initiation and growth of explosion in liquids and solids. Only a brief outline is given here since a recent monograph dealing with this has been published (Bowden and Yoffe, 1952).

Initiation of explosion

The technique of adding grits of known melting-point to the explosive, has been extended (Bowden and Williams, 1951 *a*) to the initiating explosives (such as trinitrotriazidobenzene, cyanurictriazide, and silver azide) which melt at temperatures below their decomposition temperatures. Again it is found that the minimum hot spot temperature is *c.* 500° C. This is in accord with previous work on the high explosives and on the initiating explosives such as lead azide and mercury fulminate which decompose before melting. Even in the absence of grit the low-melting initiating explosives were very sensitive to initiation by gentle impact. Recent work by Mr. A. M. Yuill has shown that this is due to the adiabatic compression of included air-bubbles so that the mechanism is similar to that responsible for the initiation of liquid and solid high explosives. The molten azides may also be initiated readily by the impact of a sharp pointed metal striker. Thermo-electric measurements show that this is due to a hot spot produced by the plastic deformation of the striker. Yoffe (1951) has studied the thermal decomposition of these molten azides. There is evidence that the transition from slow thermal decomposition to rapid explosion is due to the heat of recombination of active or atomic nitrogen at or near the surface of the explosive.

Even with a heterogeneous material such as gunpowder, which consists initially of three solid phases, the initiation can be due to hot spot formation. Blackwood and Bowden (1952) found that the impact was sensitized by grit particles and the controlling factor was again the melting-point of the grit. In this case, however, it was found that the minimum hot spot temperature required for ignition was very low (*c.* 130° C.). This and other experiments suggest that initiation requires the formation of a liquid phase which can be provided by the melted sulphur (melting-point 120° C.) and by the oxyhydrocarbons present as impurity in the charcoal.

Growth of explosion

In addition to their influence on initiation, the small gas-filled spaces which are normally present in the interstices of the crystals of a secondary explosive such as P.E.T.N. can play an important role in the growth and propagation of the low-velocity detonation. The shock wave from the first initiation centre can compress and heat these gas spaces so that they serve as new centres of initiation and allow the reaction to proceed.

High-speed camera studies of the growth of explosion from the point of initiation have shown that detonation is usually preceded by a comparatively slow burning. In the particular case of the metallic azides, which have a very simple chemical structure, e.g. PbN_6, no preliminary burning can be detected. (For work on the thermal and photochemical decomposition of azides and other explosives see the publications of Tompkins, Garner, Mott, and of Ubbelohde. Representative references to this work are given below. Eyring and collaborators (1949) have reviewed detonation theory and a monograph dealing with the stable high-velocity detonation of explosives has recently been published by J. Taylor, 1952. Further investigations in this field have been described by Bowden (1953) and size effects in the growth of explosions by Bowden and Singh (1953).)

Courtney-Pratt (1949–53) has developed a high-speed image converter camera capable of taking single pictures with exposures as short as 5×10^{-8} sec. or streak records with writing speeds up to 300,000 m./sec. at an overall aperture of f4. Detail of events occurring in 10^{-8} sec. can be resolved. He has also devised a new camera for taking a large number of 'stills'. This camera, which depends upon classical optics, uses a lenticular plate and is very simple in construction. It can take a hundred pictures at rates up to 100,000 pictures per second and, with a slightly different arrangement, twenty pictures can be recorded at intervals of a quarter of a microsecond. With these new methods it is now possible to make a more quantitative study of the growth of hot spots and of rapid reaction in the solid state.

REFERENCES

J. Blackwood and F. P. Bowden (1952), *Proc. Roy. Soc.* **A 213**, 285.
F. P. Bowden (1953), *Fourth Int. Symposium on Combustion*, 161.
—— and K. Singh (1953), *Nature*, **172**, 378.
—— and H. T. Williams (1951 *a*), ibid. **A 208**, 176.
—— —— (1951 *b*), *Research*, **4**, 339.
—— and A. D. Yoffe (1952), *Initiation and Growth of Explosion in Liquids and Solids*. Univ. Press, Cambridge.
J. S. Courtney-Pratt (1949), *Research*, **2**, 287.
—— (1952), *Photographic Journal*, **92 B**, 137.
—— *J. Photographic Science* (1953), **1**, 21.
—— Collected papers of the Fourth International Symposium on Combustion Phenomena, M.I.T. 1952.
—— Society of Motion Picture Engineers Convention on High Speed Photography. October 1952.
H. Eyring, R. E. Powell, G. H. Duffey, and R. Parlin (1949), *Chem. Rev.* **45**, 69.
W. E. Garner and J. Maggs (1939), *Proc. Roy. Soc.* **A 172**, 299.
N. F. Mott (1939), ibid. **A 172**, 325.
J. Taylor (1952), *Detonation in Condensed Explosives*. Clarendon Press, Oxford.
J. G. N. Thomas and F. C. Tompkins (1952), *J. Chem. Phys.* **20**, 662.
A. R. Ubbelohde (1948), *Phil. Trans.* **A 241**, 197.
A. D. Yoffe (1951), *Proc. Roy. Soc.* **A 208**, 188.

AUTHOR INDEX

The names of those who have worked in collaboration with the writers are printed in italics.

SUBJECT INDEX

Printed in the USA/Agawam, MA
October 8, 2013

580734.114